T0360875

Category Theory

— and — Applications

A Textbook for Beginners

Second Edition

Category Theory

—— and ——

Applications

A Textbook for Beginners

Second Edition

Marco Grandis

Università di Genova, Italy

World Scientific

W JERSEY · LONDON · SINGAPORE · BEIJING · SHANGHAI · HONG KONG · TAIPEI · CHENNAI · TOKYO

Published by

World Scientific Publishing Co. Pte. Ltd.

5 Toh Tuck Link, Singapore 596224

USA office: 27 Warren Street, Suite 401-402, Hackensack, NJ 07601

UK office: 57 Shelton Street, Covent Garden, London WC2H 9HE

Library of Congress Cataloging-in-Publication Data
Names: Grandis, Marco, author.
Title: Category theory and applications : a textbook for beginners /
    Marco Grandis, Università di Genova, Italy.
Description: Second edition. | New Jersey : World Scientific, [2021] |
    Includes bibliographical references and index.
Identifiers: LCCN 2021006491 (print) | LCCN 2021006492 (ebook) |
    ISBN 9789811236082 (hardcover) | ISBN 9789811236099 (ebook) |
    ISBN 9789811236105 (ebook other)
Subjects: LCSH: Categories (Mathematics) | Mathematical analysis. |
    Morphisms (Mathematics) | Homology theory.
Classification: LCC QA278 .G6985 2021 (print) | LCC QA278 (ebook) | DDC 512/.6--dc23
LC record available at https://lccn.loc.gov/2021006491
LC ebook record available at https://lccn.loc.gov/2021006492

British Library Cataloguing-in-Publication Data
A catalogue record for this book is available from the British Library.

For any available supplementary material, please visit
https://www.worldscientific.com/worldscibooks/10.1142/12253#t=suppl

Desk Editor: Soh Jing Wen

Printed in Singapore

To

Maria Teresa

Preface

Category Theory now permeates most of Mathematics, large parts of theoretical Computer Science and parts of theoretical Physics. Its unifying power brings together different branches, and leads to a better understanding of their roots.

This book is addressed to students and researchers of these fields and can be used as a text for a first course in Category Theory. It covers the basic tools, like universal properties, limits, adjoint functors and monads. These are presented in a concrete way, starting from examples and exercises taken from elementary Algebra, Lattice Theory and Topology, then developing the theory together with new exercises and applications.

A reader should have some elementary knowledge of these three subjects, or at least two of them, in order to be able to follow the main examples, appreciate the unifying power of the categorical approach, and discover the subterranean links brought to light and formalised by this perspective.

Applications of Category Theory form a vast and differentiated domain. This book cannot give a global picture of them, but wants to present the basic applications in Algebra and Topology, with a choice of more advanced ones, based on the interests of the author. References are given for applications in many other fields.

In this second edition, the book has been entirely reviewed, adding many applications and exercises. All non-obvious exercises have now a solution (or a reference, in the case of a relatively advanced topic); solutions are now deferred to the last chapter.

Contents

Introduction

0.1 Categories

Category Theory originated in an article of Eilenberg and Mac Lane [EiM], published in 1945. It has now developed into a branch of mathematics, with its own internal dynamics. Here we want to present its elementary part, and its strong links with the origins in Algebra and Topology.

Many mathematical theories deal with a certain kind of mathematical objects, like groups, or ordered sets, or topological spaces, etc. Each kind has its own privileged mappings, that preserve the structure in some sense, like homomorphisms of groups, or order-preserving mappings, or continuous mappings.

We have thus the 'category' Gp of groups and their homomorphisms, Ord of ordered sets and monotone mappings, Top of topological spaces and continuous mappings, and so on. More elementarily, we have the category Set of sets and their mappings.

In all these instances, the privileged mappings are called morphisms or arrows of the category; an arrow from the object X to the object Y is written as $f: X \to Y$.

Two consecutive morphisms, say $f: X \to Y$ and $g: Y \to Z$, can be composed giving a morphism $gf: X \to Z$; this partial composition law is 'as regular as it can be', which means that it is associative (when legitimate) and has a partial identity $\mathrm{id}X: X \to X$ (written also as 1_X) for every object, which acts as a unit for every legitimate composition.

'Concrete categories' are often associated with mathematical structures, in this way; but we shall see that categories are not limited to these instances, by far.

1

0.2 Universal properties

In these categories (and many 'similar' ones) we have cartesian products, constructed by forming the cartesian product of the underlying sets and putting on it the 'natural' structure of the kind we are considering, be it of algebraic character, or an ordering, or a topology, or something else.

All these procedures can be unified, so that we can better understand what we are doing: we have a family $(X_i)_{i \in I}$ of objects of a category, indexed by a set I, and we want to find an object X equipped with a family of morphisms $p_i \colon X \to X_i$ $(i \in I)$, called *cartesian projections*, which satisfies the following *universal property*:

- for every object Y and every family of morphisms $f_i \colon Y \to X_i$ $(i \in I)$ in the given category

$$
\begin{array}{ccc}
X & \xrightarrow{\ p_i\ } & X_i \\
\scriptstyle f \, \big\uparrow & \nearrow \scriptstyle f_i & \\
Y & &
\end{array}
\tag{0.1}
$$

there is a unique morphism $f \colon Y \to X$ such that $p_i f = f_i$, for all $i \in I$.

We shall see that this property determines the solution *up to isomorphism*, i.e. an invertible morphism of the given category. In fact, the proof is quite easy and some hints can be useful as of now: given two solutions $(X, (p_i))$, $(Y, (q_i))$, we can determine two morphisms $f \colon X \to Y$ and $g \colon Y \to X$ and prove that their composites coincide with the identities of X and Y.

(Let us remark, incidentally, that the product topology is much less obvious than the other product structures we have mentioned; the universal property tells us that it is indeed the 'right' choice.)

These facts bring to light a crucial aspect: a categorical definition (as the previous one) *is based on morphisms and their composition,* while the objects only step in as domains and codomains of arrows. If we want to understand what unifies the product of a family (X_i) of sets, or groups, or ordered sets, or topological spaces we should forget the nature of the objects, and think of the family of cartesian projections $p_i \colon X \to X_i$, together with the previous property. Then – in each category we are interested in – we come back to the objects in order to prove that a solution exists (and also to fix a particular solution, when this can be useful).

From a structural point of view, a category only 'knows' its objects by their morphisms and composition.

This is even more evident if we think of another procedure, which in category theory is called a 'sum', or 'coproduct'. We start again from a family $(X_i)_{i \in I}$ of objects; its (categorical) *sum* is an object X equipped with a family of morphisms $u_i \colon X_i \to X$ $(i \in I)$, called *injections*, which satisfies the following universal property:

- for every object Y and every family of morphisms $f_i \colon X_i \to Y$ $(i \in I)$ in the given category

$$
\begin{array}{ccc}
X_i & \xrightarrow{\ u_i\ } & X \\
 & \diagdown & \big\downarrow{\scriptstyle f} \\
 & {\scriptstyle f_i} & Y
\end{array}
\tag{0.2}
$$

there is a unique morphism $f \colon X \to Y$ such that $f u_i = f_i$, for all $i \in I$.

Again, the solution is determined up to isomorphism; its existence depends on the category.

The sum of a family of objects is easy to construct in the category Set of sets, by a disjoint union. We have similar solutions in Ord and Top. But in Gp the categorical sum of a family of groups is called the *free product* of the family; its construction is rather complex, and its underlying set is not the disjoint union of the sets underlying the groups. The categorical approach highlights the fact that we are 'solving the same problem' and – finally – makes clear what we are doing.

It is also important to note that the categorical definitions of product and coproduct are *dual* to each other: each of them is obtained from the other by reversing all arrows and compositions. This only makes sense – in general – within category theory, because the dual of a given category, formed by reversing its arrows and partial composition law, is a formal construction: the dual of a category of structured sets and mappings is *not* a category of the same kind (even though, *in certain cases*, it may essentially be, as a result of some duality theorem).

0.3 Diagrams in a category

In a category the objects and morphisms that we are considering are often represented by vertices and arrows in a *diagram*, as above in (0.1) and (0.2), to make evident their relationship and which compositions are legitimate.

As an important property, satisfied in the previous cases, we say that a diagram in a category is *commutative* if:

(i) whenever we have two 'paths' of consecutive arrows, from a certain vertex of the diagram to another, the two composed morphisms are the same,

(ii) whenever we have a 'loop' of consecutive arrows, from a vertex to itself, then the composed morphism is the identity of that object.

Let us consider some other instances

$$\begin{array}{ccc} X & \xrightarrow{f} & Y \\ & \searrow^{h} & \downarrow{\scriptstyle g} \\ & & Z \end{array} \qquad \begin{array}{ccc} A & \xrightarrow{f} & B \\ {\scriptstyle h}\downarrow & \searrow^{d} & \downarrow{\scriptstyle k} \\ C & \xrightarrow[g]{} & D \end{array} \qquad X \underset{v}{\overset{u}{\rightleftarrows}} Y \tag{0.3}$$

The first diagram above is commutative if and only if $gf = h$. For the second, commutativity means that $kf = d = gh$. For the third, it means that $vu = \mathrm{id}X$ and $uv = \mathrm{id}Y$; let us note that in this case the morphisms u and v are inverse to each other, and each of them is an isomorphism of our category.

Formal definitions of these notions, diagrams and commutative diagrams, will be given in 1.4.9.

0.4 Functors

It becomes now possible to view on the same level, so to say, mathematical theories of different branches, and formalise their links. A well-behaved mapping between categories is called a *functor*.

Among the simplest examples there are the *forgetful functors*, that forget the structure (or part of it), like:

$$\mathsf{Gp} \to \mathsf{Set}, \qquad \mathsf{Ord} \to \mathsf{Set}, \qquad \mathsf{Top} \to \mathsf{Set}. \tag{0.4}$$

For instance, the functor $U\colon \mathsf{Gp} \to \mathsf{Set}$ takes a group G to its underlying set $U(G)$, and a homomorphism $f\colon G \to G'$ to its underlying mapping

$$U(f)\colon U(G) \to U(G').$$

The whole procedure is well-behaved, in the sense that it preserves composition and identities.

These functors are so obvious that they are often overlooked, in mathematics; but here it will be important to keep trace of them. In particular, we shall see that they often determine other functors 'backwards', which are much less obvious: like the *free-group* functor $F\colon \mathsf{Set} \to \mathsf{Gp}$, 'left adjoint' to $U\colon \mathsf{Gp} \to \mathsf{Set}$ (of which more will be said below).

In another range of examples, a reader with some knowledge of Algebraic Topology will know that the core of this discipline is constructing *functors from a category of topological spaces to a category of algebraic structures*, and using them to reduce topological problems to simpler algebraic ones.

We have thus the sequence of *singular homology functors*, with values in the category of abelian groups

$$H_n\colon \mathsf{Top} \to \mathsf{Ab} \qquad (n \geqslant 0), \tag{0.5}$$

and the fundamental group functor $\pi_1 \colon \mathsf{Top}_\bullet \to \mathsf{Gp}$, defined on the category of pointed topological spaces and pointed continuous mappings.

0.5 Natural transformations and adjunctions

The third basic element of category theory is a *natural transformation* $\varphi \colon F \to G$ between two functors $F, G \colon \mathsf{C} \to \mathsf{D}$ with the same domain C and the same codomain D. We also write $\varphi \colon F \to G \colon \mathsf{C} \to \mathsf{D}$.

This simply amounts to a family of morphisms of D, indexed by the objects X of C

$$\varphi_X \colon F(X) \to G(X), \tag{0.6}$$

under a condition of 'naturality' which will be made explicit in the text.

Here we just give an example, based on groups and free groups (but the reader can replace groups with semigroups, or abelian groups, or R-modules, or any algebraic structure defined by 'equational axioms'). We have mentioned in the previous point the forgetful functor $U \colon \mathsf{Gp} \to \mathsf{Set}$ and the free-group functor $F \colon \mathsf{Set} \to \mathsf{Gp}$, that turns a set X into the free group generated by X.

The insertion of X in $F(X)$, as its basis, is a canonical mapping in Set

$$\eta_X \colon X \to U(F(X)). \tag{0.7}$$

All of them give a natural transformation

$$\eta \colon \mathrm{id} \to UF \colon \mathsf{Set} \to \mathsf{Set}, \tag{0.8}$$

where id is the identity functor of the category Set (turning objects and arrows into themselves) and $UF \colon \mathsf{Set} \to \mathsf{Set}$ is the composed functor (turning each set into the underlying set of the free group on it). This natural transformation is essential in linking the functors U, F and making the functor $F \colon \mathsf{Set} \to \mathsf{Gp}$ *left adjoint* to $U \colon \mathsf{Gp} \to \mathsf{Set}$.

The link is represented by the *universal property* of the insertion of the basis, namely the fact that every mapping $f \colon X \to U(G)$ with values in a group G (more precisely in its underlying set) can be uniquely extended to a homomorphism $g \colon F(X) \to G$. Formally:

- for every morphism $f \colon X \to U(G)$ in Set, there exists precisely one morphism $g \colon F(X) \to G$ in Gp such that $U(g).\eta_X = f$.

One can reformulate this link in an equivalent presentation of the adjunction, which may be more familiar to the reader: there is a (natural) bijective correspondence

$$\varphi_{XG} \colon \mathsf{Gp}(F(X), G) \to \mathsf{Set}(X, U(G)),$$
$$\varphi_{XG}(g) = U(g).\eta_X, \tag{0.9}$$

between the set of group-homomorphisms $F(X) \to G$ and the set of mappings $X \to U(G)$.

We shall see many constructions of free algebraic structures, and more generally of left (or right) adjoints of given functors (especially in Chapters 3–6). Many of them are 'real constructions', which give a good idea of the backward procedure; others are so complicated that one can doubt of their constructive character. In such cases, one can prefer to prove the existence of the adjoint, by the Adjoint Functor Theorem (in Section 3.5) or some other general statement: then the result is determined up to isomorphism, and its universal property allows its use. Much in the same way as we can define a real function as the solution of a certain differential equation with initial data, as soon as we know that the solution exists and is unique.

0.6 A brief outline

The first chapter deals with the basic tools: categories, functors and natural transformations. Ordered sets can be seen as categories of a special kind; adjunctions in this particular case are well known, as 'covariant Galois connections'; studying them (in Section 1.7) will also serve as an elementary introduction to general adjunctions.

Chapter 2 introduces the limits in a category (including products and the classical projective limits) and their dual notion, the colimits (including sums and the classical inductive limits); universal arrows with respect to a functor are a general formulation of universal properties (in Section 2.7).

Chapter 3 studies the crucial notion of adjunction: many important constructions in mathematics can be described as an adjoint functor of some obvious functor: from free algebraic structures to Stone-Čech compactification, metric completion etc. We also explore the related notion of monad and its algebras, which investigates when a category can be thought of as a category of 'algebras' over another category, in a very wide sense.

The next three chapters explore applications of category theory in Algebra, Topology and Algebraic Topology, Homological Algebra.

Chapter 7 is a brief introduction to higher dimensional category theory, with some of its applications. The starting point is the fact that 'small' categories, functors and natural transformations – with all their composition laws – form a 'two-dimensional category'. Another aspect of the power of category theory is that it is able to 'study itself'.

The second edition has four main additions, in three chapters devoted to the applications:

- Section 5.5, on singular homology,
- Section 5.6, on the smash product of pointed spaces, .
- Section 6.7, on extending homological algebra beyond pointed categories,
- Section 7.4, on local structures, from manifolds to fibre bundles.

The latter are presented as 'categories enriched on ordered categories of partial maps', in a way that blends Ehresmann's study of local structures with Lawvere's approach to enriched categories.

Several exercises and many solutions have been added. Most solutions are now deferred to a new Chapter 8, not to break the main exposition; but the solution still follows the exercise, when it is important for the sequel.

0.7 Classes of categories

It is a common feature of Mathematics to look for the natural framework where certain properties should be studied: for instance, the properties of ordinary polynomials with real coefficients are usually examined in general polynomial rings, or in more general algebras. Besides yielding more general results, the natural framework gives a clearer comprehension of what we are studying.

Category theory makes a further step in this sense. For instance, categories of modules are certainly important, but – since Buchsbaum's *Appendix* [Bu2] and Grothendieck's paper [Gt] in the 1950's – a consistent part of Homological Algebra finds its natural framework in *abelian categories* and their generalisations (see Chapter 6). Similarly, the categories of structured sets can be viewed as particular *concrete categories* (defined in 1.4.8), and the categories of equational algebras as particular *monadic categories* (see Sections 3.6 and 4.4).

In other words, when studying certain 'classes' of categories, we may (or perhaps had better) look for a *structural definition* including this class, rather than some general way of constructing the examples we want to study.

This leads again to considering categories of (small) categories, and to higher dimensional category theory.

0.8 Our approach

Notions will be presented in a concrete way, starting from examples taken from elementary mathematical theories. Then their theory is developed, with new examples and many exercises; the latter are generally endowed with a solution, or a partial solution, or suitable hints. Three chapters are devoted to applications.

We hope that a beginner can get, from these examples and applications, a concrete grasp of a theory which might risk of being quite abstract.

Many examples and applications are standard. But some of them may come out as unexpected, and hopefully intriguing, like those devoted to distributive lattices in 1.2.7, or to chains of adjunctions in 1.7.7 and 3.2.8, or to networks in 1.8.9.

The author's comments on some possibly unclear or controversial points are expressed:

- in 1.1.5, on the relationship between mathematical structures and categories,

- in 1.1.6 and 3.6.1, on varieties of algebras and artificial exclusions,

- in 1.5.9, 3.3.7 and 3.6.6, on favouring notions invariant up to categorical equivalence,

- in the introduction to Chapter 7, on the relationship between (strict and weak) 2-categories and double categories,

- in 7.3.3, about the 'basic view' of enrichment of a category over a monoidal category.

The foundational setting we use is based on standard set theory, assuming the existence of Grothendieck universes to formalise the 'smallness' conditions. This aspect, presented in 1.1.3, will be mostly left as understood.

0.9 Literature

For further study of general category theory there are excellent texts, like Mac Lane [M4], Borceux [Bo1, Bo2, Bo3], Adámek, Herrlich and Strecker [AHS], Freyd and Scedrov [FrS]. References for higher dimensional category theory will be given in Chapter 7.

A peculiar, conceptual introduction to category theory can be found in Lawvere and Schanuel [LwS].

Sheaf categories and toposes, which are not treated here, are presented in Mac Lane–Moerdijk [MaM] and [Bo3], and in more advanced texts like Johnstone's [Jo3, Jo4]. A categorical view of Set Theory can be found in Lawvere and Rosebrugh [LwR].

As already warned, the range of applications of category theory is much wider than what will be seen here, and can be presented in an elementary text.

For instance, we only explore a few categorical properties of topological vector spaces and Banach spaces, for which the reader is referred to Semadeni's book [Se]. The applications in Homological Algebra examined here follow a particular approach, discussed in Chapter 6, where other approaches are cited. Some old and new applications of categories to the theory of networks are briefly presented in 1.8.9. Morita equivalence is only mentioned, in 1.5.6.

The interactions between topology and category theory are further investigated in many articles and books, under the general heading of 'Categorical Topology'. Categories of fractions [GaZ, Bo3], introduced by Gabriel and Zisman, are important in homotopy theory; they open the way to the domain of derived categories and triangulated categories.

We do not examine the deep relationship among Category Theory, Logic and theoretical Computer Science, which is explored in texts like Makkai and Reyes [MakR], Lambek and Scott [LaS], Barr and Wells [BarW], Crole [Cr], nor the growing influence of categories and higher dimensional categories in parts of theoretical Physics, for which one can see Baez and Lauda [BaL].

Differential Geometry is studied in a 'synthetic' way in Kock [Ko], making formal use of infinitesimals. A new book by Bunge, Gago and San Luis [BunGS] extends this subject to Synthetic Differential Topology. Different forms of Galois Theory are explored in Borceux and Janelidze [BoJ].

The relationship of Category Theory with Algebraic Geometry is perhaps too complex to be simply referred to.

0.10 Notation and conventions

The symbol $\subset$ denotes *weak* inclusion.

As usual, the symbols $\mathbb{N}$, $\mathbb{Z}$, $\mathbb{Q}$, $\mathbb{R}$ and $\mathbb{C}$ denote the sets of natural, integral, rational, real and complex numbers; $\mathbb{N}^*$ is the subset of the positive integers.

Open and semi-open real intervals are denoted as $]a, b[$, $[a, b[$, and so on – a notation which distinguishes the open interval $]a, b[$ from the pair (a, b), as in Bourbaki's treatise. The standard euclidean sphere of dimension n is written as $\mathbb{S}^n$.

A singleton set can be written as $\{*\}$. The equivalence class of an element x, with respect to an assigned equivalence relation, is usually written as $[x]$. A bullet in a diagram stands for an object.

Categories, 2-categories and bicategories are generally denoted as $A, B \ldots$; strict or weak double categories as $\mathbb{A}, \mathbb{B} \ldots$

A part marked with * is out of the main line of exposition. It may refer to issues dealt with in the sequel, or be addressed to readers with some knowledge of the subject which is being discussed, or give references for higher topics.

0.11 Acknowledgements

Many points have been discussed with my colleagues and friends: in particular with Bob Paré and George Janelidze, during a long collaboration.

Diagrams and figures are composed with 'xy-pic', a free package by K.H. Rose and R. Moore.

I would also like to thank Dr Lim Swee Cheng, Ms Tan Rok Ting and Ms Soh Jing Wen, at World Scientific Publishing Co., for their kind, effective help in the publication of the two editions of this book. The covers are due to Ms Yi Ling and Mr Jimmy Low.

1

Categories, functors and natural transformations

Categories were introduced by Eilenberg and Mac Lane [EiM] in 1945, together with the other basic terms of category theory.

1.1 Categories

We start by considering concrete categories, associated with mathematical structures. But categories are not restricted to these instances, and the theory must be developed in a general way.

Given a mathematical discipline, it may not be obvious which category or categories are best suited for its study. This questionable point is discussed in 1.1.5, 1.1.6.

1.1.1 Some examples

Loosely speaking, before giving a precise definition, a category C consists of *objects* and *morphisms* together with a (partial) *composition law*: given two 'consecutive' morphisms $f \colon X \to Y$ and $g \colon Y \to Z$ we have a composed morphism $gf \colon X \to Z$. This partial operation is associative (whenever composition is legitimate) and every object X has an *identity*, written as $\mathrm{id}X \colon X \to X$ or 1_X, which acts as a unit for legitimate compositions.

The prime example is the category Set *of sets* (and mappings), where:

- an object is a set,

- the morphisms $f \colon X \to Y$ between two given sets X and Y are the (set-theoretical) mappings from X to Y,

- the composition law is the usual composition of mappings, where $(gf)(x) = g(f(x))$.

The following categories of structured sets and structure-preserving mappings (with the usual composition) will often be used and analysed:

- the category Top *of topological spaces* (and continuous mappings),

- the category Hsd *of Hausdorff spaces* (and continuous mappings),

- the category Gp *of groups* (and their homomorphisms),

- the category Ab *of abelian groups* (and homomorphisms),

- the category Mon *of monoids*, i.e. unital semigroups (and homomorphisms),

- the category Abm *of abelian monoids* (and homomorphisms),

- the category Rng *of rings*, understood to be associative and unital (and homomorphisms),

- the category CRng *of commutative rings* (and homomorphisms),

- the category R Mod *of left modules* on a fixed unital ring R (and homomorphisms),

- the category K Vct $(= K$ Mod$)$ *of vector spaces* on a commutative field K (and homomorphisms),

- the category RAlg *of unital algebras* on a fixed commutative unital ring R (and homomorphisms),

- the category Ord of *ordered sets* (and monotone mappings),

- the category pOrd of *preordered sets* (and monotone mappings),

- the category Set. *of pointed sets* (and pointed mappings),

- the category Top. *of pointed topological spaces* (and pointed continuous mappings),

- the category Ban of *Banach spaces and continuous linear mappings.*

- the category Ban$_1$ of *Banach spaces and linear weak contractions* (with norm $\leqslant 1$).

A homomorphism of a 'unital structure', like a monoid or a unital ring, is always assumed to preserve units.

For Set. we recall that a *pointed set* is a pair (X, x_0) consisting of a set X and a *base*-element $x_0 \in X$, while a pointed mapping $f \colon (X, x_0) \to (Y, y_0)$ is a mapping $f \colon X \to Y$ such that $f(x_0) = y_0$.

It may be convenient to write a pointed set as $(X, 0)$, viewed as a set equipped with a nullary operation $0 \colon \{*\} \to X$. Or simply as X, leaving the operation understood (as is generally done with algebraic structures).

Similarly, a *pointed topological space* (X, x_0) is a space with a base-point, and a pointed map $f \colon (X, x_0) \to (Y, y_0)$ is a continuous mapping from X to Y such that $f(x_0) = y_0$. The reader may know that the category Top. is important in Algebraic Topology: for instance, the fundamental group $\pi_1(X, x_0)$ is defined for a pointed topological space.

For the categories Ban and Ban$_1$ it is understood that we have chosen *either* the real *or* the complex field; when using both one can write $\mathbb{R}$Ban and $\mathbb{C}$Ban.

When a category is named after its objects alone (e.g. the 'category of groups'), this means that the morphisms are understood to be the obvious ones (in this case the homomorphisms of groups), with the obvious composition law. Of course, different categories with the same objects are given different names, like Ban and Ban$_1$ above.

1.1.2 Definition

A *category* C consists of the following data:

(a) a set Ob C, whose elements are called *objects* of C,

(b) for every pair X, Y of objects, a set $C(X, Y)$ (called a *hom-set*) whose elements are called *morphisms* (or *maps*, or *arrows*) of C from X to Y and denoted as $f: X \to Y$,

(c) for every triple X, Y, Z of objects of C, a mapping of *composition*

$$C(X, Y) \times C(Y, Z) \to C(X, Z), \qquad (f, g) \mapsto gf,$$

where gf is also written as $g.f$.

These data must satisfy the following axioms.

(i) (*Associativity*) Given three consecutive arrows, $f: X \to Y$, $g: Y \to Z$ and $h: Z \to W$, one has: $h(gf) = (hg)f$.

(ii) (*Identities*) Given an object X, there exists an endomap $e: X \to X$ which acts as an identity whenever composition makes sense; in other words if $f: X' \to X$ and $g: X \to X''$, one has: $ef = f$ and $ge = g$.

One shows, in the usual way, that e is determined by X; it is called the *identity* of X and written as 1_X or id$X$.

We generally assume that the following condition is also satisfied:

(iii) (*Separation*) For all X, X', Y, Y' objects of C, the sets $C(X, Y)$ and $C(X', Y')$ are disjoint, unless $X = X'$ and $Y = Y'$.

Therefore a map $f: X \to Y$ has a well-determined *domain* and *codomain*

$$\text{Dom}(f) = X, \qquad \text{Cod}(f) = Y.$$

Concretely, when constructing a category, one can forget about condition (iii), since one can always satisfy it by *redefining* a morphism $\hat{f}: X \to Y$ as a triple (X, Y, f) where f is a morphism from X to Y in the original sense (possibly not satisfying the Separation axiom).

Mor C denotes the set of all the morphisms of C, i.e. the disjoint union of its hom-sets. Two morphisms f, g are said to be *parallel* when they have the same domain and the same codomain.

If C is a category, the *opposite* (or *dual*) category, written as C^{op}, has the same objects as C, reversed arrows and reversed composition

$$C^{op}(X, Y) = C(Y, X),$$

$$g * f = fg, \qquad id^{op}(X) = idX. \tag{1.1}$$

Every topic of category theory has a dual instance, which comes from the opposite category (or categories). A dual notion is generally distinguished by the prefix 'co-'.

A set X can be viewed as a *discrete* category: its objects are the elements of X, and the only arrows are their (formal) identities; here $X^{op} = X$.

As usual in category theory, the term *graph* will be used to denote a simplified structure, with objects (or *vertices*) and morphisms (or arrows) $f : x \to y$, but no assigned composition nor identities. (This is called a *directed multigraph* in graph theory, or also a *quiver*.) A *morphism of graphs* preserves objects, arrows, domain and codomain. Every category has an *underlying graph*.

1.1.3 Small and large categories

We shall not insist on set-theoretical foundations. Yet some attention is necessary, to avoid involving 'the set of all sets', or requiring of a category properties of completeness that are 'too large for its size' (as we shall see in 2.2.3).

We assume the existence of a (Grothendieck) *universe* $\mathcal{U}$, which is fixed throughout; its axioms – recalled below, in 1.1.7 – say that we can perform inside it the usual operations of set theory. Its elements are called *small* sets (or $\mathcal{U}$-small sets, if necessary).

A category is understood to have objects and arrows belonging to this universe, and is said to be *small* if its set of morphisms belongs to $\mathcal{U}$, *large* if it does not (and is just a subset of $\mathcal{U}$). As a consequence, in a small category the set of objects (which is in bijective correspondence with the set of identities) also belongs to $\mathcal{U}$. A category C is said to *have small hom-sets* if all its sets $C(X, Y)$ are small; in this case C is small if and only if its set of objects is.

The categories of structured sets that we consider are generally large $\mathcal{U}$-categories, like the category Set of small sets (and mappings), or Top of small topological spaces (and continuous mappings), or Ab of small abelian groups (and homomorphisms); *in such cases, the term 'small' (referred to these structured sets) will be generally left as understood,* and we speak – as usual – of the 'category of sets', and so on.

*In fact one often needs a hierarchy of universes. For instance, Cat will denote the category of small categories and functors, introduced in 1.4.1. In order to view the (large) categories Set, Top, Ab, etc. in a similar structure we should assume the existence of a second universe $\mathcal{V}$, with $\mathcal{U} \in \mathcal{V}$, and use the category $\mathsf{Cat}_\mathcal{V}$ (also written as CAT) of $\mathcal{V}$-small categories. In a more complex situation one may need a longer chain of universes. Most of the time *these aspects will be left as understood.**

1.1.4 Isomorphisms and groupoids

In a category C a morphism $f\colon X \to Y$ is said to be *invertible*, or an *isomorphism*, if it has an inverse, i.e. a morphism $g\colon Y \to X$ such that $gf = 1_X$ and $fg = 1_Y$. The latter is uniquely determined; it is called the *inverse* of f and written as f^{-1}.

In the categories listed in 1.1.1 this definition gives the usual isomorphisms of the various structures – called 'homeomorphisms' in the case of topological spaces.

The *isomorphism relation* $X \cong Y$ between objects of C (meaning that there exists an isomorphism $X \to Y$) is an equivalence relation.

A *groupoid* is a category where every map is invertible; it is interesting to recall that this structure was introduced before categories, by H. Brandt in 1927 [Bra], if in a connected form.

The fundamental groupoid of a space X is an important structure, that contains all the fundamental groups $\pi_1(X, x)$ for $x \in X$. It will be reviewed in 5.2.9.

1.1.5 A digression on mathematical structures and categories

When studying a mathematical structure with the help of category theory, it is crucial to choose the 'right' kind of structure and the 'right' kind of morphisms, so that the result is sufficiently general and 'natural' to have good properties with respect to the goals of our study – even if we are interested in more particular situations.

(a) A first point to be verified is that the isomorphisms of the category (i.e. its invertible arrows) preserve the structure we are interested in, or we risk of studying something different from our purpose.

As a trivial example, the category T of topological spaces and *all* mappings between them has little to do with topology: an isomorphism of T is any bijection between topological spaces. Indeed T is 'equivalent' to the category of sets (as we shall see in 1.5.5), and is a 'deformed' way of looking at the latter.

Less trivially, the category M of metric spaces and continuous mappings misses crucial properties of metric spaces, since its invertible morphisms need not preserve completeness: e.g. the real line is homeomorphic to any non-empty open interval. In fact, M is equivalent to the category of *metrisable topological spaces* and continuous mappings (see 1.5.6(f)), and should be replaced with the latter.

An 'effective' category of metric spaces should be based on *Lipschitz* maps, or – more particularly – on weak contractions, so that its isomorphisms (bi-Lipschitz or isometric bijections, respectively) do preserve metric properties, like being complete or bounded (see 5.1.7).

(b) Other points will become clearer below. For instance, the category Top of topological spaces and continuous mappings is a classical framework for studying Topology. Among its good properties there is the fact that all 'categorical products' and 'categorical sums' (studied in Section 2.1, but already sketched in the Introduction) exist, and are computed *as in* Set, then equipped with a suitable topology determined by the structural maps.

More generally, this is true of all limits and colimits, and – as we shall see – is a consequence of the fact that the forgetful functor Top → Set has a left *and* a right adjoint, corresponding to discrete and chaotic topologies.

Hausdorff spaces are certainly important, but it is often better to view them *in* Top, as their category Hsd is less well behaved: colimits exist, but are not computed as in Set, and the simplest way to compute them – generally – is to take the colimit in Top and 'make it Hausdorff' (see 5.1.4(b)).

*(c) Many researchers would agree with Mac Lane, saying that even Top is not sufficiently good ([M4], Section VII.8), because it is not a cartesian closed category (see 5.1.1), and prefer – for instance – the category of compactly generated Hausdorff spaces (see 2.6.3(e)).

On the other hand, working in Homotopy Theory, one might be satisfied with the fact that the standard interval (with its cartesian powers) is exponentiable in Top, as we shall exploit in Section 5.2.

We shall also hint, in 1.6.7, at another approach called 'pointless topology', which is based on the category of locales and is favoured in topos theory.

(d) Finally we remark that artificial exclusions 'most of the time' give categories of poor properties, like the category of *non-abelian groups*, or *non-empty semigroups*. A category is a whole, and taking something out can destroy its properties and 'shape', like taking out an element of a group, or a point of a sphere.

Other comments in this sense can be found in 3.6.1; but there is also a comment *in the opposite sense*, in 5.3.8, for categories used in an 'auxiliary way'. Non-empty algebras need some further comment.

1.1.6 *Variety of algebras and* horror vacui

In Universal Algebra, a 'variety of algebras' includes all the algebraic structures of a given signature (i.e. with a certain family of operations), that satisfy a given set of equational axioms (or universally quantified identities): e.g. all groups, or all rings; but not all fields, because multiplicative inverses only exist for non-zero elements, and cannot be given by a 'general' unary operation satisfying some universal identities.

Here a *variety of algebras* will mean a category of objects defined in this way, with their homomorphisms (as made precise in Section 4.3).

We do not follow the convention that the underlying set should be non-empty, as commonly assumed in Universal Algebra – with a few exceptions like Cohn's book [Coh]. This principle has negative consequences for any theory without 0-ary operations (the *constants*), like semigroups: for instance, subsemigroups would not be closed under intersection.

For a reader with some knowledge of categorical limits, dealt with in Chapter 2, we can add that a variety of algebras, in the present sense, has all limits and colimits, while the category of non-empty semigroups lacks certain limits (like equalisers and pullbacks) and certain colimits (as an initial object), precisely because we have artificially taken out a solution. The monadicity property would also fail: see 3.6.1 and Section 4.3.

Generally speaking, the results of the traditional setting can be easily translated into the present one; the converse can be harder.

1.1.7 **Grothendieck universes*

For the interested reader, we recall the definition of a *universe* as given in [M4], Section I.6. It is a set $\mathcal{U}$ satisfying the following (redundant) properties:

(i) $x \in u \in \mathcal{U}$ implies $x \in \mathcal{U}$,

(ii) $u, v \in \mathcal{U}$ implies that $\{u, v\}$, (u, v) and $u \times v$ belong to $\mathcal{U}$,

(iii) $x \in \mathcal{U}$ implies that $\mathcal{P}x$ and $\bigcup x$ belong to $\mathcal{U}$,

(iv) the set $\mathbb{N}$ of finite ordinals belongs to $\mathcal{U}$,

(v) if $f \colon x \to y$ is a surjective mapping, $x \in \mathcal{U}$ and $y \subset \mathcal{U}$, then $y \in \mathcal{U}$.

Here $\mathcal{P}x$ is the set of subsets of x and $\bigcup x = \{y \mid y \in z \text{ for some } z \in x\}$.

1.2 Monoids and preordered sets as categories

Monoids and preordered sets can be viewed as categories of two simple kinds, providing intuition and models for some aspects of category theory.

We also review here some basic facts about the theory of lattices, to be used later on. A reader interested in this beautiful domain will find pleasure in browsing, or studying, the classical texts of G. Birkhoff and G. Grätzer [Bi, Gr2].

1.2.1 Monoids and categories

As we have seen, monoids (i.e. semigroups with unit) and their homomorphisms form a category, which we write as Mon. But we deal here with a different aspect.

A single monoid M can (and will often) be viewed as a category with one formal object $*$. The morphisms $x\colon * \to *$ are the elements of M, composed by the multiplication xy of the monoid, with identity $\mathrm{id}(*) = 1$, the unit of the monoid. If M is a group, the associated category is a groupoid.

On the other hand, in every category C, the *endomorphisms* $X \to X$ of any object form a (possibly large) monoid, under the composition law

$$\mathrm{End}(X) = \mathsf{C}(X, X), \qquad (1.2)$$

and the invertible ones form the group $\mathrm{Aut}(X)$ of *automorphisms* of X.

In this way a monoid is essentially the same as a category on a single object, while a category can be thought to be a 'many-object generalisation' of a monoid. Groups and groupoids have a similar relationship.

The theory of regular, orthodox and inverse semigroups (see [ClP, Ho, Law]) has a strong interplay with the categories of relations and their applications in Homological Algebra, which is investigated in [G9].

1.2.2 Preordered and ordered sets

We use the following terminology for orderings. A *preorder* relation $x \prec x'$ is reflexive and transitive. An *order* relation, generally written as $x \leqslant x'$, is also anti-symmetric: if $x \leqslant x' \leqslant x$ then $x = x'$.

The category of *ordered sets and monotone mappings* (the order preserving ones, also called *increasing* mappings) will be written as Ord, while we write as pOrd the category of *preordered sets and monotone mappings*.

An order relation is said to be *total* if for all x, x' we have $x \leqslant x'$ or $x' \leqslant x$.

(An ordered set is often called a '*partially* ordered set', abbreviated to 'poset', to recall that the ordering is not assumed to be total. Accordingly, the reader can find the notation Pos for the category Ord.)

It will be important to note that every hom-set $\mathsf{Ord}(X, Y)$ is canonically ordered by the *pointwise order* relation, defined as follows for $f, g \colon X \to Y$

$$f \leqslant g \quad \text{if for all } x \in X \text{ we have } f(x) \leqslant g(x) \text{ in } Y. \tag{1.3}$$

Similarly, every hom-set $\mathsf{pOrd}(X, Y)$ has a canonical preorder $f \prec g$.

A preordered set X has an associated equivalence relation $x \sim x'$ defined by the conjunction: $x \prec x'$ and $x' \prec x$. The quotient $X/\sim$ has an induced order: $[x] \leqslant [x']$ if $x \prec x'$.

If X is a preordered set, X^{op} is the opposite one (with reversed preorder), or dual one. If $a \in X$, the symbols $\downarrow a$ and $\uparrow a$ denote the downward or upward closed subsets of X generated by the element a

$$\downarrow a = \{x \in X \mid x \prec a\}, \qquad \uparrow a = \{x \in X \mid a \prec x\}. \tag{1.4}$$

In a preordered set X, the *infimum* (or *meet*) of a subset A, written as $\inf_X A$ or $\wedge A$, is the greatest lower bound of A, and is determined up to the associated equivalence relation (provided it exists). Dually, the *supremum* (or *join*) $\sup_X A = \vee A$ is the infimum of A in X^{op}. In an ordered set these results are uniquely determined – when they exist.

1.2.3 Preorders and categories

A preordered set X will often be viewed as a category, where the objects are the elements of X and the set $X(x, x')$ contains precisely one arrow if $x \prec x'$, which can be written as $(x, x') \colon x \to x'$, and no arrows otherwise. Composition and units are (necessarily) as follows

$$(x', x'').(x, x') = (x, x''), \qquad \mathrm{id}(x) = (x, x).$$

In this way a preordered set is essentially the same as a category where each hom-set has at most one element. All diagrams in these categories commute. Two elements x, x' are isomorphic objects if and only if $x \sim x'$.

In particular, each ordinal defines a category, written as $\mathbf{0}, \mathbf{1}, \mathbf{2}, \ldots$:

- $\mathbf{0}$ is the empty category,

- $\mathbf{1}$ is the *singleton category*, i.e. the discrete category on the object 0,

- $\mathbf{2}$ is the *arrow category*, with two objects, 0 and 1, and one non-identity arrow, $0 \to 1$.

Here the ordinal $\mathbf{2}$ should not be confused with the cardinal $2 = \{0, 1\}$, which is viewed as a discrete category.

1.2.4 Lattices

Classically a lattice is defined as an ordered set X where every pair x, x' of elements has a *join* $x \vee x' = \sup\{x, x'\}$ (the least element of X greater than both) and a *meet* $x \wedge x' = \inf\{x, x'\}$ (the greatest element of X smaller than both).

Here we follow a different convention, usual in category theory: *lattice* will always mean an ordered set with *finite* joins and meets; this amounts to the existence of binary joins and meets *together with* the least element $0 = \vee\emptyset$ and the greatest element $1 = \wedge\emptyset$. (This structure is called a 'bounded lattice' in Lattice Theory.)

The bounds 0 and 1 (the empty join and the empty meet) are equal in the one-point lattice $\{*\}$, and only there.

Consistently with this terminology, a *lattice homomorphism* has to preserve finite joins and meets; a *sublattice* of a lattice X is closed under such operations (and has the same bounds as X). The category of lattices and homomorphisms will be written as Lth.

Occasionally we speak of a *quasi lattice* when we only assume the existence of binary joins and meets; a homomorphism of quasi lattices only has to preserve them. A *quasi sublattice* Y of a quasi lattice X is closed under binary joins and meets in X; when X is a lattice, Y may have different bounds, or lack some of them.

For instance, if X is a lattice and $a \in X$, the downward and upward closed subsets $\downarrow a$, $\uparrow a$ of X generated by a (see (1.4)) are quasi sublattices of X, and lattices in their own right.

Let us note that the *free lattice* (see 2.7.3) generated by an element x *has three elements*: $0 < x < 1$, while the *free quasi lattice* L generated by x is just the singleton $\{x\}$: in fact every mapping $\{x\} \to X$ with values in a lattice (resp. in a quasi lattice) has a unique extension to a homomorphism $L \to X$ (resp. $\{x\} \to X$).

In the same way, the free lattice L generated by a set S can be obtained from the corresponding free quasi-lattice M by adding a (new) minimum and a (new) maximum, even when M is *already* bounded – as above.

1.2.5 Exercises and complements

The reader may know, or be interested to prove, the following results. Solutions can be found in Chapter 8.

(a) (*Lattices as equational algebras*) A lattice can be equivalently presented as a set X equipped with two binary operations $x \vee y$ and $x \wedge y$ (called *join* and *meet*) and two constants 0, 1 (or 0-ary operations), so that the following axioms are satisfied, for all $x, y, z \in X$:

(L.1) (*Associativity*) $x \vee (y \vee z) = (x \vee y) \vee z, \quad x \wedge (y \wedge z) = (x \wedge y) \wedge z,$

(L.2) (*Commutativity*) $x \vee y = y \vee x, \quad x \wedge y = y \wedge x,$

(L.3) (*Idempotence*) $x \vee x = x = x \wedge x,$

(L.4) (*Absorption*) $x \vee (x \wedge y) = x = x \wedge (x \vee y),$

(L.5) (*Identities*) $x \vee 0 = x = x \wedge 1.$

In this presentation one defines the ordering letting $x \leqslant y$ if $x \vee y = y$, or equivalently $x \wedge y = x$. The opposite lattice X^{op} is now defined by interchanging join and meet.

Again, we should be aware that $0 = 1$ is not excluded; we already know that in this case all the elements coincide, and the 'two' operations coincide as well.

(A careful reader might also observe that we are distinguishing a 'first operation' (say the join), or we cannot fix the associated order.)

(b) (*Complete lattices*) A preordered set has all meets (of its subsets) if and only if it has all joins.

An ordered set with all meets and joins is called a *complete lattice*. This agrees with the completeness of categories, as we shall see in 2.2.5(e).

*(c) Frames are particular complete lattices, related to topological spaces and 'pointless topology'; they will be briefly introduced in 1.6.7.

1.2.6 Distributive and modular lattices

A lattice is said to be *distributive* if the meet operation distributes over the join operation

(D) $(x \vee y) \wedge z = (x \wedge z) \vee (y \wedge z),$

or equivalently if the join distributes over the meet. In fact, if we assume that meets distribute over joins, we have:

$$(x \vee y) \wedge (x \vee z) = ((x \vee y) \wedge x) \vee ((x \vee y) \wedge z)$$
$$= x \vee (x \wedge z) \vee (y \wedge z) = x \vee (y \wedge z).$$

A *boolean algebra* is a distributive (bounded) lattice where every element x has a (necessarily unique) *complement* x', defined by the properties:

$$x \wedge x' = 0, \qquad x \vee x' = 1.$$

The subsets of a set X form the classical boolean algebra $\mathcal{P}X$, which is a complete lattice.

The (complete) lattice $\mathrm{Sub}A$ of subgroups of an abelian group (or submodules of a module) is not distributive, generally (see 1.2.7(d)); but one can easily check that it always satisfies a weaker, restricted form of distributivity, called modularity.

Namely, a lattice is said to be *modular* if it satisfies the following selfdual property (for all elements x, y, z)

(M)   if $x \leqslant z$ then $(x \vee y) \wedge z = x \vee (y \wedge z)$.

The category of modular (resp. distributive) lattices and their homomorphisms will be written as Mlh (resp. Dlh).

By Birkhoff's representation theorem ([Bi], III.5, Theorem 5) the free distributive lattice on n generators is finite and isomorphic to a lattice of subsets. The reader may also be interested to know that the free modular lattice on three elements is finite and (obviously!) not distributive ([Bi], III.6, Figure 10), while four generators already give an infinite free modular lattice (see the final Remark in [Bi], III.6).

1.2.7 *Exercises and complements* (Distributive lattices)

The goal of this point is to show that the (complete) lattice $X = \mathrm{Sub}(\mathbb{Z})$ of subgroups of the abelian group $\mathbb{Z}$ (or ideals of the ring $\mathbb{Z}$) is distributive – a fact that will have unexpected links with Homological Algebra, in Section 6.6.

The interested reader is invited to directly investigate the problem, before considering the layout given below. The first step is showing that X is anti-isomorphic to the divisibility lattice of natural numbers (a well-known, nearly obvious point). Then one proves that the latter is distributive, a (hopefully amusing) exercise based on our school knowledge – prime factorisation.

(a) The reader likely knows, or should prove, that each subgroup of $\mathbb{Z}$ is of the form $n\mathbb{Z}$, for a unique $n \in \mathbb{N}$ (and is an ideal of the ring of integers).

This gives us an isomorphism $X \to Y^{\mathrm{op}}$, where $Y = (\mathbb{N}, |)$ is the set of natural numbers ordered by the divisibility relation $m|n$. Therefore Y is a (complete) lattice as well, and we recognise its operations $m \vee n$ and $m \wedge n$ as fairly well-known. We also note that $1 = \min Y$ and $0 = \max Y$.

(b) Now, each $n \in \mathbb{N}^*$ has a unique decomposition $n = \Pi_p \, p^{n_p}$ as a product of powers of prime numbers p; of course the natural exponents n_p are quasi-null (i.e. all of them are 0, out of a finite number of prime indices p), so that the factorisation is essentially finite.

The reader will use this fact to prove that Y can be embedded in the cartesian product $\Pi_p \, \overline{\mathbb{N}}$ of countably many copies of the set $\overline{\mathbb{N}} = \mathbb{N} \cup \{\infty\}$, with the natural order. One can now show that Y is distributive, and X as well.

(c) As a consequence, the lattice $\mathrm{Sub}(A)$ of any cyclic group A is also distributive.

(d) The reader can easily show that this property fails for the abelian group $\mathbb{Z}^2$, or for any non-trivial abelian group $A \oplus A$. (Again, this will be of use in Section 6.6.)

On the other hand, the cyclic group $\mathbb{Z}/2 \oplus \mathbb{Z}/3 \cong \mathbb{Z}/6$ has a distributive lattice of subgroups.

(e) A reader acquainted with *principal ideal domains* may like to rethink the whole thing, for their modules.

1.3 Monomorphisms and epimorphisms

In a category, monomorphisms and epimorphisms (monos and epis for short) are defined by cancellation properties with respect to composition.

For categories of structured sets, they represent an 'approximation' to the injective and surjective mappings of the category.

1.3.1 Main definitions

In a category C a morphism $f \colon X \to Y$ is said to be a *monomorphism*, or *mono*, if it satisfies the following cancellation property:

- for every pair of maps $u, v \colon X' \to X$ (defined on an arbitrary object X') such that $fu = fv$, one has $u = v$ (see the left diagram below)

$$X' \underset{v}{\overset{u}{\rightrightarrows}} X \xrightarrow{\; f \;} Y \qquad\qquad X \xrightarrow{\; f \;} Y \underset{v}{\overset{u}{\rightrightarrows}} Y'. \qquad (1.5)$$

The morphism $f \colon X \to Y$ is said to be an *epimorphism*, or *epi*, if it satisfies the dual cancellation property:

- for every pair of maps $u, v \colon Y \to Y'$ such that $uf = vf$, one has $u = v$ (see the right diagram above).

An arrow $\rightarrowtail$ will always denote a monomorphism, while $\twoheadrightarrow$ stands for an epimorphism.

Every isomorphism is mono and epi. A category is said to be *balanced* if the converse holds: every morphism which is mono and epi is invertible.

Suppose now that we have, in a category C, two maps $m \colon A \to X$ and $p \colon X \to A$ such that $pm = \mathrm{id}A$. It follows that m is a monomorphism (called a *section*, or a *split monomorphism*), while p is an epimorphism (called a *retraction*, or a *split epi*); A is said to be a *retract* of X.

A family of morphisms $f_i \colon X \to Y_i$ $(i \in I)$ with the same domain is said to be *jointly mono* if for every pair of maps $u, v \colon X' \to X$ such that $f_i u = f_i v$ (for all indices i) one has $u = v$

$$X' \underset{v}{\overset{u}{\rightrightarrows}} X \overset{f_i}{\to} Y_i \qquad X_i \overset{f_i}{\to} Y \underset{v}{\overset{u}{\rightrightarrows}} Y'. \qquad (1.6)$$

Dually a family $f_i \colon X_i \to Y$ is *jointly epi* if for all $u, v \colon Y \to Y'$ such that $u f_i = v f_i$ (for all i) one has $u = v$.

The general properties of monos, epis and retracts will be examined in 1.3.6. Related notions, like regular monos and epis, strong monos and epis, subobjects and quotients, will be seen in the next chapter.

1.3.2 Comments

In a category of structured sets and structure-preserving mappings, an injective mapping (of the category) is obviously a monomorphism, while a surjective one is an epimorphism. The converse may require a non-trivial proof, or even fail. This can only be understood by working out the examples below.

Interestingly, a divergence appears between monos and epis: the theory of categories is selfdual, but our frameworks of structured sets are not! When we classify monos in Set, this tells us everything about the epis of Set$^\text{op}$ but nothing about the epis of Set.

In fact, in all the examples below it will be easy to prove that the monomorphisms coincide with the injective morphisms. Later we shall see, in 2.7.4(d), some conditions that ensure this fact, and hold true in all the 'usual categories of structured sets'.

On the other hand, various problems occur with epimorphisms: classifying them in various categories of algebraic structures leads to difficult problems with no elementary solution, and no real need of it: we shall see (in Section 4.4) that, in these categories, the 'regular epimorphisms' (namely the surjective homomorphisms) are more important.

1.3.3 Exercises and complements, I

A solution of these exercises is given below, but a beginner should try to give an independent one; this is easy for monos, less easy for epis.

(a) In Set a mono is the same as an injective mapping; an epi is the same as a surjective mapping. The category is balanced.

(b) In Top and Ab monos and epis coincide with the injective and surjective mappings of the category, respectively. Ab is balanced, and Top is not.

(c) In the categories Mon of monoids and Rng of rings, the monomorphisms coincide again with the injective homomorphisms. The inclusion $\mathbb{N} \to \mathbb{Z}$ (of additive monoids) is both mono and epi in Mon, which is not balanced. The same holds for the inclusion $\mathbb{Z} \to \mathbb{Q}$ in Rng.

Epimorphisms in Mon and Rng have no elementary characterisation.

(d) In a preordered set, viewed as a category, all arrows are mono and epi. The category is balanced under a precise condition on the preordering.

Solutions. (a) If $f\colon X \to Y$ is a monomorphism in Set, let us suppose that $f(x) = f(x')$, for $x, x' \in X$. We consider the mappings u, v defined on the singleton

$$\{*\} \overset{u}{\underset{v}{\rightrightarrows}} X \overset{f}{\longrightarrow} Y \qquad\qquad u(*) = x, \quad v(*) = x'. \qquad (1.7)$$

Now we have $fu = fv$, whence $u = v$ and $x = x'$, which shows that f is injective. Note that the proof works simulating an element of X with a map $\{*\} \to X$.

On the other hand, if $f\colon X \to Y$ is an epimorphism in Set, we define two mappings u, v with values in the set $\{0, 1\}$

$$X \overset{f}{\longrightarrow} Y \overset{u}{\underset{v}{\rightrightarrows}} \{0, 1\} \qquad\qquad (1.8)$$

where u is the characteristic function of the subset $f(X) \subset Y$ (with $u(y) = 1$ if and only if $y \in f(X)$) and v is the constant map $v(y) = 1$. Then $uf = vf$, whence $u = v$ and $f(X) = Y$.

(A different proof can be based on the set $Y \times \{0, 1\}$, the disjoint union of two copies of Y; or a convenient quotient of the latter.)

Since the invertible morphisms in Set are obviously the same as the bijective mappings, the category Set is balanced.

(b) For monomorphisms the proof is similar to the previous one, making use of the singleton in Top and of the group $\mathbb{Z}$ in Ab. Note that the latter allows us to simulate an element $x \in X$ by a homomorphism $u\colon \mathbb{Z} \to X$, sending the generator 1 to x.

In Top, to prove that an epi is surjective we can proceed as in (1.8), using a two-point indiscrete (or chaotic) space. In Ab we can use the quotient group $Y/f(X)$

$$X \overset{f}{\longrightarrow} Y \overset{p}{\underset{0}{\rightrightarrows}} Y/f(X) \qquad\qquad (1.9)$$

with the canonical projection p and the zero-homomorphism. We are now following a different pattern: constructing arrows is fairly free in Set, somewhat less in Top, much less in categories of algebraic structures.

We conclude here that Ab is balanced, while Top is not: a bijective continuous mapping need not be invertible in Top, i.e. a homeomorphism.

(c) For monos we proceed again as in (a), making use for Mon of the additive monoid $\mathbb{N}$ (freely generated by 1), and for Rng of the polynomial ring $\mathbb{Z}[X]$ (freely generated by the element X).

The inclusion $\mathbb{N} \to \mathbb{Z}$, of additive monoids, is injective and mono; it is also epi in Mon, because a homomorphism $f \colon \mathbb{Z} \to M$ of monoids (in additive notation) is determined by its values on $\mathbb{N}$: if $n > 0$, then

$$f(n) + f(-n) = 0_M = f(-n) + f(n),$$

whence $f(-n)$ is the additive inverse of $f(n)$ in M.

Similarly one proves that the inclusion $\mathbb{Z} \to \mathbb{Q}$ is an epimorphism in Rng: for a homomorphism $f \colon \mathbb{Q} \to R$ of unital rings, $f(1/k)$ must be inverse to $f(k)$ in R.

(d) The first assertion is obvious. Saying that the category X is balanced means here that every arrow is an isomorphism: in other words, the pre-order of X is symmetric, i.e. an equivalence relation.

1.3.4 *Exercises and complements, II*

(a) In the categories pOrd and Ord monos and epis coincide with the injective and surjective mappings of the category, respectively. These categories are not balanced.

(b) In the category Hsd of Hausdorff spaces, monomorphisms coincide with the injective continuous mappings.

Every map whose image is dense (in its codomain) is an epimorphism. We shall see in 2.3.6(b) that this condition is also necessary.

(c) In the category Gp of groups all monomorphisms are injective.

It is also true that all epimorphisms are surjective: a non-obvious fact, whose proof can be found in [M4], Section I.5, Exercise 5.

(d) Prove that the monomorphisms of Ban and Ban$_1$ coincide with their injective morphisms.

1.3.5 *Exercises and complements* (Retracts)

For the sake of simplicity, we replace a monomorphism $m \colon A \rightarrowtail X$, in a category of structured sets, by its image $m(A) \subset X$.

(a) In Set a retract of a set $X \neq \emptyset$ is any non-empty subset.

(b) In Ab retracts coincide with direct summands. More precisely, a subgroup A of an abelian group X is a retract if and only if there exists a subgroup $B \subset X$ such that $A \cap B = \{0\}$ and $A + B = X$; then X is canonically isomorphic to the direct sum $A \oplus B$.

For a reader acquainted with elementary Homological Algebra, this condition amounts to saying that the short exact sequence $A \rightarrowtail X \twoheadrightarrow X/A$ splits.

(c) In Top, the sphere $\mathbb{S}^n$ is a retract of the 'pierced space' $X = \mathbb{R}^{n+1} \setminus \{0\}$, as a subspace of $\mathbb{R}^{n+1}$.

(d) There is no elementary characterisation of retracts in Top. To prove that a subspace $A \subset X$ is a retract one 'simply' has to construct a retraction; but proving that it is not a retract can be an important, non-trivial result.

For instance, saying that the subspace $\mathbb{S}^0 = \{-1, 1\}$ of the euclidean line $\mathbb{R}$ is not a retract is a way of stating the Intermediate Value Theorem. More generally, saying that the sphere $\mathbb{S}^n$ is not a retract of $\mathbb{R}^{n+1}$ is intuitively clear (at least in low dimension), but not easy to prove. These problems are generally studied with the tools of Algebraic Topology: see 1.4.7 and 5.5.8(d).

1.3.6 General properties of monos and epis

The reader can (easily) prove the following elementary properties, that will be used throughout without reference.

We have two consecutive maps $f \colon X \to Y$ and $g \colon Y \to Z$ in a category. Property (a) is dual to (a*), and we only need to prove one of them.

(a) If f and g are mono, gf is also; if gf is mono, f is also.

(a*) If f and g are epi, gf is also; if gf is epi, g is also.

(b) If f and g are split mono, gf is also; if gf is a split mono, f is also.

(b*) If f and g are split epi, gf is also; if gf is a split epi, g is also.

(c) If f is a split mono and an epi, then it is invertible.

(c*) If f is a split epi and a mono, then it is invertible.

1.4 Functors

The structure-preserving mappings between categories are called 'functors'.

1.4.1 Functors and isomorphisms of categories

A (covariant) *functor* $F \colon \mathsf{C} \to \mathsf{D}$ consists of the following data:

(a) a mapping

$$F_0\colon \mathrm{Ob}\,\mathsf{C} \to \mathrm{Ob}\,\mathsf{D},$$

whose action is generally written as $X \mapsto F(X)$,

(b) for every pair of objects X, X' in C, a mapping

$$F_{XX'}\colon \mathsf{C}(X, X') \to \mathsf{D}(F(X), F(X')),$$

whose action is generally written as $f \mapsto F(f)$.

Composition and identities must be preserved. In other words:

(i) if f, g are consecutive maps of C then $F(gf) = F(g).F(f)$,

(ii) if X is an object of C then $F(\mathrm{id}X) = \mathrm{id}(F(X))$.

Given a second functor $G\colon \mathsf{D} \to \mathsf{E}$, one defines in the obvious way the *composed* functor $GF\colon \mathsf{C} \to \mathsf{E}$. This composition is associative and has identities: the *identity functor* of each category

$$\mathrm{id}\mathsf{C}\colon \mathsf{C} \to \mathsf{C}, \qquad\qquad X \mapsto X, \quad f \mapsto f. \tag{1.10}$$

An *isomorphism of categories* is a functor $F\colon \mathsf{C} \to \mathsf{D}$ which is invertible, i.e. admits an *inverse* $G\colon \mathsf{D} \to \mathsf{C}$; this means a functor such that $GF = \mathrm{id}\mathsf{C}$ and $FG = \mathrm{id}\mathsf{D}$. It is easy to verify that the functor F is an isomorphism if and only if all the mappings F_0 and $F_{XX'}$ considered above are bijective.

Being isomorphic categories is an equivalence relation, written as $\mathsf{C} \cong \mathsf{D}$.

A *contravariant functor* $F\colon \mathsf{C} \dashrightarrow \mathsf{D}$ is a covariant functor $\mathsf{C}^{\mathrm{op}} \to \mathsf{D}$, and takes a morphism $f\colon X \to X'$ in C to a morphism $F(f)\colon F(X') \to F(X)$ in D, preserving identities and reversing compositions

$$F(gf) = F(f).F(g),$$

with respect to the compositions of C and D, of course. By a common abuse of terminology, it can be called a *contravariant functor* $F\colon \mathsf{C}^{\mathrm{op}} \to \mathsf{D}$, a way of recalling that we are interested in C rather than in C^{op}.

Cat will denote the category of small categories and their functors. (Its 2-dimensional structure, including the natural transformations, will be examined in 1.5.3 and Section 7.1.)

1.4.2 Examples and comments

(a) A functor between two monoids, viewed as categories (see 1.2.1), is the same as a homomorphism of monoids.

(b) A functor between two preordered sets, viewed as categories (see 1.2.3), is the same as a monotone function, i.e. a preorder-preserving mapping.

(c) A functor $G \to K\,\mathsf{Vct}$ defined on a group (viewed as a category) amounts to a *representation* of G, that is a homomorphism $G \to \mathrm{Aut}(X)$ with values in the group of automorphisms of a K-vector space.

(d) Categories linked by an obvious isomorphism are often perceived as 'the same thing'.

For instance, an abelian group has a unique structure of (unitary) module on the ring $\mathbb{Z}$ of integers (where $2x = x + x$, and so on), preserved by every homomorphism of abelian groups. This fact readily shows that the category Ab is canonically isomorphic to the category $\mathbb{Z}\,\mathsf{Mod}$ of modules on the ring of integers; one generally makes no distinction between these categories.

The various equivalent ways of defining topological spaces give rise to isomorphic categories, that is rarely convenient to distinguish; the same holds for the equivalent definitions of lattices, recalled in 1.2.4 and 1.2.5.

1.4.3 Subcategories

Let C be a category. A *subcategory* D is defined by assigning:

- a subset $\mathrm{Ob}\,\mathsf{D} \subset \mathrm{Ob}\,\mathsf{C}$, whose elements are called *objects of* D,

- for every pair of objects X, Y of D, a subset $\mathsf{D}(X, Y) \subset \mathsf{C}(X, Y)$, whose elements are called *morphisms of* D, from X to Y,

so that the following conditions hold:

(i) for every pair of consecutive morphisms of D their composite in C belongs to D,

(ii) for every object of D its identity in C belongs to D.

D is thus a category, with the induced composition law. It comes with an *inclusion* functor $U \colon \mathsf{D} \to \mathsf{C}$, also written as $\mathsf{D} \subset \mathsf{C}$.

One says that D is a *full* subcategory of C if, for every pair of objects X, Y of D, we have $\mathsf{D}(X, Y) = \mathsf{C}(X, Y)$, so that D is determined by assigning the subset of its objects.

One says that D is a *wide* subcategory of C if it has the same objects, so that D is determined by assigning the subset of its maps (closed under composition and containing all identities).

For instance, Ab is a full subcategory of Gp, while Ban_1 is a wide subcategory of Ban. Of course a full and wide subcategory must be the total one.

1.4.4 Forgetful and amnestic functors

Forgetting structure, or part of it, yields various examples of functors between categories of structured sets, like the following obvious instances

$$\mathsf{Top} \to \mathsf{Set}, \qquad\qquad \mathsf{Ord} \to \mathsf{Set},$$
$$\mathsf{Rng} \to \mathsf{Ab} \to \mathsf{Set}, \qquad \mathsf{Rng} \to \mathsf{Mon} \to \mathsf{Set}. \tag{1.11}$$

These are called *forgetful* functors, and often denoted by the letter U, which refers to the *underlying* set, or *underlying* abelian group, and so on. It is also common to write $U(X)$ as $|X|$.

The term forgetful functor can also refer to the inclusion functor of a subcategory, which *forgets properties* rather than structure, like

$$\mathsf{Ab} \to \mathsf{Gp} \to \mathsf{Mon}, \qquad \mathsf{Ord} \to \mathsf{pOrd}, \tag{1.12}$$

or to functors that forget (a part of) both, like the functor $\mathsf{CRng} \to \mathsf{Mon}$ that sends a commutative ring to its underlying multiplicative monoid (forgetting the additive structure and the commutative property of the multiplication).

A functor $F\colon \mathsf{C} \to \mathsf{D}$ is said to be *amnestic* if, for every isomorphism f in C, $F(f)$ is an identity in D (if and) only if f is.

All the forgetful functors above are amnestic.

1.4.5 Quotients of categories

A *congruence* $R = (R_{XY})$ in a category C consists of a family of equivalence relations R_{XY} in each set of morphisms $\mathsf{C}(X, Y)$, that is consistent with composition:

(i) if $f\,R_{XY}\,f'$ and $g\,R_{YZ}\,g'$, then $gf\,R_{XZ}\,g'f'$.

Equivalently (by transitivity) it is sufficient to verify that:

(ii) if $f\,R\,f'$, then $fh\,R\,f'h$ and $kf\,R\,kf'$ (for all legitimate compositions).

The *quotient category* $\mathsf{D} = \mathsf{C}/R$ has the same objects of C and

$$\mathsf{D}(X, Y) \;=\; \mathsf{C}(X, Y)/R_{XY}.$$

In other words, a morphism $[f]\colon X \to Y$ in D is an equivalence class of morphisms $X \to Y$ in C. The composition is induced by that of C, which is legitimate because of condition (i):

$$[g].[f] = [gf]. \tag{1.13}$$

The *projection functor* $P\colon \mathsf{C} \to \mathsf{C}/R$ is the identity on objects and sends a morphism f to its equivalence class $[f]$.

In Top the homotopy relation $f \simeq f'$ is a congruence of categories; the quotient category

$$\mathsf{hoTop} = \mathsf{Top}/\simeq$$

is called the *homotopy category of topological spaces*, and is important in Algebraic Topology. Plainly, a continuous mapping $f \colon X \to Y$ is a homotopy equivalence (i.e. there exists a continuous mapping $g \colon Y \to X$ such that $gf \simeq \mathrm{id}X$ and $fg \simeq \mathrm{id}Y$) if and only if its homotopy class $[f]$ is an isomorphism of hoTop. All this will be examined in a more detailed way in Section 5.2.

*The *weak homotopy category* HoTop is obtained by a more complex procedure, formally inverting all the continuous mappings that – loosely speaking – induce isomorphisms in all homotopy sets and groups [GaZ]. It can also be realised as a full subcategory of hoTop.*

Every category C has a finest congruence $f = g$, whose quotient is trivially isomorphic to C, and a coarsest congruence $f \sim g$, meaning that f and g are parallel arrows. In the second case, the quotient category is the (possibly large) set Ob C preordered by the relation $X \prec Y$ whenever there is some arrow $X \to Y$ in C; it will be called the *preordered set associated to* C (in a universal way, as we shall see in Exercise 3.1.4(f)), and denoted as po(C).

Let us note that the relation of isomorphism is wider in a quotient category, but may happen to coincide.

1.4.6 Products of categories

If C and D are categories, one defines the *product category* C × D.

An object is a pair (X, Y) where X is in C and Y in D. A morphism is a pair of morphisms

$$(f, g) \colon (X, Y) \to (X', Y'), \tag{1.14}$$

for $f \in \mathsf{C}(X, X')$ and $g \in \mathsf{D}(Y, Y')$.

The composition of (f, g) with $(f', g') \colon (X', Y') \to (X'', Y'')$ is (obviously) component-wise:

$$(f', g').(f, g) = (f'f, g'g).$$

Similarly one defines the product $\mathsf{C} = \Pi \mathsf{C}_i$ of a family of categories indexed by a small set. It has (cartesian) *projection functors* $P_i \colon \mathsf{C} \to \mathsf{C}_i$.

A *functor in two variables* is an ordinary functor $F \colon \mathsf{C} \times \mathsf{D} \to \mathsf{E}$ defined on the product of two categories. Fixing an object X_0 in C we have a functor $F(X_0, -) \colon \mathsf{D} \to \mathsf{E}$; and symmetrically.

If the category C has small hom-sets $C(X, Y)$ (see 1.1.3), there is a functor of morphisms, or *hom-functor*

$$\text{Mor}: C^{op} \times C \to \text{Set},$$

$$(X, Y) \mapsto C(X, Y), \qquad (f, g) \mapsto g. - .f.$$

(1.15)

Here, for $f: X' \to X$ and $g: Y \to Y'$ in C, the mapping $g. - .f$ acts as:

$$g. - .f: C(X, Y) \to C(X', Y'), \qquad u \mapsto guf.$$

This quite important functor will reappear in various 'enriched' forms. It is described as 'a functor in two variables, contravariant in the first and covariant in the second'.

Fixing one of these variables one gets a 'representable functor' (covariant or contravariant on C), as we shall see in 1.6.4.

1.4.7 Faithful and full functors

For a functor $F: C \to D$ let us consider again the mappings (of sets):

$$F_{XX'}: C(X, X') \to D(F(X), F(X')), \qquad f \mapsto F(f).$$

(1.16)

F is said to be *faithful* if all these mappings are injective (for X, X' in C); F is said to be *full* if all of them are surjective.

An isomorphism of categories is always full and faithful. All the forgetful functors we have seen in 1.4.4 are faithful. The inclusion functor $D \to C$ of a subcategory is always faithful; it is full if and only if D is a full subcategory of C. A projection functor $P: C \to C/R$ on a quotient category is always full and bijective on objects.

An *embedding* of categories is a faithful functor, injective on the objects.

There are some (obvious) preservation and reflection properties of functors:

(a) every functor preserves commutative diagrams, isomorphisms, retracts, split monos and split epis (commutative diagrams were presented in Section 0.3 of the Introduction and are formalised below, in 1.4.9),

(b) a faithful functor *reflects* monos and epis (i.e. if $F(f)$ is mono or epi, then f is also) and commutative diagrams,

(c) a full and faithful functor reflects isomorphisms, split monos and split epis.

Applying point (a), a standard way of proving that a topological subspace $A \subset X$ is not a retract (in Top) is to find a functor $F: \text{Top} \to \text{Ab}$ such that the associated homomorphism $F(A) \to F(X)$ is not a split mono in Ab (see 5.5.8(d)).

1.4.8 Concrete categories

The notion of 'category of structured sets' is either vague or has different, complicated formalisations, which can hardly cover all the cases we may want.

As a formal and simple alternative, a *concrete category* is defined as a category A *equipped* with a faithful functor $U\colon A \to$ Set, called its *forgetful functor*. As a consequence U *reflects monos and epis*, but need not preserve them. Concrete categories are extensively studied in [AHS].

It is not easy to find categories that can *not* be made concrete: even Set^{op} can be, by the *contravariant functor of subsets*

$$\mathcal{P}^*\colon \text{Set}^{op} \to \text{Set},$$
$$X \mapsto \mathcal{P}X, \qquad (f\colon X \to Y) \mapsto (f^*\colon \mathcal{P}Y \to \mathcal{P}X), \tag{1.17}$$

where $f^*(B) = f^{-1}(B)$ is the pre-image of $B \subset Y$.

As a consequence, if the category C is concrete via $U\colon C \to$ Set, then C^{op} *can* be made concrete via the composed functor $\mathcal{P}^*U^{op}\colon C^{op} \to$ Set (letting apart whether this is of interest).

It is thus important to note that the homotopy category hoTop (introduced in 1.4.5) cannot be made concrete: this was proved by P. Freyd [Fr3, Fr4], in 1970.

More generally a category A equipped with a faithful functor $U\colon A \to C$ is said to be *concrete over* C. (We shall see in Lemma 1.5.7 how this notion is related to subcategories.)

1.4.9 Examples and complements (Functors as diagrams)

Diagrams in a category have already been presented in Section 0.3 of the Introduction, in an informal way. We can now be more precise.

(a) Let S be a small category. A *diagram* in the category C, *of type* S (or *based on* S) will be any functor $X\colon S \to C$.

It can be written in *index notation*, with indices $i \in$ ObS and $a\colon i \to j$ in MorS:

$$X\colon S \to C, \qquad i \mapsto X_i, \quad a \mapsto (X_a\colon X_i \to X_j), \tag{1.18}$$

or as a system $((X_i), (u_a))$, with $u_a = X_a$.

(b) The diagram $X\colon S \to C$ is said to be *commutative* if it factorises through the canonical projection $P\colon S \to$ po(S) with values in the preordered set po(S) defined in 1.4.5. This means that X coincides on every pair of parallel morphisms of S.

Therefore, every functor defined on a preordered set S is automatically a commutative diagram.

In particular, a *coherent system of isomorphisms* is a functor $X\colon S \to \mathsf{C}$ where the preorder of S is an equivalence relation $i \sim j$ (i.e. a symmetric preorder). The system amounts thus to a family (X_i) of objects of C equipped with a family of (iso)morphisms $u_{ij}\colon X_i \to X_j$ for $i \sim j$, such that

$$u_{jk}.u_{ij} = u_{ik}, \quad u_{ii} = \mathrm{id}(X_i) \qquad \text{(for } i \sim j \sim k \text{ in } S). \tag{1.19}$$

Since the well-known Coherence Theorem of monoidal categories, by Mac Lane [M3], a 'coherence theorem' in category theory states that, under suitable hypotheses, all the isomorphisms of a certain kind form a coherent system. The extended version for bicategories is in [MaP].

(c) A functor $\mathbf{1} \to \mathsf{C}$ amounts to an object of C, while a functor $\mathbf{2} \to \mathsf{C}$ amounts to a morphism of C.

(d) Let $E = \{1, \underline{e}\}$ be the idempotent monoid on two elements, with $\underline{e}^2 = \underline{e}$. A functor $F\colon E \to \mathsf{C}$ amounts to an object X of C equipped with an idempotent endomorphism $e\colon X \to X$; or – more simply – to an idempotent endomorphism e (which determines its domain-codomain).

(The monoid E is isomorphic to the join-semilattice $(\mathbf{2}, \vee)$, with unit 0 and $1 \vee 1 = 1$. But of course we should not confuse the ordinal category $\mathbf{2}$, which has two objects, with the monoid-category E, which has only one.)

(e) A functor $X\colon (\mathbb{N}, \leqslant) \to \mathsf{C}$, defined on the ordered set of natural numbers (i.e. the ordinal ω), amounts to a sequence of consecutive morphisms of C

$$X_0 \longrightarrow X_1 \longrightarrow X_2 \longrightarrow \ \dots \ \longrightarrow X_n \longrightarrow \ \dots \tag{1.20}$$

while a functor $X\colon (\mathbb{N}, +) \to \mathsf{C}$, defined on the additive monoid of natural numbers, amounts to an endomorphism $u_1\colon X_* \to X_*$ (and its powers, $u_n = (u_1)^n$, including $u_0 = \mathrm{id}(X_*)$).

(f) The product $\mathbf{2} \times \mathbf{2}$ has four objects and five non-identity arrows, as in the left diagram below

$$
\begin{array}{ccc}
00 \longrightarrow 10 & \qquad & X_{00} \longrightarrow X_{10} \\[-2pt]
\downarrow \searrow \downarrow & & \downarrow \quad = \quad \downarrow \\[-2pt]
01 \longrightarrow 11 & & X_{01} \longrightarrow X_{11}
\end{array}
\tag{1.21}
$$

A functor $X\colon \mathbf{2} \times \mathbf{2} \to \mathsf{C}$ amounts to a commutative square in C, as in the right diagram above.

(g) A diagram in C can also be defined as a morphism of graphs $X\colon \Gamma \to$ C, where Γ is a small graph. Commutativity is then expressed by conditions (i), (ii) of Section 0.3.

This is not more general, because every graph generates a free category, by finite paths of consecutive arrows, as we shall see in 3.7.3. In fact, this second approach is *less expressive*: for instance, commutative squares or idempotent endomorphisms cannot be formalised as arbitrary diagrams on a given graph, but as diagrams satisfying certain conditions.

1.5 Natural transformations and equivalence of categories

Given two *parallel* functors $F, G\colon$ C $\to$ D (between the same categories) there can be 'higher arrows' $\varphi\colon F \to G$, called 'natural transformations' (and functorial isomorphisms, or natural isomorphisms, when they are invertible).

'Category', 'functor' and 'natural transformation' are the three basic terms of category theory, since the very beginning of the theory in [EiM].

It is interesting to note that only the last term is taken from the common language: one can say that Eilenberg and Mac Lane introduced categories and functors because they wanted to formalise the natural transformations that they were encountering in Algebra and Algebraic Topology (as Mac Lane remarks in [M4], at the end of Section I.6). Much in the same way as a general theory of *continuity* (a familiar term for a familiar notion) requires the introduction of *topological spaces* (denoted by a theoretical term).

A reader acquainted with basic homotopy theory can take advantage of a formal parallelism, where *spaces* correspond to *categories, continuous mappings* to *functors, homotopies* of mappings to *invertible natural transformations*, and *homotopy equivalence* of spaces to *equivalence* of categories.

This analogy is even deeper in the domain of Directed Algebraic Topology, where *directed homotopies* need not be reversible and correspond to general *natural transformations*: see [G8].

1.5.1 *Natural transformations*

Given two functors $F, G\colon$ C $\to$ D between the same categories, a *natural transformation* $\varphi\colon F \to G$ (or $\varphi\colon F \to G\colon$ C $\to$ D) consists of the following data:

- for each object X of C a morphism $\varphi X\colon FX \to GX$ in D, called the *component* of φ on X (also written as φ_X, or just φ),

so that, for every arrow $f\colon X \to X'$ in C, we have a commutative square in

D (*naturality condition* of φ on f), whose diagonal will be written as $\varphi(f)$
when useful

$$
\begin{array}{ccc}
FX & \xrightarrow{\varphi X} & GX \\
{\scriptstyle Ff}\downarrow & & \downarrow{\scriptstyle Gf} \\
FX' & \xrightarrow{\varphi X'} & GX'
\end{array}
\qquad \varphi X'.F(f) = G(f).\varphi X = \varphi(f). \qquad (1.22)
$$

In particular, *the identity of a functor* $F: \mathsf{C} \to \mathsf{D}$ is the natural transformation $\mathrm{id}F: F \to F$, with components $(\mathrm{id}F)_X = \mathrm{id}(FX): FX \to FX$.

A natural transformation $\varphi: F \to G$ often arises from a 'canonical choice' of a family $(\varphi X: FX \to GX)_X$ of morphisms, *but these two aspects should not be confused*. There are canonical choices of such families that are not natural (see 1.5.2(e)), and natural transformations 'defined' by means of the axiom of choice (see 1.5.8).

Replacing the category C with a graph Γ (see 1.1.2), one can consider a *natural transformation* $\varphi: F \to G: \Gamma \to \mathsf{D}$ between two morphisms of graphs defined on Γ, *with values in a category*.

1.5.2 Exercises and complements

(a) Prove that the natural transformations $\varphi: U \to U$ of the forgetful functor $U: R\,\mathsf{Mod} \to \mathsf{Set}$ are in bijective correspondence with the scalars $\lambda \in R$. *Hints:* use the ring R as a left module on itself.

(b) Determine the natural endotransformations of the forgetful functor $U: R\,\mathsf{Mod} \to \mathsf{Ab}$.

(c) Prove that the natural transformations $\varphi: F \to F$ of the identity functor $F: R\,\mathsf{Mod} \to R\,\mathsf{Mod}$ correspond to the scalars λ of the centre of the ring R, i.e. the subring of the elements that commute with all the others.

This bijective correspondence shows that the category $R\,\mathsf{Mod}$ determines the centre of the ring R, in a structural way, and leads to the relation of Morita equivalence of rings (see 1.5.6(b)).

(d) Determine the natural endotransformations of the *contravariant functor of subsets* $\mathcal{P}^*: \mathsf{Set}^{\mathrm{op}} \to \mathsf{Set}$ (defined in (1.17)).

(e) Consider the *covariant functor of subsets*

$$
\mathcal{P}_*: \mathsf{Set} \to \mathsf{Set},
$$
$$
X \mapsto \mathcal{P}X, \qquad (f: X \to Y) \mapsto (f_*: \mathcal{P}X \to \mathcal{P}Y), \qquad (1.23)
$$

where $f_*(A) = f(A)$ is the image of $A \subset X$. Show that the canonical choices $\varphi_X(A) = X$ and $\psi_X(A) = X \setminus A$ do not give natural transformations $\mathcal{P}_* \to \mathcal{P}_*$.

(f) A natural transformation $f \to g \colon X \to Y$ between monotone mappings of preordered sets amounts to the preordering $f \prec g$ introduced in (1.3), and is determined by f, g.

(g) Characterise the natural transformations $f \to g \colon X \to Y$ between homomorphisms of monoids.

1.5.3 *Operations*

Two natural transformations $\varphi \colon F \to G$ and $\psi \colon G \to H$ have a *vertical composition* $\psi\varphi \colon F \to H$ (also written as $\psi.\varphi$)

$$\mathsf{C} \; \begin{array}{c} \xrightarrow{F} \\ \xrightarrow{\downarrow\varphi} \\ \xrightarrow{\downarrow\psi} \\ \xrightarrow{H} \end{array} \; \mathsf{D} \qquad (\psi\varphi)(X) = \psi X.\varphi X \colon FX \to HX. \qquad (1.24)$$

Moreover there is a *whisker composition* of natural transformations with functors, or *reduced horizontal composition*, written as $K\varphi H$ (or $K \circ \varphi \circ H$, when useful to distinguish it)

$$\mathsf{C}' \xrightarrow{\;H\;} \mathsf{C} \; \begin{array}{c} \xrightarrow{F} \\ \xrightarrow{\downarrow\varphi} \\ \xrightarrow{G} \end{array} \; \mathsf{D} \xrightarrow{\;K\;} \mathsf{D}' \qquad (1.25)$$

$$K\varphi H \colon KFH \to KGH, \qquad (K\varphi H)(X') = K(\varphi(HX')).$$

(The binary operations φH and $K\varphi$ are also called whisker compositions; they can be obtained from the ternary operation, letting $K = \mathrm{id}\mathsf{D}$ or $H = \mathrm{id}\mathsf{C}$.)

This 2-dimensional structure of the framework of categories, where natural transformations play the role of 2-dimensional arrows between functors, will be further analysed in Section 7.1 (including the *full* horizontal composition of natural transformations).

An *isomorphism of functors*, or functorial isomorphism, or natural isomorphism, is a natural transformation $\varphi \colon F \to G \colon \mathsf{C} \to \mathsf{D}$ which is invertible with respect to vertical composition. It is easy to see that this happens if and only if all the components φX are invertible in D.

(The original term of 'natural equivalence', in [EiM], is now obsolete and will not be used, because it can be confused with an equivalence of categories.)

Isomorphism of (parallel) functors is an equivalence relation, written as $F \cong G$.

1.5.4 Exercises (The calculus of natural transformations)

The reader should prove the following properties of the compositions of natural transformations, which will often be used in computations.

All the verifications are trivial, except point (d) which is easy (and can be found in Chapter 8).

(a) $$\chi(\psi\varphi) = (\chi\psi)\varphi,$$

$$K'(K\varphi H)H' = (K'K)\varphi(HH') \qquad (associativities),$$

(b) $$\varphi.\mathrm{id}F = \varphi = \mathrm{id}G.\varphi,$$

$$1_Y \circ \varphi \circ 1_X = \varphi, \qquad K \circ \mathrm{id}F \circ H = \mathrm{id}(KFH) \qquad (identities),$$

(c) $$K(\psi\varphi)H = (K\psi H).(K\varphi H) \qquad (distributivity),$$

(d) $$(\psi G).(H\varphi) = (K\varphi).(\psi F) \qquad (interchange).$$

The last, more precisely a 'reduced interchange property', is about two 'horizontally consecutive' natural transformations, as in the following diagram

$$\mathsf{C} \ \underset{G}{\overset{F}{\underset{\downarrow\varphi}{\rightrightarrows}}} \ \mathsf{D} \ \underset{K}{\overset{H}{\underset{\downarrow\psi}{\rightrightarrows}}} \ \mathsf{E} \tag{1.26}$$

(e) For a natural transformation $\varphi\colon F \to G\colon \mathsf{C} \to \mathsf{C}$ of *endofunctors*, taking $\varphi = \psi$ we have a property that will often be used (for adjunctions and monads)

$$(\varphi G).(F\varphi) = (G\varphi).(\varphi F) \qquad (self\text{-}interchange). \tag{1.27}$$

1.5.5 Equivalence of categories

Isomorphisms of categories have been defined in 1.4.1.

More generally, an *equivalence of categories* is a functor $F\colon \mathsf{C} \to \mathsf{D}$ which is invertible up to isomorphism of functors (defined in 1.5.3), i.e. there exists a functor $G\colon \mathsf{D} \to \mathsf{C}$ such that $GF \cong \mathrm{id}\mathsf{C}$ and $FG \cong \mathrm{id}\mathsf{D}$. The functor G can be called a *weak inverse* of F.

An *adjoint equivalence of categories* is a coherent version of this notion; namely it is a four-tuple $(F, G, \eta, \varepsilon)$ where

- $F\colon \mathsf{C} \to \mathsf{D}$ and $G\colon \mathsf{D} \to \mathsf{C}$ are functors,

- $\eta\colon \mathrm{id}\mathsf{C} \to GF$ and $\varepsilon\colon FG \to \mathrm{id}\mathsf{D}$ are functorial isomorphisms,

- $F\eta = (\varepsilon F)^{-1}\colon F \to FGF, \qquad \eta G = (G\varepsilon)^{-1}\colon G \to GFG.$

(The direction of η and ε is written above as in the more general case of an adjunction, where these transformations need not be invertible: see Section 3.1.)

The following conditions on a functor $F \colon \mathsf{C} \to \mathsf{D}$ are equivalent, forming a very useful *characterisation of the equivalence of categories*:

(i) F is an equivalence of categories,

(ii) F can be completed to an adjoint equivalence $(F, G, \eta, \varepsilon)$,

(iii) F is faithful, full and *essentially surjective on objects*.

The last property means that: for every object Y of D there exists some object X in C such that $F(X)$ is isomorphic to Y in D. The proof of the equivalence of these three conditions requires the axiom of choice and is given below, in Theorem 1.5.8.

One says that the categories C, D are *equivalent*, written as $\mathsf{C} \simeq \mathsf{D}$, if there exists an equivalence of categories between them (or, equivalently, an *adjoint* equivalence of categories). This is indeed an equivalence relation, as is easily proved using condition (iii) above, or can be proved directly using the operations we have seen above (in 1.5.3).

The role of equivalences in category theory is fundamental. Like everything, it should not be viewed as 'absolute'. Comments on this point can be found in 1.5.9, 3.3.7 and 3.6.6.

1.5.6 Exercises and complements

(a) The category of finite sets (and mappings between them) is equivalent to its full subcategory of finite cardinals, which is small and therefore cannot be isomorphic to the former.

*(b) The reader may know that two rings R, S are said to be *Morita equivalent* if their categories of modules are equivalent [Mo, AnF].

This is a crucial notion in ring theory, that becomes trivial in the domain of commutative rings: in fact, if two rings are Morita equivalent one can prove that their centres are isomorphic (using what we have seen in 1.5.2); but – quite interestingly – commutative rings can be Morita equivalent to non-commutative ones, like their rings of square matrices. Thus, studying left modules on the ring $M_n(\mathbb{R})$ is equivalent to studying real vector spaces.*

(c) A category is said to be *skeletal* if it has no pair of distinct isomorphic objects. It is easy to show, by the previous characterisation of equivalences, that every category has a *skeleton*, i.e. an equivalent skeletal category. The latter can be obtained as a full subcategory, by *choosing* precisely one object in every class of isomorphic objects.

(d) We have described in (a) a skeleton of the category of finite sets that can be constructed without any choice, even though we do need the axiom of choice to prove that the inclusion of this skeleton has a weak inverse.

In a different way, a preordered set X has a natural skeleton *formed by a quotient*, the associated ordered set $X/\!\!\sim$ (see 1.2.2); but again we need the axiom of choice to prove that the projection $X \to X/\!\!\sim$ has a weak inverse.

(e) The reader can easily prove that two categories are equivalent if and only if they have isomorphic skeletons.

Loosely speaking, this says that an equivalence of categories amounts to multiplying or deleting isomorphic copies of objects, even though there can be no canonical way of doing this.

(f) The category M of metric spaces and continuous mappings (already considered in 1.1.5) is equivalent to the category MTop of metrisable topological spaces and continuous mappings, by the forgetful functor M $\to$ MTop.

We also note that the forgetful functor MTop $\to$ Set is amnestic, while the forgetful functor M $\to$ Set is not.

(g) A category C is equivalent to the singleton category **1** if and only if it is non-empty and indiscrete, i.e. each hom-set $C(X, Y)$ has a unique element.

1.5.7 Lemma (Modifying functors and categories)

(a) Starting from a functor $F\colon C \to D$ and an arbitrary family of isomorphisms $\varphi_X\colon F(X) \to G(X)$ in D (for X in C), there is precisely one way of extending these data to a functor $G\colon C \to D$ and a natural transformation $\varphi\colon F \to G$ (obviously invertible).

(b) Given a faithful functor $U\colon A \to C$, one can embed A as a subcategory of a suitable category C′ equivalent to C.

(c) If the faithful functor U is essentially surjective on the objects, one can make this embedding wide (and even the identity on objects).

For instance, the forgetful functor Top $\to$ Set is surjective on objects, and Top is a wide subcategory of the category T $\simeq$ Set introduced in 1.1.5(a).

Proof (a) We define G on a morphism $f\colon X \to Y$ of C as

$$G(f) = \varphi_Y.F(f).(\varphi_X)^{-1}\colon G(X) \to G(Y).$$

Besides preserving identities and composition, this is necessary and sufficient to make φ natural.

(b) We construct C′ with objects C_A where A belongs to A and $U(A) = C$, and other (distinct) objects C_* where C belongs to C. Then we let

$C'(C_X, D_Y) = C(C, D)$, where X (resp. Y) is either an object of A or the symbol $*$. The composition is that of C.

Now the functor $U'\colon A \to C'$ that takes A to $U(A)_A$ and f to $U(f)$ is injective on objects and arrows, and identifies A with a subcategory of C'. On the other hand, the obvious full embedding $C \to C'$ taking C to C_* is an equivalence of categories, since ever object C_A is isomorphic to C_*.

(c) If U is essentially surjective on the objects, we can omit the objects of type C_* in the construction of C', and rewrite the object C_A as A, so that $Ob\,C' = Ob\,A$. $\qquad\square$

1.5.8 Theorem (Characterisation of equivalences)

The conditions (i)–(iii) of 1.5.5 are equivalent.

Proof Part of the proof is a consequence of general properties of adjunctions, but it is useful to have now this important theorem, much before Chapter 3.

(i) $\Rightarrow$ (iii). We assume that $F\colon C \to D$ has a weak inverse $G\colon D \to C$, with a functorial isomorphism $\eta\colon 1 \to GF$.

Then F is faithful: given two parallel morphisms $f, f'\colon X \to X'$, we have two commutative squares with horizontal isomorphisms

$$\begin{array}{ccc} X & \xrightarrow{\ \eta X\ } & GFX \\ {\scriptstyle f}\big\|\big\|{\scriptstyle f'} & {\scriptstyle GFf}\big\|\big\|{\scriptstyle GFf'} & \\ X' & \xrightarrow[\ \eta X'\ ]{} & GFX' \end{array} \qquad (1.28)$$

so that $Ff = Ff'$ implies $f = f'$. In the same way G is faithful (because $FG \cong 1$).

To show that F is full, we take a morphism $g\colon FX \to FX'$ and define $f = (\eta X')^{-1}.Gg.\eta X\colon X \to X'$. Now we have the parallel morphisms $g, Ff\colon FX \to FX'$ in D. Both $G(g)$ and $G(Ff)$ make the following square commute in C

$$\begin{array}{ccc} X & \xrightarrow{\ \eta X\ } & GFX \\ {\scriptstyle f}\big\downarrow & {\scriptstyle Gg}\big\downarrow{\scriptstyle GFf} & \\ X' & \xrightarrow[\ \eta X'\ ]{} & GFX' \end{array} \qquad (1.29)$$

Since the horizontal arrows are isomorphisms, $G(g) = G(Ff)$; since G is faithful, $g = F(f)$.

Finally it is obvious that F is essentially surjective on objects, because each object Y of D is isomorphic to $F(GY)$.

(iii) $\Rightarrow$ (ii). Assuming that F is faithful, full and essentially surjective on objects, we 'construct' with the axiom of choice a weak inverse G.

For every object Y of D we choose an object $G(Y)$ in C *and* an isomorphism $\varepsilon Y \colon F(G(Y)) \to Y$ in D. For every morphism $g \colon Y \to Y'$ in D, we define $G(g) \colon GY \to GY'$ as the unique morphism of C such that

$$F(G(g)) = (\varepsilon Y')^{-1}.g.\varepsilon Y \colon F(GY) \to F(GY'),$$

which makes the following square commutative in D

$$
\begin{array}{ccc}
FGY & \xrightarrow{\ \varepsilon Y\ } & Y \\
{\scriptstyle FGg}\downarrow & & \downarrow{\scriptstyle g} \\
FGY' & \xrightarrow[\varepsilon Y']{} & Y'
\end{array}
\qquad g.\varepsilon Y = \varepsilon Y'.FG(g). \qquad (1.30)
$$

It is easy to verify that G is a functor. For instance, for a composed morphism $g'g \colon Y \to Y' \to Y''$, we have:

$$F(G(g'g)) = (\varepsilon Y'')^{-1}.g'g.\varepsilon Y = ((\varepsilon Y'')^{-1}.g'.\varepsilon Y').((\varepsilon Y')^{-1}.g.\varepsilon Y)$$
$$= FG(g').FG(g) = F(Gg'.Gg),$$

and $G(g'g) = Gg'.Gg$, because F is faithful.

The family $\varepsilon = (\varepsilon Y)_Y$ of isomorphisms is a natural transformation $\varepsilon \colon FG \to \mathrm{id}D$, by (1.30), whence a functorial isomorphism.

Moreover, for every X in C there is a unique morphism $\eta X \colon X \to GFX$ such that

$$F(\eta X) = (\varepsilon FX)^{-1} \colon FX \to FG(FX),$$

and ηX is invertible (by 1.4.7(c)).

Using again the faithfulness of F, one shows that $\eta = (\eta X)_X$ is natural, whence a functorial isomorphism.

We already have one coherence condition: $F\eta = (\varepsilon F)^{-1}$. For the other, the self-interchange property (1.27) of $\varepsilon \colon FG \to 1 \colon D \to D$ gives $\varepsilon.FG\varepsilon = \varepsilon.\varepsilon FG$. Here ε is invertible, and $FG\varepsilon = \varepsilon FG$. Therefore:

$$F(\eta GY) = (\varepsilon FGY)^{-1} = (FG\varepsilon Y)^{-1} = F(G\varepsilon Y)^{-1},$$

and the faithfulness of F gives $\eta GY = (G\varepsilon Y)^{-1}$, for all Y in D.

(ii) $\Rightarrow$ (i). Obvious. $\qquad\qquad\qquad\qquad\qquad\qquad\qquad\qquad\quad$ $\square$

1.5.9 A first digression on categorical equivalence

There is now a tendency in category theory to prefer notions invariant up to equivalence of categories (let us say an 'E-notion', for short), rather than just invariant up to isomorphism of categories (let us say an 'I-notion'), or not even invariant in this sense (let us say an 'S-notion', for 'strict').

In the author's opinion invariance up to categorical equivalence is certainly an important property. Yet, it should be put in balance with other aspects, like simplicity, effectiveness and adherence to well-founded practice in Mathematics.

(a) The product $C \times D$ of categories has been defined in a strict way (in 1.4.6). Replacing this notion by a weaker one, invariant up to equivalence, would be particularly strange and ineffective: we no longer have a functor $Cat^2 \to Cat$. A reader familiar with the cartesian closedness of Cat (to be seen in 3.4.7(a)) may wonder what this may mean, with respect to an E-notion of product.

(b) A *faithful functor* can be viewed as an 'E-replacement' of the S-notion of *subcategory* (as shown by Lemma 1.5.7). We can agree that a faithful functor is theoretically a more important issue; nevertheless the notion of subcategory keeps its concrete importance.

(c) Some instances where an I-notion can be preferable to the corresponding E-notion will be seen in 3.3.7 (for comma categories), in 3.6.6 (for monadicity, following Mac Lane) and in 7.1.5 (for flexible limits, following [BKPS]).

1.6 Categories of functors, representable functors

This section is about categories $Cat(S, C)$ of functors $S \to C$, which includes the well-known case of simplicial sets. Representable functors are also introduced here.

S is always a small category.

1.6.1 Categories of functors

For any category C we write as $Cat(S, C)$, or C^S, the category whose objects are the functors $F \colon S \to C$ and whose morphisms are the natural transformations $\varphi \colon F \to G$, with vertical composition.

These objects can be viewed as diagrams in C, based on S, as in 1.4.9. The reader will note that we are extending the notation $Cat(S, C)$ to possibly large categories C.

In particular the arrow category **2**, with two objects (0 and 1) and one non-identity arrow, $0 \to 1$ (see 1.2.3) gives the *category of morphisms* C^2 of C, where a map $f = (f', f'') \colon x \to y$ is a commutative square of C, as in the left diagram below

$$
\begin{array}{ccc}
X' & \xrightarrow{f'} & Y' \\
x\downarrow & & \downarrow y \\
X'' & \xrightarrow{g'} & Y''
\end{array}
\qquad
\begin{array}{ccccc}
X' & \xrightarrow{f'} & Y' & \xrightarrow{f''} & Z' \\
x\downarrow & & \downarrow y & & \downarrow z \\
X'' & \xrightarrow{g'} & Y'' & \xrightarrow{g''} & Z''
\end{array}
\qquad (1.31)
$$

These maps are composed by pasting commutative squares, as in the right diagram above. The identity of x is: $\mathrm{id}(x) = (\mathrm{id}X', \mathrm{id}X'') \colon x \to x$.

Writing an object of C^S in index notation, as in (1.18), the *diagonal functor*

$$
D \colon \mathsf{C} \to \mathsf{C}^\mathsf{S}
$$
$$
(DA)_i = A, \quad (DA)_a = \mathrm{id}A \qquad (i \in \mathrm{Ob}\,\mathsf{S},\ a \in \mathrm{Mor}\,\mathsf{S}),
\qquad (1.32)
$$

sends an object A to the constant functor at A, and a morphism $f \colon A \to B$ to the natural transformation $Df \colon DA \to DB \colon \mathsf{S} \to \mathsf{C}$ whose components $(Df)_i$ are constant at f.

1.6.2 Exercises and complements

(a) C^0 'is' the singleton category **1** (including the case $\mathsf{C} = \mathbf{0}$). C^1 is isomorphic to C. C^3 is the category of pairs of consecutive arrows of C.

(b) If $\mathsf{S} \neq \mathbf{0}$ the diagonal functor $D \colon \mathsf{C} \to \mathsf{C}^\mathsf{S}$ is an embedding.

(c) Prove that a natural transformation $\varphi \colon F \to G \colon \mathsf{C} \to \mathsf{D}$ can be viewed as a functor $\mathsf{C} \to \mathsf{D}^2$, or equivalently as a functor $\mathsf{C} \times \mathbf{2} \to \mathsf{D}$.

(d) Analyse the category

$$
G\mathsf{Set} = \mathsf{Cat}(G, \mathsf{Set}), \qquad (1.33)
$$

where G is a group (viewed as a category). An object, called a *G-set*, is a set equipped with an action of G, and an arrow, called a *G-morphism*, is a G-equivariant mapping – notions that the reader may already know or can deduce from (1.33). Similarly we have the category $G\mathsf{Top} = \mathsf{Cat}(G, \mathsf{Top})$ of *G-spaces*. All this can be extended to a monoid G.

*(e) With a basic knowledge of Homological Algebra, one can describe the category $\mathrm{Ch}^+\mathsf{Ab}$ of positive cochain complexes of abelian groups as a full subcategory of the category $\mathsf{Cat}(\mathbb{N}, \mathsf{Ab})$ that we have seen in 1.4.9(e), and give a similar description of the category $\mathrm{Ch}_+\mathsf{Ab}$ of positive chain complexes (that will be used in 5.5.2).

1.6.3 Categories of presheaves

A functor $S^{op} \to C$, defined on the opposite category S^{op}, is also called a *presheaf* of C on the (small) category S. They form the *presheaf category* $Psh(S, C) = C^{S^{op}}$.

We are particularly interested in the category $Psh(S, Set) = Set^{S^{op}}$ of *presheaves of sets* on S.

The category S is canonically embedded in the latter, by the *Yoneda embedding*

$$y \colon S \to Set^{S^{op}}, \qquad y(i) = S(-, i) \colon S^{op} \to Set, \qquad (1.34)$$

which sends every object i to the corresponding presheaf $y(i)$, *represented by i* (according to the definition below, in 1.6.4).

Taking as S the category $\underline{\Delta}$ of positive finite ordinals (and monotone maps), one gets the category $Smp(C) = C^{\underline{\Delta}^{op}}$ of *simplicial objects* in C, and – in particular – the well-known category SmpSet of *simplicial sets*, that will be studied in Section 5.3. Here the Yoneda embedding sends the positive ordinal $\{0, ..., n\}$ to the simplicial set Δ^n, freely generated by one simplex of dimension n.

Simplicial groups, in the category SmpGp, are also important.

Cubical sets, another category of presheaves of interest in Topology, will be studied in Section 5.4.

A reader interested in categories of *sheaves* is referred to [MaM, Bo3].

1.6.4 Representable functors

Assume now that C has small hom-sets. A functor $F \colon C \to Set$ is said to be *representable* if it is isomorphic to a functor

$$C(X_0, -) \colon C \to Set, \qquad X \mapsto C(X_0, X),$$
$$(f \colon X \to Y) \mapsto (f.- \colon C(X_0, X) \to C(X_0, Y)), \qquad (1.35)$$

for some object X_0 in C; this object is said *to represent F*.

Then the Yoneda Lemma (in 1.6.6) describes the natural transformations $F \to G$, for every functor $G \colon C \to Set$; it also proves that the object which represents a functor is determined up to isomorphism (see 1.6.6(b)).

A contravariant functor $F \colon C \dashrightarrow Set$ is in fact a covariant functor $F \colon C^{op} \to Set$; therefore it is *representable* if it is isomorphic to a functor

$$C(-, X_0) \colon C^{op} \to Set, \qquad X \mapsto C(X, X_0),$$
$$(f \colon X \to Y) \mapsto (-.f \colon C(Y, X_0) \to C(X, X_0)), \qquad (1.36)$$

for some object X_0 in C.

1.6.5 Exercises and complements

(a) Show that the forgetful functors

$$\mathsf{Top} \to \mathsf{Set}, \qquad R\,\mathsf{Mod} \to \mathsf{Set}, \qquad \mathsf{Rng} \to \mathsf{Set},$$

are representable.

(b) By definition, the canonical forgetful functors of Ban and Ban_1 are represented by the scalar field K ($\mathbb{R}$ or $\mathbb{C}$):

$$U = \mathsf{Ban}(K, -) \colon \mathsf{Ban} \to \mathsf{Set}, \qquad U(X) = |X|,$$
$$B_1 = \mathsf{Ban}_1(K, -) \colon \mathsf{Ban}_1 \to \mathsf{Set}, \quad B_1(X) = \{x \in X \mid \|x\| \leqslant 1\}, \tag{1.37}$$

and give, respectively, the underlying set $|X|$ and the *unit ball* $B_1(X)$ of a Banach space (up to functorial isomorphism).

(c) Show that the contravariant functor of subsets $\mathcal{P}^* \colon \mathsf{Set}^{\mathrm{op}} \to \mathsf{Set}$ (defined in (1.17)) is representable, making use of the characteristic function of a subset.

(d) A representable covariant functor preserves monomorphisms. A representable contravariant functor takes epimorphisms of its domain to injective mappings.

(e) Review the exercises 1.5.2(a) and (d) at the light of the Yoneda Lemma, below.

1.6.6 Yoneda Lemma

(a) Let $F, G \colon \mathsf{C} \to \mathsf{Set}$ be two functors, with $F = \mathsf{C}(X_0, -)$. The canonical mapping

$$y \colon \mathrm{Nat}(F, G) \to G(X_0), \qquad y(\varphi) = \varphi_{X_0}(\mathrm{id}X_0) \in G(X_0), \tag{1.38}$$

from the set of natural transformations $\varphi \colon F \to G$ to the set $G(X_0)$ is a bijection.

(b) The functor $G \colon \mathsf{C} \to \mathsf{Set}$ is represented by the object X_0 of C if and only if there exists some $x_0 \in G(X_0)$ such that:

(*) for every X in C and every $x \in G(X)$ there is a unique morphism $f \colon X_0 \to X$ such that $G(f)(x_0) = x$.

This morphism f is an isomorphism if and only if G is also represented by X, via $x \in G(X)$.

Proof (a) We construct a backward mapping

$$y' \colon G(X_0) \to \mathrm{Nat}(F, G),$$
$$(y'(x))_X \colon \mathsf{C}(X_0, X) \to GX, \quad (f \colon X_0 \to X) \mapsto (Gf)(x), \tag{1.39}$$

where the family $\psi = y'(x)$ is indeed a natural transformation $F \to G$. In fact, for $g \colon X \to Y$ in C

$$
\begin{array}{ccc}
\mathsf{C}(X_0, X) & \xrightarrow{\psi X} & GX \\
{\scriptstyle g.-}\downarrow & & \downarrow{\scriptstyle Gg} \\
\mathsf{C}(X_0, Y) & \xrightarrow[\psi Y]{} & GY
\end{array}
\tag{1.40}
$$

$$(G(g).\psi X)(f) = G(g)((Gf)(x)) = G(gf)(x) = \psi_Y(gf).$$

Then we have to show that the mappings y, y' are inverse to each other.

First, a natural transformation $\varphi \colon F \to G$ is taken to the element $x = y(\varphi) = \varphi_{X_0}(\mathrm{id}X_0) \in G(X_0)$, and the latter to the natural transformation $y'(x) \colon F \to G$. The component $(y'(x))_X$ takes the morphism $f \colon X_0 \to X$ to $(Gf)(x)$, and it is sufficient to prove that $\varphi_X(f) = (Gf)(x)$; this follows from the naturality of φ on f

$$
\begin{array}{ccc}
\mathsf{C}(X_0, X_0) & \xrightarrow{\varphi X_0} & GX_0 \\
{\scriptstyle f.-}\downarrow & & \downarrow{\scriptstyle Gf} \\
\mathsf{C}(X_0, X) & \xrightarrow[\varphi X]{} & GX
\end{array}
\tag{1.41}
$$

$$\varphi_X(f) = \varphi_X(f.\mathrm{id}X_0) = (Gf)(\varphi_{X_0}(\mathrm{id}X_0)) = (Gf)(x).$$

The other way round, an element $x \in G(X_0)$ is taken to the natural transformation $\varphi = y'(x) \colon F \to G$ and then to

$$y(\varphi) = \varphi_{X_0}(\mathrm{id}X_0) = (G(\mathrm{id}X_0))(x) = x.$$

(b) Giving a natural transformation $\varphi \colon F \to G$ is equivalent to giving an element $x_0 \in G(X_0)$, linked by the previous bijection: $x_0 = y(\varphi)$ and $\varphi = y'(x_0)$.

The natural transformation φ is an isomorphism if and only if each component $\varphi X \colon FX \to GX$ is a bijection of sets, which is equivalent to condition (*) – taking into account formula (1.41). $\qquad\square$

1.6.7 Frames and locales

There is a natural functor

$$\mathcal{O} \colon \mathsf{Top}^{\mathrm{op}} \to \mathsf{Lth}, \tag{1.42}$$

that assigns to a space X the lattice $\mathcal{O}(X)$ of its open subsets, and to a continuous mapping $f\colon X \to Y$ the preimage homomorphism

$$f^*\colon \mathcal{O}(Y) \to \mathcal{O}(X).$$

In fact, $\mathcal{O}(X)$ is a *frame*, i.e. a complete lattice where arbitrary joins distribute over binary meets

$$(\vee\, U_i) \wedge V = \vee\, (U_i \wedge V),$$

since these operations are unions and intersections in $\mathcal{P}(X)$. In the same way $f^*\colon \mathcal{O}(Y) \to \mathcal{O}(X)$ is a *homomorphism of frames*, which means that it preserves arbitrary joins and binary meets.

As a surrogate of Top, we now introduce the formal *category of locales* Loc, namely the opposite of the category of frames and their homomorphisms. The previous functor (1.42) is thus reinterpreted as

$$\mathcal{O}\colon \mathsf{Top} \to \mathsf{Loc}. \tag{1.43}$$

The main topological notions can be studied in Loc, although the relationship with topological spaces is complex. Loc is the domain of 'pointless topology', which is also important in Topos Theory. An interested reader can see [Jo1, Jo2, Bo3].

Exercises and complements. (a) Composing the functor $\mathcal{O}\colon \mathsf{Top}^{\mathrm{op}} \to \mathsf{Lth}$ with the forgetful functor $U\colon \mathsf{Lth} \to \mathsf{Set}$, we get a representable functor

$$U\mathcal{O}\colon \mathsf{Top}^{\mathrm{op}} \to \mathsf{Set}.$$

(b) This functor is faithful on the full subcategory $T_0\mathsf{Top}^{\mathrm{op}}$ of T_0-spaces.

1.7 A category of Galois connections

After the category Ord of ordered sets and monotone mappings, we now study the category AdjOrd of ordered sets and (covariant) Galois connections between them – or adjunctions between ordered sets; its importance will appear in the comments and examples below.

All this will also serve as an introduction to general adjunctions, dealt with in Section 3.1.

The classical *contravariant* form of Galois connections can be found in 1.7.8.

1.7.1 Definition

Given a pair X, Y of ordered sets, a (covariant) *Galois connection* (f, g) between them can be presented in the following equivalent ways.

(i) We assign two monotone mappings $f: X \to Y$ and $g: Y \to X$ such that:

$$f(x) \leqslant y \text{ in } Y \quad \Leftrightarrow \quad x \leqslant g(y) \text{ in } X. \tag{1.44}$$

(ii) We assign a monotone mapping $g: Y \to X$ such that for every $x \in X$ there exists in Y:

$$f(x) = \min\{y \in Y \mid x \leqslant g(y)\}. \tag{1.45}$$

(ii*) We assign a monotone mapping $f: X \to Y$ such that for every $y \in Y$ there exists in X:

$$g(y) = \max\{x \in X \mid f(x) \leqslant y\}. \tag{1.46}$$

(iii) We assign two monotone mappings $f: X \to Y$ and $g: Y \to X$ such that

$$\mathrm{id}X \leqslant gf, \qquad\qquad fg \leqslant \mathrm{id}Y. \tag{1.47}$$

By these formulas g determines f (called its *left adjoint*) and f determines g (its *right adjoint*). One writes $f \dashv g$ (as in the general notation of adjoints in category theory), and the connection (f, g) is also called an *adjunction* between ordered sets.

The relations (1.47) imply that

$$f = fgf, \qquad\qquad g = gfg. \tag{1.48}$$

More generally we can consider Galois connections between preordered sets, with the obvious modifications. For instance, in (1.48) we only get $f \sim fgf$ (meaning that $f(x) \sim fgf(x)$, for all $x \in X$) and $g \sim gfg$.

1.7.2 Exercises and complements

(a) Prove that the four conditions of 1.7.1 are indeed equivalent.

(b) Of course an isomorphism of ordered sets is, at the same time, left and right adjoint to its inverse. Conversely, if $f \dashv g \dashv f$ then f and g are isomorphisms, inverse to each other.

(c) More generally, a monotone mapping $f: X \to Y$ between ordered sets *may* have adjoints $h \dashv f \dashv g$ (one or both), which can be viewed as 'best approximations', of different kinds, to an inverse which may not exist

$$fh \geqslant 1, \quad hf \leqslant 1, \qquad\qquad gf \geqslant 1, \quad fg \leqslant 1. \tag{1.49}$$

For instance, the embedding of ordered sets $i: \mathbb{Z} \to \mathbb{R}$ has both adjoints, which are well known!

(d) The embedding $\mathbb{Q} \to \mathbb{R}$ has no left nor right adjoint: an irrational number has no distinguished rational approximation.

(e) Prove that two mappings $f \colon X \rightleftarrows Y \colon g$ between ordered sets, that satisfy condition (1.44), are necessarily monotone.

1.7.3 Properties

Let us come back to a general Galois connection $f \dashv g$ between ordered sets X, Y.

The mapping f *preserves all the existing joins* (also infinite), while g *preserves all the existing meets*. In fact, if $x = \vee x_i$ in X then $f(x_i) \leqslant f(x)$ (for all indices i). Supposing that $f(x_i) \leqslant y$ in Y (for all i), it follows that $x_i \leqslant g(y)$ (for all i); but then $x \leqslant g(y)$ and $f(x) \leqslant y$.

From the relations $f = fgf$ and $g = gfg$ it follows that:

(a) $gf = \mathrm{id} \Leftrightarrow f$ is injective $\Leftrightarrow f$ is a split mono $\Leftrightarrow g$ is surjective
$\qquad\qquad \Leftrightarrow g$ is a split epi $\Leftrightarrow f$ reflects the order relation,

(a*) $fg = \mathrm{id} \Leftrightarrow f$ is surjective $\Leftrightarrow f$ is a split epi $\Leftrightarrow g$ is injective
$\qquad\qquad \Leftrightarrow g$ is a split mono $\Leftrightarrow g$ reflects the order relation.

Moreover the connection restricts to an isomorphism (of ordered sets) between the sets of *fixed* (or *closed*) *elements* of X and Y, defined as follows

$$\begin{aligned} \mathrm{Fix}(X) &= g(Y) = \{x \in X \mid x = gf(x)\}, \\ \mathrm{Fix}(Y) &= f(X) = \{y \in Y \mid y = fg(y)\}. \end{aligned} \qquad (1.50)$$

Terminology. (a) The term 'fixed element' refers to a fixed element of the operator $gf \colon X \to X$ or $fg \colon Y \to Y$, and is appropriate in both cases.

The fixed elements of X are often called 'closed elements'; this is also appropriate, because $gf \colon X \to X$ is a *closure* operator: monotone, idempotent and *inflationary* (in the sense that $gf \geqslant \mathrm{id}$). On the other hand, the operator $fg \colon Y \to Y$ is *deflationary* ($fg \leqslant \mathrm{id}$), and there is some incongruence in saying that an element $y \in Y$ is closed if $y = fg(y)$; these elements are also called 'kernel elements'.

(b) In the contravariant form this disparity vanishes: both gf and fg are closure operators, and the name of 'closed elements', in X and Y, is always appropriate (see 1.7.8).

1.7.4 A category of adjunctions

An adjunction $f \dashv g$ will also be written as an arrow $(f, g) \colon X \dashrightarrow Y$, often dot-marked and conventionally directed as the left adjoint $f \colon X \to Y$. Such arrows have an obvious composition

$$(f', g').(f, g) = (f'f, gg') \qquad \text{(for } (f', g'): Y \rightarrowtail Z), \qquad (1.51)$$

and form the category AdjOrd *of ordered sets and Galois connections.*

Each hom-set $\mathsf{AdjOrd}(X, Y)$ is canonically ordered: for two adjunctions $(f, g), (f', g'): X \rightarrowtail Y$ we let $(f, g) \leqslant (f', g')$ if the following equivalent conditions hold

$$f \leqslant f', \qquad\qquad g' \leqslant g, \qquad (1.52)$$

since $f \leqslant f'$ gives $g' \leqslant gfg' \leqslant gf'g' \leqslant g$.

The relationship between AdjOrd and the usual category Ord (on the same objects) will become clearer in Chapter 7, where we shall amalgamate them to form the double category AdjOrd, with horizontal arrows in Ord, vertical arrows in AdjOrd and convenient double cells (in 7.2.7). Other similar double categories of lattices will also be of interest (see 7.5.2).

1.7.5 Direct and inverse images of subsets

The transfer of subobjects along morphisms is an important feature, that we examine here in Set and will be developed in Chapter 6 in other frameworks related to Homological Algebra.

Every set A has an ordered set $\mathrm{Sub}A = \mathcal{P}A$ of subsets, which actually is a complete boolean algebra (as recalled in 1.2.6). A mapping $f: A \to B$ gives two monotone mappings, of *direct* and *inverse image*, or image and preimage

$$f_*: \mathrm{Sub}A \;\rightleftarrows\; \mathrm{Sub}B : f^*,$$

$$f_*(X) = f(X), \quad f^*(Y) = f^{-1}(Y) \qquad\qquad (X \subset A,\ Y \subset B). \qquad (1.53)$$

These mappings form a Galois connection $f_* \dashv f^*$, since

$$X \subset f^*f_*(X), \qquad f_*f^*(Y) = Y \cap f(A) \subset Y, \qquad (1.54)$$

which implies, in particular, that f_* preserves unions and f^* preserves intersections.

(In fact f^* is a homomorphism of complete boolean algebras; it also has a right adjoint, related to the universal quantifier, while the left adjoint f_* is related to the existential one; we shall not use these facts.)

All this defines a *transfer functor* for subobjects of Set

$$\mathrm{Sub}: \mathsf{Set} \to \mathsf{AdjOrd}, \qquad (1.55)$$

with values in the category of ordered sets and Galois connections defined above – a sort of 'amalgamation' of the covariant and contravariant functors of subsets, considered in 1.5.2.

1.7.6 The graph of a Galois connection

The *graph* of the adjunction $(f, g) \colon X \rightarrowtail Y$ will be the following ordered set

$$
\begin{aligned}
G(f, g) &= \{(x, y) \in X \times Y \mid f(x) \leqslant y\} \\
&= \{(x, y) \in X \times Y \mid x \leqslant g(y)\},
\end{aligned} \tag{1.56}
$$

with the order induced by the cartesian product $X \times Y$ (that is a categorical product *in* Ord, as defined in 2.1.1).

The given adjunction has a canonical factorisation in two adjunctions

$$
X \underset{g'}{\overset{f'}{\rightleftarrows}} G(f, g) \underset{g''}{\overset{f''}{\rightleftarrows}} Y \tag{1.57}
$$

$$
\begin{aligned}
f'(x) &= (x, f(x)), & g'(x, y) &= x, \\
f''(x, y) &= y, & g''(y) &= (g(y), y).
\end{aligned}
$$

This forms a *natural weak factorisation system* in AdjOrd, in the sense of [GT] (see 2.5.5). Note that the first morphism is a monomorphism of AdjOrd (in fact its 'covariant part' f' is an injective mapping, because $g'f' = \mathrm{id}$) and the second is an epimorphism (in fact f'' is surjective, as $f''g'' = \mathrm{id}$).

1.7.7 Exercises and complements (Chains of adjunctions)

Non-trivial *chains of Galois connections* between two ordered sets X, Y

$$
\ldots f_{-2} \dashv f_{-1} \dashv f_0 \dashv f_1 \dashv f_2 \ldots \qquad (f_{2i} \colon X \rightleftarrows Y \colon f_{2i+1}), \tag{1.58}
$$

seem not to be frequent, 'in nature'.

Yet, as an interesting, unexpected fact, a Galois connection $\mathbb{Z} \rightleftarrows \mathbb{Z}$ on the integral (ordered) line *always* produces an unbounded chains of adjunctions.

The reader is invited to investigate this intriguing situation, construct examples and study variations, replacing $\mathbb{Z}$ with other ordered sets, like $\mathbb{N}$ or the real line. Then the results can be compared with the following outline – which has probably the fault of being too explicit.

(a) The following conditions on a monotone function $f \colon \mathbb{Z} \to \mathbb{Z}$ are equivalent:

(i) $f(\mathbb{Z})$ is lower and upper unbounded,

(ii) f has a right adjoint,

(ii*) f has a left adjoint.

In fact, by form (ii) of Definition 1.7.1, f has a right adjoint if and only if every subset $S(y) = \{x \in X \mid f(x) \leqslant y\}$ is non-empty and upper bounded, which is equivalent to condition (i); the latter is selfdual.

(b) It follows that each adjunction $f_0 \dashv f_1$ on $\mathbb{Z}$ can be extended to a unique chain of Galois connections, as in (1.58).

(c) As a *trivial case*, if one of these mappings is invertible, all of them are and we get a chain of adjunctions of period 2 (or possibly of period 1)

$$... \; f \dashv f^{-1} \dashv f \dashv f^{-1} \; ... \qquad (1.59)$$

Now, a bijective monotone mapping $f \colon \mathbb{Z} \to \mathbb{Z}$ is necessarily a translation $f(x) = x + k$. Therefore the only chain of period 1 is that of $f = \mathrm{id}\mathbb{Z}$.

On the other hand, the ordered set $\mathbb{Z}^2$ has involutive symmetries, like $f(x, y) = (y, x)$, and various chains $f \dashv f$ of period 1.

(d) In the chain (1.58), f_0 is injective if and only if f_1 is surjective (by 1.7.3(a)). *In such a case* all f_{2i} are split mono, all f_{2i+1} are split epi, and the adjunctions are expressed by the following conditions (for $i \in \mathbb{Z}$)

$$f_{2i+1} f_{2i} = \mathrm{id} = f_{2i-1} f_{2i}, \qquad f_{2i} f_{2i+1} \leqslant \mathrm{id}, \qquad f_{2i} f_{2i-1} \geqslant \mathrm{id}. \qquad (1.60)$$

Moreover $f_{2i} \leqslant f_{2i+2}$ and $f_{2i-1} \geqslant f_{2i+1}$, because:

$$f_{2i} \leqslant f_{2i+2} f_{2i+1} f_{2i} = f_{2i+2}, \qquad f_{2i-1} = f_{2i+1} f_{2i} f_{2i-1} \geqslant f_{2i+1}.$$

Once we exclude the trivial case (1.59), the sequence (f_{2i}) is strictly increasing while (f_{2i+1}) is strictly decreasing: in fact we know that the relation $f \dashv g \dashv f$ gives $g = f^{-1}$. Since no f_{2i} can be surjective, all mappings f_i are different.

The reader can compute the chain of adjunctions produced by the injective mapping $f_0(x) = 2x$.

(e) There is an interesting case of this kind, where all d_i are injective and all e_i are surjective

$$d_i, e_i \colon \mathbb{Z} \to \mathbb{Z}, \qquad ... \; d_{i+1} \dashv e_i \dashv d_i \dashv e_{i-1} \dashv d_{i-1} \; ... \qquad (1.61)$$

$$d_i(x) = x \text{ if } x < i, \qquad d_i(x) = x + 1 \text{ if } x \geqslant i,$$

$$e_i(x) = x \text{ if } x \leqslant i, \qquad e_i(x) = x - 1 \text{ if } x > i,$$

$$e_i d_i = \mathrm{id} = e_{i-1} d_i, \qquad d_{i+1} e_i \leqslant \mathrm{id}, \qquad d_i e_i \geqslant \mathrm{id},$$

$$d_i > d_{i+1}, \qquad e_i < e_{i+1}.$$

These mappings (or better their restrictions to finite ordinals) are related to the faces and degeneracies of simplicial sets (see 5.3.6(a)).

(f) On the other hand, if we start from an increasing mapping $f \colon \mathbb{Z} \to \mathbb{Z}$

that satisfies condition (i) and is neither injective nor surjective, we get a chain (1.58) of adjunctions, with $f_0 = f$, where no f_i is injective or surjective.

For instance, from the chain (1.61) one can readily compute the chain of Galois connections of the mapping $f = d_i e_i$, by composing adjunctions (as in 1.7.4).

(g) Replacing $\mathbb{Z}$ with the real line $\mathbb{R}$ *one cannot have non-trivial chains of adjunctions of length higher than 3*.

In fact in an adjunction $f \dashv g$ of increasing functions $\mathbb{R} \to \mathbb{R}$, the left adjoint f is lower and upper unbounded (as above) and preserves all sup (hence it is left-continuous) while the right adjoint g is lower and upper unbounded and preserves all inf (hence it is right-continuous).

Therefore, in a chain $f \dashv g \dashv h$, the function g is lower and upper unbounded and continuous. If g is strictly increasing, then it is invertible and we fall in the trivial case of period 2 or 1. Otherwise g is constant on some non-trivial interval $[a, b]$, but then both f and h are not continuous at the point $g([a, b])$ and the chain $f \dashv g \dashv h$ cannot be extended.

(h) Other variations on the ordered sets X, Y of diagram (1.58) are left to the reader; of course, these sets need not be the same. Chains of adjunctions between categories (that are not ordered sets) will be examined in 3.2.8.

1.7.8 Contravariant Galois connections

A *contravariant Galois connection* between the ordered sets X, Y is a pair of *decreasing* (or *antitone*) mappings $f \colon X \to Y$ and $g \colon Y \to X$ such that

$$gf \geqslant \mathrm{id}X, \qquad\qquad fg \geqslant \mathrm{id}Y. \qquad\qquad (1.62)$$

This *symmetric notion* is equivalent to a covariant Galois connection:

- replacing Y with Y^{op} we have two monotone functions $f \dashv g$,

- replacing X with X^{op} we have two monotone functions $g \dashv f$.

Classically, the term 'Galois connection' used to refer to the antitone case, which covers important concrete issues – some of which are recalled below. This terminology is still used in [M4]. Presently, the covariant setting is often privileged because only in this case composition is possible and we have a category.

The properties of a contravariant Galois connection are deduced from the covariant case, in an obvious way. Here both endomappings $gf \colon X \to X$ and $fg \colon Y \to Y$ are closure operators, and we can freely denote the sets of fixed elements as $\mathrm{cl}(X)$ and $\mathrm{cl}(Y)$ (in (1.50)).

1.7.9 Exercises and complements

(a) We are given an arbitrary subset $\perp \subset X \times Y$ of a cartesian product of sets, which we view as an 'orthogonality relation': $x \perp y$ means that $(x, y) \in \perp$.

There is an associated contravariant Galois connection

$$R: \mathcal{P}X \rightleftarrows \mathcal{P}Y : L,$$

$$R(A) = A\perp = \{y \in Y \mid \text{for all } x \in A, \ x \perp y\}, \tag{1.63}$$

$$L(B) = \perp B = \{x \in X \mid \text{for all } y \in B, \ x \perp y\},$$

that establishes a bijective correspondence between the closed subsets of X and those of Y

$$\mathrm{cl}(X) = \{A \subset X \mid A = \perp(A\perp)\} = \{\perp B \mid B \subset Y\},$$

$$\mathrm{cl}(Y) = \{B \subset Y \mid B = (\perp B)\perp\} = \{\perp A \mid A \subset X\}. \tag{1.64}$$

(b) For a commutative unital ring R, let us take the affine space $X = R^n$ and the polynomial ring $Y = R[x_1, ..., x_n]$ (for some $n \geqslant 1$). Every polynomial $p \in Y$ can be evaluated at any point $x \in X$, giving $p(x) \in R$. We write $x \perp p$ to mean that $p(x) = 0$. The associated contravariant Galois connection is well known in Algebraic Geometry, and sends:

- a subset $A \subset R^n$ to the ideal $\mathcal{R}(A)$ of all polynomial which annihilate at every point of A,

- a subset $B \subset R[x_1, ..., x_n]$ to the *affine algebraic set* $\mathcal{L}(B)$ of all points of R^n annihilated by every polynomial of B.

*If R is an algebraically closed field, Hilbert's *Nullstellensatz* characterises the closed subsets of $R[x_1, ..., x_n]$ as the radical ideals of this ring ([ZS], Section VII.3). On the other hand, the closed subsets of R^n are its affine algebraic sets, by definition.*

(c) An orthogonality (endo)relation $f \perp g$ between morphisms of a category will be introduced in 2.4.3.

(d) Let X be a preordered set. Applying point (a) to the relation $x \prec y$ we obtain a contravariant Galois connection between the mappings of upper and lower bounds

$$U, L: \mathcal{P}X \to \mathcal{P}X,$$

$$U(A) = \{x \in X \mid a \prec x, \text{ for all } a \in A\}, \tag{1.65}$$

$$L(A) = \{x \in X \mid x \prec a, \text{ for all } a \in A\}.$$

(e) One can easily deduce from these formulas the fact that, in a preordered set X

$$\vee A = \wedge(U(A)), \qquad \wedge A = \vee(L(A)). \tag{1.66}$$

More precisely, the first formula means that, for any subset A of a pre-ordered set, $\vee A$ exists if and only if $\wedge(U(A))$ exists, and – in this case – they coincide. (Exercise 1.2.5(b) is now a consequence.)

*(f) A contravariant Galois connection will be used in Exercise 2.4.7(a).

1.8 Elementary categories of relations and partial mappings

We now review the elementary construction of some categories of relations and partial mappings, for sets, or abelian groups, etc.

Various extensions of relations (to regular categories, abelian categories, etc.) will be examined in Chapters 4 and 6; in the latter we also show how categories of relations are important in Homological Algebra.

Relations and partial mappings will often (but not necessarily) be written as dot-marked arrows $X \rightarrowtail Y$, to distinguish them from ordinary mappings. Categories of partial mappings will be denoted by calligraphic letters like $\mathcal{S}, \mathcal{I}, \mathcal{C}, \ldots$

1.8.1 Categories and involutions

We begin by fixing our terminology for involutive endofunctors, starting from two examples related to ordered sets.

(i) The category Ord, of ordered sets and monotone mappings, has an involutive *covariant* endofunctor:

$$X \mapsto X^{\mathrm{op}}, \qquad (f \colon X \to Y) \mapsto (f^{\mathrm{op}} \colon X^{\mathrm{op}} \to Y^{\mathrm{op}}), \tag{1.67}$$

where X^{op} has reversed order and f^{op} is the 'same' mapping as f.

(ii) The category AdjOrd, of ordered sets and Galois connections (introduced in 1.7.4), has an involutive *contravariant* endofunctor:

$$X \mapsto X^{\mathrm{op}}, \qquad (u \colon X \rightarrowtail Y) \mapsto (u^{\mathrm{op}} \colon Y^{\mathrm{op}} \rightarrowtail X^{\mathrm{op}}), \tag{1.68}$$

where $u^{\mathrm{op}} = (g^{\mathrm{op}}, f^{\mathrm{op}})$, if $u = (f, g)$ with $f \dashv g$.

This situation will become clearer in the domain of double categories: the two procedures (i) and (ii) – which have the same action on the objects – can be combined to give a *vertical involution* of the double category AdjOrd, horizontally covariant (on monotone mappings) and vertically contravariant (on adjunctions): see 7.2.9.

We shall use the following terminology.

(a) A *category with reversor* will be a category C equipped with an involutive *covariant* endofunctor $R\colon \mathsf{C} \to \mathsf{C}$, often denoted as in (1.67), and said to be *trivial* when $R = \mathrm{id}\mathsf{C}$.

Obvious non-trivial examples are Ord and Cat. The name of 'reversor' is taken from Directed Algebraic Topology [G8], where the categories of 'directed spaces' in some sense (like preordered topological spaces, 'd-spaces', and small categories) have such an endofunctor, that turns a directed structure into the opposite one. In the trivial case, when $R = \mathrm{id}\mathsf{C}$, we go back to classical, non-directed 'spaces'.

(b) A *category with involution* will be a category C equipped with an involutive *contravariant* endofunctor $J\colon \mathsf{C} \dashrightarrow \mathsf{C}$, often denoted as in (1.68). It follows that C is selfdual, i.e. isomorphic to C^{op}.

AdjOrd is an obvious example; the same holds for AdjCat, that will be introduced in 3.2.1.

(c) More particularly, an *involutive category* will be a category C equipped with a (contravariant) involution *which is the identity on objects*. In other words every morphism $f\colon X \to Y$ has an associated *opposite morphism* $f^\sharp\colon Y \to X$ so that

$$f^{\sharp\sharp} = f, \qquad (gf)^\sharp = f^\sharp g^\sharp, \qquad (1_X)^\sharp = 1_X. \qquad (1.69)$$

Classical examples are the categories of relations introduced below.

The involution is said to be *regular* (in the sense of von Neumann) if $ff^\sharp f = f$ (for all maps f), as happens with relations of groups or modules. In this case all the semigroups of endomorphisms $\mathsf{C}(X, X)$ are regular, in the well-known sense of semigroup theory. The reader is warned that this has nothing to do with the notion of 'regular category', that will be seen in 4.2.1.

1.8.2 Ordered categories

In this book an *ordered category* is a category C equipped with an order relation $f \leqslant g$ between *parallel morphisms*, which is consistent with composition: for $f, g\colon X \to Y$ and $f', g'\colon Y \to Z$

(i) if $f \leqslant g$ and $f' \leqslant g'$ then $f'f \leqslant g'g$.

(This notion is related to 'enrichment', see 7.3.3; a different 'internal' notion is recalled in 7.3.6.)

A functor $F\colon \mathsf{C} \to \mathsf{D}$ between ordered categories is *monotone*, or *increasing*, if it preserves the order. Any subcategory of an ordered category inherits an ordering.

For instance the category Ord of ordered sets is canonically ordered by the *pointwise order* $f \leqslant g$, defined in (1.3). This ordering is inherited by the subcategories Lth, Mlh and Dlh of lattices, modular lattices and distributive lattices (see 1.2.4–1.2.6).

Also AdjOrd has a canonical order, defined in 1.7.4.

1.8.3 *Relations of sets*

The category Rel Set of *sets and relations* (or *correspondences*) is well known. A relation $u \colon X \dashrightarrow Y$, often denoted by a dot-marked arrow, is a subset of the cartesian product $X \times Y$. It can be viewed as a 'partially defined, multi-valued mapping', that sends an element $x \in X$ to the subset $\{y \in Y \mid (x, y) \in u\}$ of Y.

The identity relations and the composite of u with $v \colon Y \dashrightarrow Z$ are

$$\mathrm{id}X = \{(x, x) \mid x \in X\},$$
$$vu = \{(x, z) \in X \times Z \mid \exists y \in Y \colon (x, y) \in u \text{ and } (y, z) \in v\}. \tag{1.70}$$

It is easy to verify that Rel Set is a category (Exercise 1.8.4(b)). It is actually an *involutive ordered category*, in the following sense. First it is an involutive category as defined in 1.8.1: the *opposite relation* $u^\sharp \colon Y \dashrightarrow X$ is obtained by reversing pairs. Second, it is an ordered category, in the sense of 1.8.2: given two parallel relations $u, u' \colon X \dashrightarrow Y$ we say that $u \leqslant u'$ if $u \subset u'$ as subsets of $X \times Y$. Third, these structures are consistent: if $u \leqslant v$ then $u^\sharp \leqslant v^\sharp$ (and conversely, as a consequence).

The involution is *not* regular (in the sense of 1.8.1): see 1.8.4(c).

The category Set is embedded in the category of relations Rel Set, identifying a mapping $f \colon X \to Y$ with its graph $\{(x, y) \in X \times Y \mid y = f(x)\}$; the condition $f \leqslant g$ for parallel mappings amounts to $f = g$.

1.8.4 *Exercises and complements*

(a) It may help to draw here a synopsis of future investigations of Rel Set, as endowed with important structures – some of them perhaps unexpected:

(i) as a semiadditive category with arbitrary biproducts, in 2.2.6(c) and 6.4.5(d),

(ii) as a symmetric monoidal closed category (with a 'peculiar' internal hom), in 3.2.8(d) and 3.4.7(h),

(iii) as a Kleisli category for a monad on Set, in 3.7.2(c),

(iv) as a locally ordered 2-category in 4.5.1 (modified as a locally pre-ordered bicategory, in 4.5.5).

The reader is warned that the points (i)–(iii) have no counterpart in many other categories of relations – for instance in Rel Ab, introduced below.

The interpretation of Rel Set as a category of partially defined, multi-valued mappings will be better understood at the light of point (iii). Bicommutative squares in Rel Set are analysed in 2.3.8.

(b) Verify that Rel Set is an involutive ordered category.

(c) Show that the involution is not regular. *Hints:* use an endo-relation of a two-element set.

(d) A relation $f\colon X \to Y$ of sets is a mapping if and only if it is everywhere defined and single-valued. These conditions can be respectively characterised as follows

$$f^{\sharp}f \geqslant \mathrm{id}X, \qquad ff^{\sharp} \leqslant \mathrm{id}Y. \qquad (1.71)$$

This means that $f \dashv f^{\sharp}$ in the ordered category Rel Set, viewed as a 2-category: see 7.1.2.

1.8.5 Relations of abelian groups and modules

Abelian groups have a similar category of relations, also called *additive relations* or *additive correspondences* (see [M1, M2, Hi]).

A *relation of abelian groups* $u\colon X \to Y$ is a subgroup of the cartesian product $X \times Y$. Composition, order and involution are defined as above, giving the involutive ordered category Rel Ab.

Ab is embedded in the former as the wide subcategory of everywhere defined, single-valued relations.

It is important to note that the involution of Rel Ab is regular, i.e.

$$uu^{\sharp}u = u, \qquad \text{for all } u \text{ in Rel Ab}. \qquad (1.72)$$

In fact the inclusion $u \leqslant uu^{\sharp}u$ is obvious (and also holds for relations of sets). The other way round, if (x,y), (x',y) and (x',y') are in u, we have

$$(x,y') = (x,y) - (x',y) + (x',y') \in u. \qquad (1.73)$$

More generally for any (unital) ring R we have the involutive ordered category Rel $(R\,\mathsf{Mod})$ of left R-modules, where a relation is a submodule of the cartesian product.

We also have the involutive ordered category Rel Gp of groups, where a relation is a subgroup of the cartesian product. The involution is still regular, with the same proof as in (1.73), since the commutativity of addition has not been used.

1.8.6 Partial mappings

A partial mapping (of sets) $f\colon X \nrightarrow Y$ is a mapping $\mathrm{Def}\, f \to Y$ defined on a subset of X, its subset of *definition*. Equivalently, it is a single-valued relation.

For a partial mapping $g\colon Y \nrightarrow Z$, the composite $gf\colon X \nrightarrow Z$ is obvious (as in the well-known case of partially defined functions $\mathbb{R} \nrightarrow \mathbb{R}$):

$$\mathrm{Def}\,(gf) = f^{-1}(\mathrm{Def}\, g), \qquad (gf)(x) = g(f(x)), \qquad (1.74)$$

for $x \in \mathrm{Def}\,(gf)$.

We have thus a wide subcategory $\mathcal{S} \subset \mathsf{Rel\,Set}$ that contains Set. As a subcategory of $\mathsf{Rel\,Set}$, $\mathcal{S}$ is an ordered category: $f' \leqslant f$ means that f' is a restriction of f, defined on a subset of X contained in $\mathrm{Def}\,(f)$.

The category $\mathcal{S}$ *should not be viewed as essentially different* from the usual categories of structured sets and 'total' mappings, since *it is equivalent* to the category $\mathsf{Set}_\bullet$ of pointed sets, via the functor

$$R\colon \mathsf{Set}_\bullet \to \mathcal{S}, \qquad (X, x_0) \mapsto X \setminus \{x_0\}, \qquad (1.75)$$

that sends a pointed mapping $f\colon (X, x_0) \to (Y, y_0)$ to its restriction

$$R(f)\colon X \setminus \{x_0\} \nrightarrow Y \setminus \{y_0\},$$

$$\mathrm{Def}\, R(f) = Y \setminus f^{-1}(\{y_0\}).$$

In fact this functor is plainly faithful, full and surjective on objects (see 1.5.5).

The weak inverse $S\colon \mathcal{S} \to \mathsf{Set}_\bullet$ is obtained by choosing, for every set X, a pointed set

$$S(X) = (X \cup \{x_0\}, x_0) \qquad (x_0 \notin X),$$

and defining S on partial mappings in the obvious (and unique) way that gives $RS = 1$. Note that SR is only isomorphic to the identity.

> Set theory can give a canonical, perhaps confusing, choice for S: one can take $x_0 = X$, since $X \notin X$. A better choice might be the empty partial mapping $0_X\colon \{*\} \nrightarrow X$; see the solution of Exercise 3.4.2(f).

This equivalence will be used in 2.2.5(d) to transfer to $\mathcal{S}$ categorical constructions that are easily computed and well known in $\mathsf{Set}_\bullet$, like categorical and tensor products.

1.8.7 Partial bijections

More particularly a *partial bijection* (or partial isomorphism) of sets $f\colon X \nrightarrow Y$ is a bijection between a subset $\mathrm{Def}\, f$ of X and a subset $\mathrm{Val}\, f$ of Y, or equivalently a single-valued, injective relation.

Partial bijections form a wide subcategory

$$\mathcal{I} \subset \mathcal{S} \subset \mathrm{Rel\,(Set)}. \tag{1.76}$$

It is equivalent to the wide subcategory of Set. formed by the pointed mappings $f \colon (X, x_0) \to (Y, y_0)$ that are injective out of the preimage $f^{-1}\{y_0\}$.

$\mathcal{I}$ is used as a basic tool for graphic representations of spectral sequences, in [G9]. We shall see in 6.3.3(d) that it is a pointed category with good exactness properties, but lacks products and sums.

1.8.8 *Partial continuous mappings

In many cases one should consider partial morphisms defined on subobjects of a particular kind.

For topological spaces, for instance, partial continuous mappings should be 'reasonably' *defined on subspaces* (i.e. 'regular subobjects', as we shall see); but in fact the important structures of partial continuous mappings are even more restricted.

While studying topological manifolds, it can be useful to use the (ordered) category $\mathcal{C}$ of spaces and partial continuous mappings *defined on open subspaces*, together with its subcategory of spaces and partial homeomorphisms *between open subspaces*. This approach, introduced in [G2] and exposed in [G12], will be briefly reviewed in Section 7.4.

Exercises and complements. *(a) The interested reader can verify that the category $\mathcal{C}$ is equivalent to a full subcategory $\mathsf{Top}'_\bullet$ of the category $\mathsf{Top}_\bullet$ of pointed topological spaces, consisting of those objects (X, x_0) where the base point x_0 is closed and adherent to every other point (i.e. the only neighbourhood of x_0 is X). (A solution can be found in [G12], 2.1.5(d).)

*(b) The category $\mathcal{D}$ of topological spaces and partial continuous mappings *defined on closed subspaces* is also of some interest. It is equivalent to another full subcategory $\mathsf{Top}''_\bullet$ of $\mathsf{Top}_\bullet$, consisting of those objects (X, x_0) where the base point x_0 is open and dense (cf. [G12], 2.1.5(e)).

1.8.9 *Corelations and networks

The involutive ordered category $\mathsf{C} = \mathrm{Cor\,Set}$ *of sets and corelations* can be constructed in a dual way with respect to $\mathrm{Rel\,Set}$, letting $\mathsf{C}(X, Y)$ be the ordered set of the quotients of the sum $X + Y$; more explicitly, this is the set of the equivalence relations on the disjoint union of X and Y.

The reader can easily find out how corelations compose, by using the equivalence relation generated by a relation. (The pictures below tell everything about that!)

Early in the history of categories, since the late 1960's, the full subcategory $\mathcal{T}$ *of finite sets and corelations* was used in the theory of networks, by G. Darbo [Da1, Da2] and some of his collaborators like F. Parodi and G. Testa. The subject was apparently ahead of its time, and later forgotten; but recently J. Baez has independently reinvented it, with his students, in a similar perspective [BaP, Fo].

In fact, a corelation $\varphi\colon \alpha \rightarrowtail \beta$ between finite sets (called a 'transducer' in [Da1] and related articles) can be viewed as an electric circuit made of perfectly conductive wires, connecting the 'terminals' belonging to α to those belonging to β, as in the following picture

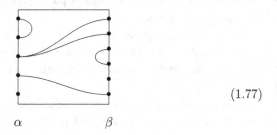

$$(1.77)$$

Transducers compose as if we were connecting wires

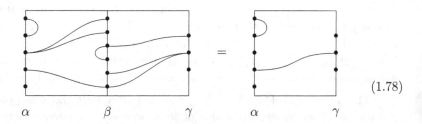

$$(1.78)$$

In [Da1] a 'universe of devices' is defined as a pair $(\mathcal{D}, \sigma)$, formed of a functor and a natural transformation

$$\mathcal{D}\colon \mathcal{T} \to \mathsf{Set},$$
$$\sigma_{\alpha\beta}\colon \mathcal{D}\alpha \times \mathcal{D}\beta \;\to\; \mathcal{D}(\alpha + \beta),$$

$$(1.79)$$

satisfying some natural axioms.

(It would now be called a unitary lax monoidal functor, from the symmetric monoidal category $(\mathcal{T}, +, \emptyset)$ to the cartesian monoidal category of sets, according to definitions that we shall see in Sections 3.4 and 7.2.4.)

In a model of the theory, $\mathcal{D}\alpha$ represents the set of devices with terminals belonging to the finite set α. A corelation $\varphi\colon \alpha \rightarrowtail \beta$ takes a device $A \in \mathcal{D}\alpha$ to a device $\varphi_*(A) \in \mathcal{D}\beta$, as suggested in the diagram below

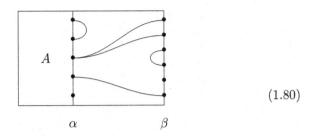

$$(1.80)$$

Finally, the mapping $\sigma_{\alpha\beta}$ takes a pair (A, B) of devices to a device $A + B$ which, in a concrete model, is obtained by putting A and B in a box and letting out their terminals

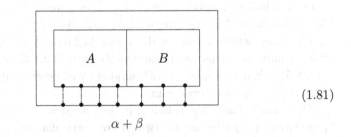

$$(1.81)$$

The theory is quite general, and not limited at all to electric devices.

2

Limits and colimits

Limits and colimits in a category C are defined by a universal property, which determines a solution up to isomorphism.

The general notion of a limit (introduced in Section 2.2) includes cartesian products, equalisers and pullbacks (analysed in Sections 2.1 and 2.3). Dually, colimits comprise sums, coequalisers, pushouts and the classical 'inductive limits' (analysed in Sections 2.1, 2.3 and 2.6).

The construction of limits is dealt with in Theorem 2.2.4. Their description by universal arrows and representability is in 2.7.2, by Kan extensions in 2.8.5. Their relationship with adjoints will be examined in 3.2.2 and 3.2.4, with comma categories in 3.3.5.

The name of limits and colimits, in the categorical sense, comes from the preexisting 'projective limits' (which are particular limits) and 'inductive limits' (particular colimits), for algebraic structures and topological spaces. In the usual categories of structured sets, inductive limits have important peculiar properties, while projective limits need no separate study.

2.1 Basic limits and colimits

We begin by defining *products* and *equalisers* in a category C, and computing them in various concrete categories. Then we do the same for the dual notions, *sums* and *coequalisers*. We end this section with a brief introduction to biproducts and additive categories.

The general definition of limits and colimits is deferred to the next section, where we shall see that all limits and colimits can be constructed with the basic ones, dealt with here (Theorem 2.2.4).

Various examples are presented as easy exercises, generally left to the reader. It will be useful to complete the missing details, and work out similar results for other categories of structured sets.

Other results and computations of basic limits and colimits will be given later: for Ban and Ban_1 in 2.2.6; for ordered sets in 3.2.3(d); for varieties of algebras in Chapter 4; for various categories of topological or metric spaces in Chapter 5; for abelian categories and their generalisations in Chapter 6.

2.1.1 Products

The simplest case of a limit, in a category C, is the *product* of a family $(X_i)_{i \in I}$ of objects, indexed by a *small set I*.

This is defined as an object X equipped with a family of morphisms $p_i \colon X \to X_i$ $(i \in I)$, called (cartesian) *projections*, which satisfies the following universal property

$$
\begin{array}{ccc}
X & \xrightarrow{\ p_i\ } & X_i \\
{\scriptstyle f}\uparrow\ & \nearrow{\scriptstyle f_i} & \\
Y & &
\end{array}
\qquad (2.1)
$$

(i) for every object Y and every family of morphisms $f_i \colon Y \to X_i$ there is a unique morphism $f \colon Y \to X$ such that, for all $i \in I$, $p_i f = f$.

The morphisms f_i are called the *components* of f, and the latter is often written as (f_i), even though it is not the same as the family of its components, of course.

The product of a family of objects need not exist. If it does, it is determined up to a unique *coherent* isomorphism, in the sense that if also Y is a product of the family $(X_i)_{i \in I}$ with projections $q_i \colon Y \to X_i$, then the unique morphism $f \colon X \to Y$ which commutes with all projections (i.e. $q_i f = p_i$, for all indices i) is an isomorphism.

Indeed there is also a unique morphism $g \colon Y \to X$ such that $p_i g = q_i$ (for all i); moreover $gf = \mathrm{id}X$ (because $p_i(gf) = p_i(\mathrm{id}X)$, for all i) and $fg = \mathrm{id}Y$.

Therefore one speaks of *the* product of the family $(X_i)_{i \in I}$, denoted as $\Pi_i X_i$.

We say that a category C *has products* (resp. *finite products*) if every family of objects indexed by a small set (resp. a finite set) has a product in C.

In particular the product of the empty family of objects $\emptyset \to \mathrm{Ob}\,C$ means an object X (equipped with no projections) such that for every object Y (equipped with no maps) there is a unique morphism $f \colon Y \to X$ (satisfying no condition). The solution is called the *terminal* object of C; again, it need not exist, but is determined up to a unique isomorphism. It can be written as $\top$.

2.1.2 Exercises and complements (Products)

(a) In Set, Top, Ab, Ord all products exist and are the usual cartesian ones (constructed on the product of the underlying sets). The terminal object is the singleton, with the corresponding structure.

Let us recall that a topological product X with projections $p_i \colon X \to X_i$ has the coarsest topology making all projections continuous. Similarly, in Ord, the product has the coarsest order relation making all projections monotone: $(x_i) \leqslant (y_i)$ if, for all indices i, $x_i \leqslant y_i$ in X_i.

(b) Products in the categories Set. and Top. are also obvious

$$\Pi_i(X_i, \overline{x}_i) = (\Pi_i X_i, (\overline{x}_i)). \tag{2.2}$$

(c) In the category X associated to a preordered set, the categorical product of a family of points $x_i \in X$ amounts to their inf (the greatest lower bound), and the terminal object amounts to the greatest element of X. These elements need not exist; they are determined up to the equivalence relation associated to our preorder, and uniquely determined in an ordered set.

(d) Products in Cat have been considered in 1.4.6; the terminal object is the singleton category **1**.

(e) It is easy to prove that a category has finite products if and only if it has binary products $X_1 \times X_2$ and a terminal object.

Solutions. All verifications are straightforward, except Exercise (e), which requires some work.

The empty product and binary products are accounted for, while unary products always exist, trivially. The reader can begin to prove that a ternary product $X_1 \times X_2 \times X_3$ can be obtained as $(X_1 \times X_2) \times X_3$, with projections

$$(X_1 \times X_2) \times X_3 \to X_1 \times X_2 \to X_1,$$
$$(X_1 \times X_2) \times X_3 \to X_1 \times X_2 \to X_2, \tag{2.3}$$
$$(X_1 \times X_2) \times X_3 \to X_3,$$

and write the general argument by induction on the number of factors.

2.1.3 Equalisers and regular monomorphisms

Another basic limit is the *equaliser* of a pair $f, g \colon X \to Y$ of parallel maps of C. This is an object E with a map $m \colon E \to X$ such that $fm = gm$ and the following universal property holds:

$$E \xrightarrow{\ m\ } X \underset{g}{\overset{f}{\rightrightarrows}} Y \qquad\qquad (m = \mathrm{eq}(f,g)) \qquad\qquad (2.4)$$

with w, h, Z in the diagram below.

(ii) every map $h\colon Z \to X$ such that $fh = gh$ factorises uniquely through m (i.e. there is a unique map $w\colon Z \to E$ such that $mw = h$).

The equaliser morphism is necessarily a monomorphism, and is called a *regular monomorphism*. It is easy to see that a regular mono which is epi is an isomorphism.

Products and equalisers are particular instances of the limit of a functor, that will be introduced in 2.2.1.

2.1.4 Exercises and complements (Equalisers)

(a) In Set the natural solution for the equaliser of a pair of mappings $f, g\colon X \to Y$ is the subset

$$E = \{x \in X \mid f(x) = g(x)\}, \qquad\qquad (2.5)$$

with the inclusion $m\colon E \to X$.

(b) In Top, Ab, Ord, Set$_\bullet$ and Top$_\bullet$ (and many other concrete categories) we construct the equaliser in the same way, putting on E the structure induced by X as a subspace, or a subgroup, etc.

(c) In Cat, the equaliser of two functors $F, G\colon \mathsf{C} \to \mathsf{D}$ is the subcategory of the objects and arrows of C on which F and G coincide.

(d) In any category, every split monomorphism is regular.

(e) The reader will note that in Top a monomorphism $m\colon X' \to X$ is any injective map (see 1.3.3(b)), while a regular mono must be the inclusion of a subspace $E \subset X$, up to homeomorphism; this means a *topological embedding* $m\colon X' \to X$, i.e. an injective continuous mapping where X' has the coarsest topology making m continuous: the open subsets of X' are the preimages of those of X. (The converse is also true: every embedding of a subspace is a regular mono, see 2.4.7(d).)

(f) We have a similar situation in Ord: a monomorphism is any injective monotone mapping, and is a regular mono if and only if it *reflects* the ordering (inducing an isomorphism between the domain and an ordered subset of the codomain).

2.1.5 Sums

As already mentioned in Section 1.1, every notion of category theory has a dual notion.

The *sum*, or *coproduct*, of a family $(X_i)_{i \in I}$ of objects of C is dual to its product. Explicitly, it is an object X equipped with a family of morphisms $u_i \colon X_i \to X$ $(i \in I)$, called *injections*, which satisfy the following universal property:

$$X_i \xrightarrow{\ u_i\ } X \atop \searrow_{f_i} \quad \downarrow f \atop Y \qquad (2.6)$$

(i*) for every object Y and every family of morphisms $f_i \colon X_i \to Y$, there is a unique morphism $f \colon X \to Y$ such that, for all $i \in I$, $f u_i = f_i$.

The map f will be written as $[f_i]$, by its *co-components*.

Again, if the sum of the family (X_i) exists, it is determined up to a unique coherent isomorphism, and denoted as $\sum_i X_i$, or $X_1 + \ldots + X_n$ in a finite case.

The sum of the empty family is the *initial* object $\bot$: this means that every object X has precisely one map $\bot \to X$.

2.1.6 Exercises and complements (Sums)

All points are easy or referred to, except (e), whose solution can be found in Chapter 8.

(a) Sums in Set are realised as 'disjoint unions'. There is a rather natural way of doing this

$$X = \Sigma_i X_i = \bigcup_i X_i \times \{i\}, \qquad u_i(x) = (x, i) \quad (\text{for } x \in X_i). \qquad (2.7)$$

The initial object is the empty set.

(b) In Top we do the same, equipping the set X with the finest topology that makes all the injections continuous; in other words, the open subsets of X are arbitrary unions of open subsets of the summands. The initial object is the empty space.

Similarly, in Ord we endow the set X with the finest ordering that makes all the injections monotone: $(x, i) \leqslant (y, j)$ if and only if $i = j$ and $x \leqslant y$ in X_i.

(c) In Abm, Ab and R Mod categorical sums are realised as 'direct sums'

$$\bigoplus A_i = \{(a_i) \in \Pi_i A_i \mid a_i = 0 \text{ except for a finite set of indices}\}, \qquad (2.8)$$

where the family (a_i) is said to be *quasi-null*. The initial object is the singleton.

(d) In Set. and Top. a sum $\Sigma_i(X_i, \bar{x}_i)$ can be built as a quotient of the unpointed sum $\Sigma_i X_i$, identifying all the base points $\bar{x}_i$; their class gives the new base point.

In Set. the sum can also be realised – more simply – as a subset of the cartesian product

$$\{(x_i) \in \Pi_i X_i \mid x_i = \bar{x}_i \text{ for all indices } i \text{ except one at most}\}, \qquad (2.9)$$

namely the union of all the 'cartesian axes' of the product, pointed again at the 'origin' $\bar{x} = (\bar{x}_i)$.

(e) In Top. this second construction works in the finite case: a finite sum, can be constructed as a subspace of the cartesian product. The operation is generally denoted by the join-symbol $\vee$

$$(X, x_0) \vee (Y, y_0) = (X \times \{y_0\}) \cup (\{x_0\} \times Y) \subset X \times Y, \qquad (2.10)$$

and the space is pointed at (x_0, y_0). For instance, $(\mathbb{R}, 0) \vee (\mathbb{R}, 0)$ can be realised as the subspace of the cartesian plane formed of the two cartesian axes, pointed at the origin.

(f) In the category X associated to a preordered set, the categorical sum of a family of points $x_i \in X$ amounts to its sup (the least upper bound), while the initial object amounts to the least element of X.

(g) In Cat a sum ΣC_i of categories is their obvious disjoint union. The initial object is the empty category **0**.

*(h) In Gp categorical sums are classically known as 'free products': see 4.1.3(c). Finite sums in CRng are constructed in 3.4.7(j).

2.1.7 Coequalisers and regular epimorphisms

The *coequaliser* of a pair $f, g: X \to Y$ of parallel maps of C is a map $p: Y \to C$ such that $pf = pg$ and

$$X \overset{f}{\underset{g}{\rightrightarrows}} Y \overset{p}{\longrightarrow} C \qquad (p = \operatorname{coeq}(f, g)) \qquad (2.11)$$

(ii*) every map $h: Y \to Z$ such that $hf = hg$ factorises uniquely through p (i.e. there is a unique map $w: C \to Z$ such that $wp = h$).

A coequaliser morphism is necessarily an epimorphism, and is called a *regular epimorphism*. A regular epi which is mono is an isomorphism.

Sums and coequalisers are particular instances of the *colimit* of a functor, that will be introduced in 2.2.1.

2.1.8 Exercises and complements (Coequalisers)

(a) In Set the natural solution for the coequaliser of two mappings $f, g\colon X \to Y$ is the projection

$$p\colon Y \to Y/R, \tag{2.12}$$

on the quotient modulo the equivalence relation of Y spanned by the pairs $(f(x), g(x)) \in Y^2$, for $x \in X$.

(b) In Top we do the same, putting on Y/R the quotient topology, namely the finest that makes the mapping p continuous. In pOrd we do the same, putting on Y/R the induced preordering (the finest that makes p monotone). In Ord we first compute the preorder Y/R in pOrd, and then take the associated ordered set (as in 1.2.2); this fact will be extended to all colimits, in 3.2.3(d).

(c) In Ab we take the quotient Y/H, modulo the subgroup

$$H = \{f(x) - g(x) \mid x \in X\},$$

which amounts to the quotient Y/R' modulo the congruence of abelian groups spanned by the previous equivalence relation R. The reader will adapt this construction to Gp.

(d) Coequalisers in Set$_\bullet$ and Top$_\bullet$ are easy, and left to the reader.

(e) In Top an epimorphism $p\colon Y \to Y'$ is any surjective map (see 1.3.3(b)), while a regular epi must be the projection on a quotient space $p\colon Y \to Y/R$, up to homeomorphism; this means a *topological projection* $p\colon Y \to Y'$, i.e. a surjective continuous mapping where Y' has the finest topology making p continuous: a subset of Y is open if and only if its preimage is open in Y. (The converse is also true: every topological projection is a regular epi, see 2.4.5(c).)

(f) In Cat, the coequaliser of the distinct embeddings $f, g\colon \mathbf{1} \to \mathbf{2}$ of the singleton category in the arrow category is the free monoid $(\mathbb{N}, +)$, as a category on one object (see Exercise 1.4.9(e)).

*(g) In general, the coequaliser in Cat of two functors $F, G\colon \mathsf{C} \to \mathsf{D}$ is the quotient of D modulo the *generalised congruence* generated by this pair; the latter, as defined in [BBP], also involves equivalent objects. One can avoid giving a 'construction' of the coequaliser category (necessarily complicated) and just prove its existence by the Adjoint Functor Theorem 3.5.2.

2.1.9 Zero objects, biproducts and additive categories

We end this section by briefly introducing some important notions, that will be more completely analysed in Section 6.4.

First, a *zero object* of a category, often written as 0, is both initial and terminal. This exists in Abm, Ab, R Mod, Gp, Set$_\bullet$, Top$_\bullet$,... but not in Set, Top, Cat,... where the initial and terminal object are distinct.

A category with zero object is said to be *pointed*; then each pair of objects A, B has a zero-morphism $0_{AB} \colon A \to B$ (also written as 0), which is given by the composite $A \to 0 \to B$.

As a stronger fact, the reader may have remarked that, in Abm, Ab and R Mod, a *finite* product ΠA_i and the corresponding finite sum ΣA_i are realised as the same object $A = \bigoplus A_i$, which satisfies:

- the property of the product, by a family of projections $p_i \colon A \to A_i$,

- the property of the sum, by a family of injections $u_i \colon A_i \to A$,

- the equations $p_i u_i = \mathrm{id} A_i$ and $p_j u_i = 0 \colon A_i \to A_j$ (for $i \neq j$).

Such an object (in a pointed category) is called a *biproduct*. The empty biproduct is the zero object.

Other examples, including Ban and Rel Set (but not Ban$_1$), will be seen in 2.2.6.

As a related notion, a *preadditive*, or $\mathbb{Z}$-*linear*, category is a category C where every hom-set $C(A, B)$ is equipped with a structure of abelian group, generally written as $f + g$, so that composition is additive in each variable (i.e. bilinear over $\mathbb{Z}$)

$$(f + g)h = fh + gh, \qquad k(f + g) = kf + kg, \qquad (2.13)$$

where $h \colon A' \to A$ and $k \colon B \to B'$. A preadditive category on a single object 'is' a (unital) ring.

An *additive category* C is a preadditive category with finite products, or equivalently with finite sums, which are then biproducts. The equivalence will be proved in Theorem 6.4.3, where we also see that, *in this case*, the sum of parallel maps is determined by the categorical structure.

Typical examples are Ab, R Mod, Ban (while Abm and Rel Set are just $\mathbb{N}$-linear and 'semiadditive': see 6.4.3, 6.4.5).

Exercises and complements. (a) In every pointed category with binary sums and products, there is a canonical morphism $f \colon X + Y \to X \times Y$.

(b) We have used this morphism in 2.1.6(e), and an obvious extension to arbitrary sums and products in 2.1.6(d).

2.2 General limits and completeness

This section contains the main definitions and basic results about limits and colimits in a category C. The main theorem proves that all limits can be constructed from products and equalisers.

2.2.1 *Limits and colimits*

Let S be a small category and $X: S \to C$ a functor, written in index notation (as in 1.4.9(a)), for $i \in S = Ob\,S$ and $a: i \to j$ in S:

$$X: S \to C, \qquad i \mapsto X_i, \quad a \mapsto (X_a: X_i \to X_j). \qquad (2.14)$$

A *cone* for the functor X is an object A of C (the *vertex* of the cone) equipped with a family of maps $(f_i: A \to X_i)_{i \in S}$ in C such that all the following triangles commute

$$
\begin{array}{ccc}
A \xrightarrow{\; f_i \;} X_i & \qquad & X_a.f_i = f_j \\
 \searrow_{f_j} \;\; \downarrow{X_a} & & \\
 X_j & & \text{(for } a: i \to j \text{ in S).}
\end{array}
\qquad (2.15)
$$

The *limit* of $X: S \to C$ is a universal cone $(L, (u_i: L \to X_i)_{i \in S})$. This means a cone of X such that every cone $(A, (f_i: A \to X_i)_{i \in S})$ factorises uniquely through the former; in other words, there is a unique map $f: A \to L$ such that, for all $i \in S$, $u_i f = f_i$. The solution need not exist; when it does, it is determined up to a unique coherent isomorphism and the limit object L is denoted as $\mathrm{Lim}(X)$.

The uniqueness part of the universal property amounts to saying that *the family* $u_i: L \to X_i$ $(i \in S)$ *is jointly mono* (see 1.3.1). A cone of X that satisfies the existence part of the universal property is called a *weak limit*; of course such a solution is 'determined' in a (very) weak way.

Dually, a *cocone* $(A, (f_i: X_i \to A)_{i \in S})$ of X satisfies the condition $f_j.X_a = f_i$ for every $a: i \to j$ in S. The *colimit* of the functor X is a universal cocone $(L', (u_i: X_i \to L')_{i \in S})$: the universal property says now that for every cocone $(A, (f_i))$ of X there is a unique map $f: L' \to A$ such that $f u_i = f_i$ (for all i). The colimit object is denoted as $\mathrm{Colim}(X)$.

The uniqueness part of the universal property means that the family $u_i: X_i \to L'$ is jointly epi. A cocone that satisfies the existence part of the universal property is called a *weak colimit*.

2.2.2 *Examples and complements*

(a) The product ΠX_i of a family $(X_i)_{i \in S}$ of objects of C is the limit of the corresponding functor $X \colon S \to C$, defined on the *discrete* category whose objects are the elements $i \in S$ (and whose morphisms reduce to formal identities of such objects). The sum ΣX_i is the colimit of this functor X.

We recall that the family of cartesian projections (p_i) of a product is *jointly mono*. These projections are 'often' epi but *not necessarily*: for instance in Set (and Top) this fails whenever some of the factors X_i are empty and others are not. However, one can easily show that, if C has a zero object (as defined in 2.1.9), all cartesian projections are split epi.

(b) The equaliser in C of a pair of parallel morphisms $f, g \colon X_0 \to X_1$ is the limit of the obvious functor defined on the category $0 \rightrightarrows 1$. The coequaliser is the colimit of this functor.

(c) Pullbacks and pushouts will be studied in the next section. Filtered colimits in Section 2.6.

(d) The existence of all limits in Set will follow from the existence of products and equalisers, in Theorem 2.2.4. However, as an interesting exercise, the reader can prove that the limit of any functor $X \colon S \to$ Set can be constructed as the set of its cones with vertex at the singleton set $\{*\}$:

$$\mathrm{Lim}(X) = \{(x_i) \in \Pi X_i \mid X_a(x_i) = x_j, \text{ for all } a \colon i \to j \text{ in S}\}. \qquad (2.16)$$

(e) In Set every non-empty set is weakly terminal; $\emptyset$ is the unique weakly initial object. In a pointed category every object is weakly terminal and weakly initial.

(f) Weak pullbacks of sets are used in 2.3.8. A weakly initial object will be used in the proof of the Initial Object Theorem, in 3.5.1.

*(g) Weak limits and colimits are studied in [G4, G6]. They are important in Homotopy Theory. A reader familiar with the notion of homotopy pullback (resp. pushout) of topological spaces [Mat] will likely know, or easily prove, that it gives a weak pullback (resp. pushout) in the homotopy category hoTop.

2.2.3 *Complete categories and the preservation of limits*

A category C is said to be *complete* (resp. *finitely complete*) if it has a limit for every functor $S \to C$ defined on a *small* category (resp. a *finite* category).

More precisely one speaks – in the first case – of a *small-complete* category. Of course we cannot expect Set to have products indexed by a large set, or limits for all functors $S \to$ Set defined on a large category. In other words, the basis S should be of a 'smaller size' than the category C where we construct limits, or we force C to be just a preordered set (where products are infima, and are not influenced by the size of the set of indices).

The question is settled by a neat result of P. Freyd: a small category C has the limit of every functor defined on a small category (if and) only if it is a preordered set with all infima (see 2.2.5(f)). This is why, *in an arbitrary category, the existence of all (small) limits does not imply the existence of colimits*, as is the case of preordered sets (see 2.2.5(e)).

A functor $F: \mathsf{C} \to \mathsf{D}$ is said to *preserve the limit* $(L, (u_i: L \to X_i)_{i \in S})$ of a functor $X: \mathsf{S} \to \mathsf{C}$ if the transformed cone

$$(FL, (Fu_i: FL \to FX_i)_{i \in S})$$

is the limit of the composed functor $FX: \mathsf{S} \to \mathsf{D}$. One says that F *preserves limits* if it preserves all the limits which exist in C. Analogously for the preservation of products, equalisers, etc. Saying that a functor *does not preserve limits* simply means that it does not preserve *some* of them.

Dually we have *cocomplete* categories and the property of *preservation of colimits*.

2.2.4 Theorem (Construction and preservation of limits)

A category is complete (resp. finitely complete) if and only if it has equalisers and products (resp. finite products). Moreover, if C is complete (resp. finitely complete), a functor $F: \mathsf{C} \to \mathsf{D}$ preserves all limits (resp. all finite limits) if and only if it preserves equalisers and products (resp. finite products).

Dual results hold for colimits. In particular, a category is cocomplete (resp. finitely cocomplete) if and only if it has coequalisers and sums (resp. finite sums).

Proof The reader can complete this outline. (A similar, more detailed argument can be found in [M4], Section V.2, Theorem 1.)

Let the functor $X: \mathsf{S} \to \mathsf{C}$ be written in index notation, as in 2.2.1, and consider the small products

$$\prod_i X_i \qquad \text{indexed by the objects } i \in \mathrm{Ob}\,\mathsf{S}, \text{ with projections } p_i,$$
$$\prod_a X_{j(a)} \quad \text{indexed by the arrows } a \text{ in } \mathsf{S}, \text{ with projections } q_a, \tag{2.17}$$

where it is understood that $a: i(a) \to j(a)$ in S.

A cone $(A, (f_i))$ of X amounts to a map $f: A \to \prod_i X_i$ that equalises the following two maps u, v (and gives back the original cone letting $f_i = p_i f$)

$$u, v: \prod_i X_i \to \prod_a X_{j(a)},$$
$$q_a u = p_{j(a)}, \qquad q_a v = X_a.p_{i(a)} \qquad \text{(for } a \text{ in S).} \tag{2.18}$$

Taking the equaliser $m \colon L \to \prod_i X_i$ of the pair u, v, we have a cone $(L, (u_i))$ of X, with $u_i = p_i m \colon L \to X_i$ (for $i \in \mathrm{Ob}\, \mathsf{S}$). This cone is the limit.

The second part of the statement, on the preservation of limits, is an obvious consequence, once we assume that products and equalisers exist, so that all limits can be obtained as above. $\qquad\square$

2.2.5 Exercises and complements, I

(a) Taking into account the computations of Section 2.1 it follows that the categories Set, Top, Ab, Ord, Set, and Top, are complete and cocomplete.

The forgetful functor Top $\to$ Set preserves limits and colimits, while Ab $\to$ Set only preserves limits; Ord $\to$ Set preserves limits and sums, but it does not preserve coequalisers (and general colimits, as a consequence).

(b) The reader will note that all limits in Top are constructed as in Set and equipped with the coarsest topology making all the structural mappings $u_i \colon L \to X_i$ continuous. Similarly, all colimits in Top are constructed as in Set, and equipped with the finest topology making all the structural mappings $u_i \colon X_i \to L'$ continuous.

(c) One can proceed in a similar way in the category pOrd of preordered sets. In Ord this procedure works for limits and sums; for coequalisers and general colimits it must be 'corrected', replacing the colimit in pOrd with the associated ordered set (see 3.2.3(d)).

(d) *The equivalence between* Set, *and* $\mathcal{S}$ (in 1.8.6) *proves that also the latter has all limits and colimits.* It is interesting to work out the case of products, showing that the product 1×1 in $\mathcal{S}$ has three elements, while 2×3 has eleven.

Sums in $\mathcal{S}$ can also be constructed in this way, but they simply amount to disjoint unions of sets.

(e) In a preordered set X, viewed as a category, equalisers and coequalisers are (obviously) trivial, while we have seen in Section 2.1 that products and sums amount to infima and suprema, respectively.

Therefore X is a complete category if and only if it has all infima, which is equivalent to the existence of all suprema (as recalled in 1.2.5(b)): in other words, the category X is complete if and only if it is cocomplete. In the ordered case this means that X is a complete lattice.

Similarly, an ordered set is a (bounded) lattice if and only if, as a category, it is finitely complete and cocomplete.

*(f) Freyd's result recalled in 2.2.3 can be written in this form: a small category C with all small products is a preordered set (with all infima).

2.2.6 Exercises and complements, II

(a) It is interesting to consider limits and colimits in the categories Ban_1 and Ban (introduced in 1.1.1).

The reader can prove that Ban_1 has binary products and sums, and Ban has binary biproducts (see 2.1.9). After this, it is easy to show that both Ban_1 and Ban are finitely complete.

*(b) In fact Ban_1 is even complete and cocomplete, while Ban is finitely complete and cocomplete. The construction of coequalisers (as quotient linear spaces modulo closed linear subspaces, with the induced norm) is less elementary than the previous points above; the construction of arbitrary products and sums in Ban_1 is even more delicate.

All this will not be proved here. A reader interested in the non-elementary aspects of these categories is referred to Semadeni's book [Se]. A fundamental result that simplifies this topic is the fact that a normed linear space is complete if and only if every family of elements of summable norm is summable.

(c) $\mathsf{Rel\,Set}$ has arbitrary biproducts. *Hints:* use a disjoint union of sets.

(d) $\mathsf{Rel\,Ab}$ lacks terminal and initial object.

*(e) $\mathsf{Rel\,Set}$ and $\mathsf{Rel\,Ab}$ lack equalisers and coequalisers. *Hints:* use Exercises 2.8.2(e) and 6.6.2(h).

(f) If the category C is complete (resp. finitely complete), so is the category C^S (for any small category S), with limits computed pointwise on each object of S. Similarly for colimits.

2.2.7 Representable functors detect limits and colimits

Let C be a category with small hom-sets, and $X\colon \mathsf{S} \to \mathsf{C}$ a functor defined on a small category. The universal property of the limit of X can be rewritten as follows:

- a cone $(L, (u_i\colon L \to X_i))$ of X is a limit in C if and only if, for every A in C, the set $\mathsf{C}(A, L)$ is the limit in Set of the sets $\mathsf{C}(A, X_i)$, with limit-cone $u_i.-\colon \mathsf{C}(A, L) \to \mathsf{C}(A, X_i)$.

This rewriting tells us that:

(i) each representable covariant functor $F_A = \mathsf{C}(A, -)\colon \mathsf{C} \to \mathsf{Set}$ preserves all (the existing) limits,

(ii) the family (F_A), indexed by the objects $A \in \mathrm{Ob}\,\mathsf{C}$, reflects them.

Colimits in C are limits in C^op. Thus, a representable contravariant functor $\mathsf{C}(-, A)\colon \mathsf{C}^\mathrm{op} \to \mathsf{Set}$ takes *colimits* of C to *limits* of sets; globally, their family reflects limits of Set into colimits of C.

For instance, for the contravariant functor of subsets $\mathsf{Set}^{\mathrm{op}} \to \mathsf{Set}$ of 1.6.5(c), we (plainly) have $\mathcal{P}(\sum_i X_i) \cong \prod_i \mathcal{P}(X_i)$, since a subset $A \subset \sum_i X_i$ amounts to a family of subsets $A_i \subset X_i$.

2.2.8 Creating limits

There is another interesting relationship between functors and limits (or colimits); again, it does not assume their existence (as defined in [M4], Section V.1).

One says that a functor $F\colon \mathsf{C} \to \mathsf{D}$ *creates limits* for a functor $X\colon \mathsf{S} \to \mathsf{C}$ if:

(i) for every limit cone $(L', (v_i\colon L' \to FX_i)_{i \in S})$ of the composed functor $FX\colon \mathsf{S} \to \mathsf{D}$, there is a unique cone $(L, (u_i\colon L \to X_i)_{i \in S})$ of X in C taken by F to $(L', (v_i))$,

(ii) this cone $(L, (u_i))$ is a limit of X in C (preserved by F).

The relationship of this notion with the reflection and preservation of limits is explored below, in 2.2.9. Saying that F creates products obviously means that F creates limits for every functor defined on a small discrete category.

For instance the forgetful functor $\mathsf{Ab} \to \mathsf{Set}$ creates all small limits (which do exist). Indeed, a limit of abelian groups can be constructed by taking the limit L' of the underlying sets and putting on it the unique structure that makes the mappings v_i into homomorphisms.

We shall see that the same happens for all varieties of algebras (in Section 4.4). On the other hand, these forgetful functors do not create general *colimits*: they do not even preserve them – with an important exception, filtered colimits, dealt with in Section 2.6.

As another example, the forgetful functor $\mathsf{Top} \to \mathsf{Set}$ preserves all limits and colimits but does not create them: for instance the product topology is the coarsest topology on the product set that makes all cartesian projections continuous, and is not the only one having this outcome (generally): any finer topology would do.

Of course we can restrict the limits that we are considering to the S-based limits, for a given S, or for some set of small categories; e.g. finite limits, or products, etc.

2.2.9 Exercises and complements (Acting on limits)

Let $F\colon \mathsf{C} \to \mathsf{D}$ be a functor and $X\colon \mathsf{S} \to \mathsf{C}$ a functor defined on a small category.

(a) If F creates limits for X, then it reflects its limit.

(b) If F creates limits for X, and FX has a limit in D, the limit of X exists and F preserves it.

(c) F creates unary products if and only if it reflects identities; then, F is injective on the objects.

(d) If X has a limit L preserved by F, and F reflects identities, then F creates the limit of X.

(e) The forgetful functor Top $\to$ Set does not reflect identities (but is amnestic, see 1.4.4).

(f) The reader likely knows that a differentiable manifold can be defined in two (slightly) different ways: as a convenient topological space equipped with a differentiable atlas, *or* equipped with an equivalence class of them. After defining the differentiable maps, these two approaches give equivalent categories of differentiable manifolds; in each of them the forgetful functor with values in Top preserves finite products, but only in the second case it reflects identities, and creates finite products.

2.3 Pullbacks and pushouts

Pullbacks and pushouts are an important pair of limit and colimit, dual to each other. We work in an arbitrary category C.

2.3.1 Pullbacks

The *pullback* of a pair of morphisms (f, g) of C with the same codomain

$$f \colon X_1 \to X_0 \leftarrow X_2 \colon g$$

is the limit of the corresponding functor defined on the category $1 \to 0 \leftarrow 2$.

This amounts to the usual definition: an object A equipped with two maps $u_i \colon A \to X_i$ $(i = 1, 2)$ which form a commutative square $fu_1 = gu_2$

$$(2.19)$$

in a universal way:

- for every triple (B, v_1, v_2) such that $fv_1 = gv_2$, there is a unique map $w \colon B \to A$ such that $u_1 w = v_1$, $u_2 w = v_2$.

The pair (u_1, u_2) is jointly mono. The pullback-object A is also called a *fibred product* over X_0 and written as $X_1 \times_{X_0} X_2$. The square diagram $f u_1 = g u_2$ in (2.19) is also called a *pullback*, or a *pullback diagram*; it can be marked in the pullback corner, as above.

In the ordered set X the pullback of the diagram $x_1 \leqslant x_0 \geqslant x_2$ is the meet $x_1 \wedge x_2$. Saying that X has pullbacks means that every *upper-bounded* pair of elements has a meet.

In Set (resp. Top, Ab) the pullback-object can be realised as a subset (resp. subspace, subgroup) of the product $X_1 \times X_2$:

$$A = \{(x_1, x_2) \in X_1 \times X_2 \mid f(x_1) = g(x_2)\}. \tag{2.20}$$

2.3.2 Exercises and complements

(a) Generalising the last construction, prove that a category that has binary products and equalisers also has pullbacks: A is constructed as the equaliser of two maps $X_1 \times X_2 \to X_0$. Pullbacks in concrete categories are usually computed in this way.

(b) This result does not need the terminal object. When the latter exists, the product $X_1 \times X_2$ amounts to the pullback $X_1 \times_\top X_2$ over the terminal object.

(c) Note that the pullback of the symmetric pair $X \to \top \leftarrow X$ is the product $(X \times X, p, q)$, with *different* projections (generally). When a fixed choice of pullbacks is used, *one cannot assume that the choice is symmetric*; even though it certainly is, up to isomorphism.

2.3.3 Pushouts

Dually, the *pushout* of a pair (f, g) of morphisms with the same domain is the colimit of the obvious functor defined on the category $1 \leftarrow 0 \to 2$.

This amounts to an object A equipped with two maps $u_i \colon X_i \to A$ $(i = 1, 2)$ which form a commutative square with $u_1 f = u_2 g$, in a universal way:

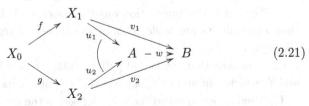

$$\tag{2.21}$$

- for every triple (B, v_1, v_2) such that $v_1 f = v_2 g$ there is a unique map $w \colon A \to B$ such that $w u_1 = v_1$, $w u_2 = v_2$.

The pushout square can also be marked in the appropriate corner.

A category that has binary sums and coequalisers also has pushouts: A is constructed as the coequaliser of the maps $j_1 f, j_2 g \colon X_0 \to X_1 + X_2$ given by the injections $j_i \colon X_i \to X_1 + X_2$.

The pushout-object A is also called a *pasting* over X_0 and written as $X_1 +_{X_0} X_2$. In Set this means a quotient of the sum $X_1 + X_2$ modulo the equivalence relation generated by all the pairs $(f(x), g(x))$ (viewed in the sum), for $x \in X_0$.

In the category Top of topological spaces, a pasting $X_1 +_{X_0} X_2$ is constructed as in Set and equipped with the quotient topology of the sum. It is a useful way of constructing new spaces. The reader can note that the usual process of verifying the continuity of a mapping $f \colon X \to Y$ on the subsets of a *closed* cover (X_1, X_2) of X amounts to saying that X is the pushout of X_1 and X_2 over $X_1 \cap X_2$. This fact can be easily generalised to a *finite closed cover* (or 'transferred' to an arbitrary open cover).

The well known Seifert–van Kampen Theorem exhibits the fundamental group of a space as a pushout of groups. R. Brown's version generalises this result, using fundamental groupoids [Bro1, Bro2].

2.3.4 Kernel pairs and cokernel pairs

Given a morphism $f \colon A \to B$, the pullback $R_f \rightrightarrows A$ of the pair (f, f) is called the *kernel pair* of f

$$
\begin{array}{ccc}
A & \xrightarrow{\ f\ } & B \\
\uparrow{\scriptstyle u} & \searrow & \uparrow{\scriptstyle f} \\
R_f & \xrightarrow{\ v\ } & A
\end{array}
\qquad (2.22)
$$

In Set it can be realised as $R_f = \{(x, x') \in A \times A \mid f(x) = f(x')\}$, the equivalence relation R_f associated to f. Then, the coequaliser of the pair $u, v \colon R_f \rightrightarrows A$ is the (projection on) the quotient A/R_f. Kernel pairs and their coequalisers are basic tools of the theory of regular categories: see Chapter 4.

One can note that the kernel pair has little to do with the equaliser of f and f, which – in any category – is the identity of its domain.

The dual notion is called the *cokernel pair* of the morphism $f \colon A \to B$. In Set it can be realised as a quotient of the sum $B + B$; a similar construction works in any category with binary sums and coequalisers.

2.3.5 Exercises and complements, I

The following properties of pullbacks and pushouts are important; the verification is straightforward, and left to the reader.

(a) *Characterising monos and epis.* The morphism m is mono if and only if the left square below is a pullback, if and only if the kernel pair of m consists of two equal arrows

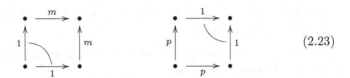

$$(2.23)$$

(a bullet stands for an object).

Dually the morphism p is epi if and only if the right square is a pushout, if and only if the cokernel pair of p consists of two equal arrows.

Note that these characterisations do not require the existence of pullbacks and pushouts in our category.

(b) A functor which preserves pullbacks (resp. pushouts) also preserves monomorphisms (resp. epimorphisms). This proves again that a representable functor preserves monomorphisms, as we have already seen in 1.6.5(d).

(c) *Pasting property.* If the two squares below are pullbacks, so is their 'pasting', i.e. the outer rectangle

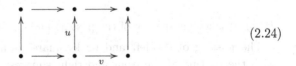

$$(2.24)$$

(d) *Depasting property.* If in the commutative diagram above the outer rectangle is a pullback and the pair (u, v) is jointly mono then the left square is a pullback.

The hypothesis on the pair (u, v) is automatically satisfied in two cases frequently met: when the right square is also a pullback, or one of the morphisms u, v is mono.

(e) Pushouts have dual properties with respect to (b) and (c). In an abelian category pullbacks and pushouts are characterised by exactness properties of an associated sequence of morphisms: see 6.4.8.

2.3.6 Exercises and complements, II

(a) A category with pullbacks and terminal object is finitely complete. *Hints.* The equaliser of two maps $A \rightrightarrows B$ can be constructed by a pullback, using the diagonal morphism $B \to B^2$.

(b) Epimorphisms in Hsd are precisely the maps whose image is dense in their codomain. *Hints.* We have already seen that this condition is sufficient (in 1.3.4(b)); conversely, we can consider the embedding $C \to Y$ of a closed subspace into a Hausdorff space, and show that its cokernel pair in Top also 'works' in Hsd.

*(c) The class of limits generated by the existence of pullbacks, in a category, is characterised in [Pa].

2.3.7 Lemma

In a category C we suppose that the following diagram is commutative.

If the two lateral squares and the back square are pullbacks (with vertices P, Q and R), then the front square is also a pullback

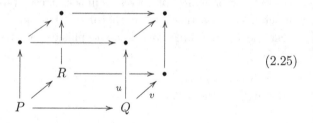

$$(2.25)$$

Proof It is a consequence of the pasting and depasting properties, in 2.3.5.

The pasting of the left and back squares is a pullback, and coincides with the pasting of the front and right squares; but the right square is a pullback, whence the front square is also.

Also here one can drop the hypothesis that the right square be a pullback and replace it with: the pair (u, v) is jointly mono.   □

2.3.8 *Bicommutative squares and exact squares

This point is about categories of relations of sets and modules, and will be used in Chapter 6.

(a) First we work in Set.

It is easy to see that a square of mappings

$$
\begin{array}{ccc}
A & \xrightarrow{\ f\ } & X \\
{\scriptstyle g}\downarrow & & \downarrow{\scriptstyle h} \\
Y & \xrightarrow[\ k\ ]{} & B
\end{array}
\qquad (2.26)
$$

is commutative if and only if we have $gf^\sharp \leqslant k^\sharp h$ in $\mathrm{Rel\,Set}$ (or $fg^\sharp \leqslant h^\sharp k$, equivalently). In fact

$$
hf = kg \quad \Rightarrow \quad gf^\sharp \leqslant k^\sharp kgf^\sharp = k^\sharp hff^\sharp \leqslant k^\sharp h,
$$

$$
gf^\sharp \leqslant k^\sharp h \quad \Rightarrow \quad kg \leqslant kgf^\sharp f \leqslant kk^\sharp hf \leqslant hf \quad \Rightarrow \quad kg = hf.
$$

We say that the square (2.26) is *bicommutative* in $\mathrm{Rel\,Set}$ if it satisfies the stronger condition

(i) $gf^\sharp = k^\sharp h\colon X \rightarrowtail Y \qquad (\Leftrightarrow\ fg^\sharp = h^\sharp k\colon Y \rightarrowtail X)$,

i.e. if the original square (of mappings) commutes in $\mathrm{Rel\,Set}$ when we 're-verse' two parallel edges.

Plainly, condition (i) is equivalent to:

(ii) for $(x, y) \in X \times Y$, $h(x) = k(y)$ if and only if there exists some $a \in A$ such that $f(a) = x$, $g(a) = y$.

Equivalently again, the square (2.26) is a *weak pullback* in Set (see 2.2.1), which means that:

(iii) $hf = kg$ and for every commutative square $hf' = kg'$ there is *some* map u such that $f' = fu$, $g' = gu$

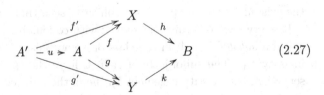

$$(2.27)$$

The sufficiency of (iii) follows from taking $A' = \{*\}$, its necessity from the axiom of choice.

(b) Similarly a square (2.26) of homomorphisms of $R\,\mathrm{Mod}$ (in particular of Ab) is said to be *exact* if it becomes *bicommutative* in the category of relations, i.e. if it satisfies (i).

This is still equivalent to (ii), since the forgetful functor $\mathrm{Rel}\,(R\,\mathrm{Mod}) \to \mathrm{Rel\,Set}$ is faithful; this condition can also be expressed by an exact sequence due to Hilton [Hi] (see 6.4.8). Characterisation (iii) holds here *if* we take $A' = R$ (or require A' to be a free module).

2.4 Subobjects and quotients, regular and strong

In a category, a subobject of an object A is defined as a 'selected' mono-morphism $m\colon X \rightarrowtail A$, in a suitable equivalence class; dually, a quotient of A is a selected epimorphism $p\colon A \twoheadrightarrow X$.

Nevertheless, in many categories it is convenient to restrict the monos or epis we are considering, in order to get the 'good' notion of subobject or quotient: strong monomorphisms and strong epimorphisms (defined in 2.4.4) are important in categories of topological spaces, while strong epimorphisms and ordinary monos are important in varieties of algebras. This is discussed in 2.4.7.

In these cases, strong can be equivalently replaced with regular. But strong monos and epis have the advantage of being closed under composition, in every category.

Categories of structured sets, usually, have a natural choice for the 'good' notion of subobject and quotient.

(In the present edition, this section has been extended.)

2.4.1 Subobjects and quotients

Let A be an object of the category C. A subobject of A is defined as an equivalence class of monomorphisms, or better as a *selected representative* of such a class.

More precisely, given two monos m, n *with values in A*, we say that $m \prec n$ if there is a (uniquely determined) morphism u such that $m = nu$. We say that m and n *are equivalent,* or $m \sim n$, if $m \prec n \prec m$, which amounts to the existence of a (unique) isomorphism u such that $m = nu$.

In every class of equivalent monos with codomain A, one is *selected* and called a *subobject* of A; in the class of isomorphisms we always choose the identity 1_A. The subobjects of A in C form the (possibly large) *ordered* set Sub$A$, with greatest element 1_A; here, the induced *order* $m \prec n$ is also written as $m \leqslant n$.

Epimorphisms with a fixed domain A are dealt with in a dual way. Their preorder and equivalence relation are also written as $p \prec q$ (meaning that p factorises through q) and $p \sim q$. A *quotient* of A is a *selected* representative of an equivalence class of epimorphisms with domain A; they form the ordered set Quo$A$, with greatest element 1_A; again the induced order is also written as $p \leqslant q$.

The category C is said to be *well powered* (resp. *well copowered*) if all its sets Sub$A$ (resp. Quo$A$) are small, as is often the case with categories of (small) structured sets.

Duality of categories turns subobjects of C into quotients of C^{op}, *preserving* their order.

2.4.2 Induced morphisms

The following commutative diagrams in C

$$
\begin{array}{ccc}
A & \xrightarrow{\ f\ } & B \\
m \uparrow & & \uparrow n \\
A' & \xrightarrow[f']{} & B'
\end{array}
\qquad\qquad
\begin{array}{ccc}
A & \xrightarrow{\ f\ } & B \\
p \downarrow & & \downarrow q \\
A'' & \xrightarrow[f'']{} & B''
\end{array}
\qquad (2.28)
$$

respectively express the fact that:

(i) the morphism f' is *induced* by f, on the monomorphisms (or subobjects) m, n,

(ii) the morphism f'' is *induced* by f, on the epimorphisms (or quotients) p, q.

Both kinds of induction are consistent with composition. In case (i) we also speak of a *restricted* morphism.

Remarks. (a) In a concrete category of structured sets there is generally a natural choice of subobjects based on subsets, and a natural choice of quotients based on quotient sets.

(b) The definition of a subobject (or a quotient) as an equivalence class is often preferred in category theory (cf. [M4]). However, when working with induced morphisms as in (2.28), one is forced to shift to selected representatives – even in the general theory.

(c) From a formal point of view, we are considering the (generally large) set $M(A)$ of monomorphisms $\bullet \rightarrowtail A$ in C, preordered as above, and we replace it with its skeleton $Sub A$, a (possibly small) ordered set. This is often convenient, for theoretical and concrete reasons.

As in 1.5.6(d), the skeleton can be simply obtained as the associated ordered set, whose elements are equivalence classes of monos. Yet, in this shortcut the interplay of subobjects and general morphisms of C becomes complicated. Much in the same way as the skeleton of a category is better defined by a choice, rather than by a quotient.

2.4.3 Orthogonal morphisms

The following relationship between morphisms of C will be useful. We say that f *is orthogonal to* g, and write $f \perp g$, if for all arrows u, v that make a commutative square $vf = gu$

$$
\begin{array}{ccc}
\bullet & \xrightarrow{\ f\ } & \bullet \\
{\scriptstyle u}\downarrow & \nearrow{\scriptstyle w} & \downarrow{\scriptstyle v} \\
\bullet & \xrightarrow[\ g\ ]{} & \bullet
\end{array}
\tag{2.29}
$$

there is precisely one morphism w such that $wf = u$ and $gw = v$.

Plainly, this relation is not symmetric. For a (generally large) subset $H \subset \mathrm{Mor}\,\mathsf{C}$ we have the following *orthogonal classes*

$$
\begin{aligned}
\perp H &= \{f \in \mathrm{Mor}\,\mathsf{C} \mid \forall h \in H,\ f \perp h\}, \\
H\perp &= \{f \in \mathrm{Mor}\,\mathsf{C} \mid \forall h \in H,\ h \perp f\}.
\end{aligned}
\tag{2.30}
$$

Applying 1.7.9(a), the orthogonality relation gives a contravariant Galois connection

$$
\begin{aligned}
L\colon \mathcal{P}(\mathrm{Mor}\,\mathsf{C}) &\rightleftarrows \mathcal{P}(\mathrm{Mor}\,\mathsf{C}) : R, \\
L(H) = \perp H, &\qquad R(H) = H\perp.
\end{aligned}
\tag{2.31}
$$

2.4.4 Regular and strong subobjects and quotients

Let us recall that an epimorphism $p\colon A \to X$ of C is said to be *regular* if it is the coequaliser of a pair of morphisms $\bullet \rightrightarrows A$ (see 2.1.7).

As a more general property, an epimorphism $p\colon A \to X$ is said to be *strong* if it is orthogonal to every monomorphism: for every commutative square $mu = vp$ where m is mono, there exists a (unique) morphism w such that $wp = u$ and $mw = v$ (the last two conditions are actually equivalent)

$$
\begin{array}{ccc}
A & \xrightarrow{\ p\ } & X \\
{\scriptstyle u}\downarrow & \nearrow{\scriptstyle w} & \downarrow{\scriptstyle v} \\
Y & \xrightarrowtail[\ m\ ]{} & B
\end{array}
\tag{2.32}
$$

In fact, every regular epimorphism is strong, as proved below (in 2.4.5(a)). Plainly, both notions are invariant up to the equivalence relation $p \sim q$ (of epimorphisms defined on A). A quotient $p\colon A \to X$ is said to be *regular* (resp. *strong*) if it is, as an epimorphism.

Dually a monomorphism $m\colon Y \to B$ is said to be *strong* if every epimorphism is orthogonal to it: for every commutative square as above, where p is epi, there exists a morphism w such that $wp = u$ and $mw = v$. Every regular monomorphism is a strong mono. A subobject is said to be *regular* (resp. *strong*) if it is, as a monomorphism.

2.4.5 Exercises and complements

We begin with a solved exercise. The others have a solution in Chapter 8.

(a) Every regular epimorphism is strong.

In fact, in the diagram below (with a commutative square), let p be the coequaliser of f and g

$$
\begin{array}{ccccc}
A' & \underset{g}{\overset{f}{\rightrightarrows}} & A & \overset{p}{\longrightarrow} & X \\
& & \downarrow u & \overset{w}{\nearrow} & \downarrow v \\
& & Y & \underset{m}{\rightarrowtail} & B
\end{array}
\tag{2.33}
$$

Then $uf = ug$ (cancelling m), and there is a morphism w such that $wp = u$. We already noted that $mw = v$ is a consequence.

(b) Strong epis are closed under composition.

(c) If a composite gf is a strong epimorphism, g is also.

(d) A strong epimorphism which is mono is an isomorphism.

2.4.6 Proposition

(a) In a category C *with equalisers and cokernel pairs, a monomorphism is regular if and only if it is the equaliser of its cokernel pair.*

(a*) Dually, in a category with coequalisers and kernel pairs, an epimorphism is regular if and only if it is the coequaliser of its kernel pair.

Proof (a) We give a direct, analytic argument. A clearer, more complex, proof will be the subject of the next exercise.

Let $m\colon E \rightarrowtail A$ be the equaliser of $f, g\colon A \rightrightarrows B$ and $f', g'\colon A \rightrightarrows C$ the cokernel pair of m

$$
\begin{array}{ccc}
& C & \\
{\scriptstyle f'}\uparrow\uparrow{\scriptstyle g'} & & \searrow{\scriptstyle u} \\
E \overset{m}{\rightarrowtail} & A & \underset{g}{\overset{f}{\rightrightarrows}} B \\
{\scriptstyle v}\nwarrow & \uparrow{\scriptstyle h} & \\
& H &
\end{array}
\tag{2.34}
$$

We take the equaliser $h\colon H \rightarrowtail A$ of (f', g'). The pair (f, g) coequalises m, and factorises through the cokernel pair (f', g'), giving a morphism $u\colon C \to B$ such that $uf' = f$ and $ug' = g$.

Now $fh = gh$, and $h = mv$ for a unique $v\colon H \to E$. Then $f'm = g'm$, and also m factorises through h. Finally v is invertible, and m is also the equaliser of (f', g'). $\qquad\square$

2.4.7 Exercises and complements, II

(a) Extend Proposition 2.4.6(a) proving that, in a category C with equalisers and cokernel pairs:

(i) an equaliser morphism is the equaliser of its cokernel pair,

(ii) a cokernel pair is the cokernel pair of its equaliser.

Hints. Build a contravariant Galois connection, inspired by the proof of the previous proposition. It may be simpler to use preordered sets.

(b) In Set every monomorphism (resp. epimorphism) is regular.

The reader will show that every subset $E \subset X$ is the equaliser of its cokernel pair, and every quotient set X/R is the coequaliser of its kernel pair.

Therefore all subobjects are regular (and strong), and can be identified with subsets. All quotients are regular (and strong), and can be identified with set-theoretical quotients modulo an equivalence relation.

(c) In Ab and R Mod every monomorphism (resp. epimorphism) is regular (and even 'normal', as we shall see in 6.1.3). All subobjects are regular and can be identified with subgroups or submodules; all quotients are regular and can be identified with ordinary quotients modulo a subobject.

(d) In Top every subspace $E \subset X$ is the equaliser of its cokernel pair, and every quotient space X/R is the coequaliser of its kernel pair.

Here every injective continuous mapping is a monomorphism, while the regular and strong subobjects coincide and amount to *inclusions of subspaces*: the 'general subobjects' are less important than the strong (or regular) ones. Similarly, every surjective continuous mapping is an epi, while the regular and strong quotients coincide and amount to projections on quotient spaces, in the usual sense.

(e) We shall see in Theorem 4.4.4 that, in a variety of algebras, regular epimorphisms coincide with the strong ones, and give the usual quotients of algebras, modulo a congruence. On the other hand, general subobjects amount to the usual subalgebras.

(f) In a variety of algebras, general epimorphisms on the one hand, regular and strong monomorphisms on the other, can be difficult to determine and – finally – of a limited interest.

Knowing a few cases should be enough. We already said that epimorphisms have no elementary classification in Mon and Rng. All subgroups are regular subobjects in Gp (a non-trivial fact, see [AHS], Exercise 7H), while a subsemigroup need not be a regular subobject in the category of semigroups ([AHS], Exercise 7I).

2.4.8 Exercises and complements (Preimages and images)

The following issues are neither obvious nor too hard. A reader may find it interesting, or amusing, to work out the proof.

(a) If the left square below is a pullback and n is mono, or a strong mono, or a regular mono, so is m (the first point is already known)

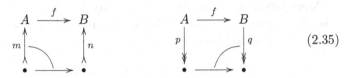

$$(2.35)$$

The monomorphism $m : \bullet \rightarrowtail A$ is determined up to the equivalence relation of monos; it is called the *preimage*, or *inverse image*, of n along f and written as $f^*(n)$. When we work with subobjects, $f^*(n)$ is strictly determined by their choice.

(a*) Dually, if the right square above is a pushout and p is epi, or a strong epi, or a regular epi, so is q. (Again, the first point is already known.)

The epimorphism $q : B \twoheadrightarrow \bullet$ is determined up to the equivalence relation of epis; it is called the *image* of p along f and written as $f_\circ(p)$.

(b) For the sake of simplicity, we generally adopt the following conventions.

(i) *The unit constraint for pullbacks.* The pullback of an identity along any morphism (always exists and) is chosen to be an identity: $f^*(1_B) = 1_A$.

(i*) *The unit constraint for pushouts.* The pushout of an identity along any morphism (always exists and) is chosen to be an identity: $f_\circ(1_A) = 1_B$.

2.5 Factorisation systems

Many categories have a privileged factorisation system (E, M), for instance the 'epi-mono' factorisation in Set and R Mod, or 'regular epi - general mono' in any variety of algebras. Yet the general notion (in 2.5.2) does not require that the system be *proper*, i.e. that E be a set of epimorphisms and M a set of monomorphisms; there are important cases where this does not hold.

We briefly sketch this subject, by a sequence of exercises with detailed hints. The reader can find further information in [AHS, Bo1, FrK, CaJKP] and in the papers cited below. The (non-elementary) relationship of factorisation systems with monads will be outlined in Section 7.7.

We work in a category C. The subsets of monomorphisms, epimorphisms and isomorphisms in Mor C are written as Mono, Epi and Iso, respectively.

2.5.1 Examples

Two examples will be of help to introduce the subject.

(a) As an obvious, well known fact, every mapping $f\colon A \to B$ in Set has an *epi-mono factorisation* $f = me$, which is uniquely determined up to a unique 'central' isomorphism.

More precisely, given two factorisations $f = me = m'e'$ of this kind, there is a unique morphism w that makes the diagram below commutative, and it is an isomorphism

$$
\begin{array}{ccccc}
A & \xrightarrow{\ e\ } & X & \xrightarrow{\ m\ } & B \\[2pt]
\Big\| & & \Big\downarrow{\scriptstyle w} & & \Big\| \\[2pt]
A & \xrightarrow[\ e'\ ]{} & X & \xrightarrow[\ m'\ ]{} & B
\end{array}
\tag{2.36}
$$

This decomposition $f = me$ will also be called *the EM-factorisation* of f, where $E = $ Epi and $M = $ Mono are the (large) subsets of Mor Set of epis and monos, respectively. The pair (E, M) is called a factorisation system for Set. It is a *proper* one – which in general just means that $E \subset$ Epi and $M \subset$ Mono.

The reader likely knows various similar factorisations, in concrete categories; some of them will be recalled below.

(b) The category of morphisms C^2 of an arbitrary category C, described in 1.6.1, has a canonical factorisation which plays an important role in the theory of factorisations.

In fact every morphism $f = (f', f'')\colon x \to y$ can be factorised as in the right diagram below

$$
\begin{array}{ccc}
X' & \xrightarrow{\ f'\ } & Y' \\[2pt]
{\scriptstyle x}\Big\downarrow & & \Big\downarrow{\scriptstyle y} \\[2pt]
X'' & \xrightarrow[\ f''\ ]{} & Y''
\end{array}
\qquad\qquad
\begin{array}{ccccc}
X' & \xrightarrow{\ 1\ } & X' & \xrightarrow{\ f'\ } & Y' \\[2pt]
{\scriptstyle x}\Big\downarrow & & \Big\downarrow{\scriptstyle \overline{f}} & & \Big\downarrow{\scriptstyle y} \\[2pt]
X'' & \xrightarrow[\ f''\ ]{} & Y'' & \xrightarrow[\ 1\ ]{} & Y''
\end{array}
\tag{2.37}
$$

where $\overline{f} = yf' = f''x$ is the diagonal of the commutative square f.

If we let E be the set of morphisms (i, g) where i is an isomorphism, as in the left commutative square below, and M be the set of morphisms (f, j) where j is an isomorphism, as in the right square

$$
\begin{array}{ccc}
\bullet & \xrightarrow{\ i\ } & \bullet \\[2pt]
\Big\downarrow & & \Big\downarrow \\[2pt]
\bullet & \xrightarrow[\ g\ ]{} & \bullet
\end{array}
\qquad\qquad
\begin{array}{ccc}
\bullet & \xrightarrow{\ f\ } & \bullet \\[2pt]
\Big\downarrow & & \Big\downarrow \\[2pt]
\bullet & \xrightarrow[\ j\ ]{} & \bullet
\end{array}
\tag{2.38}
$$

it is easy to see that the EM-factorisation considered above is unique up to isomorphism, in C^2. The reader will note that:

- this factorisation system (E, M) is not proper,

- restricting E to the set E_0 where i is an identity and M to the set M_0 where j is an identity, we even have a *strict factorisation system* (E_0, M_0), in the sense that the factorisation (2.37) is now strictly determined,

- such a restriction can also be made in case (a), taking – for instance – $E_0 = $ Epi and M_0 the set of inclusions of subsets.

2.5.2 Factorisation systems

We now define a *factorisation system* for C as a pair (E, M) of (generally large) subsets of Mor C such that:

(i) E and M are closed under composition and contain all the isomorphisms (forming two wide subcategories E, M of C),

(ii) every morphism f of C has a factorisation $f = me$, where $e \in E$ and $m \in M$,

(iii) (*uniqueness*) for any two factorisations $f = me = m'e'$ of this kind, there is a unique morphism w that makes the diagram (2.36) commute, and it is an isomorphism.

In this situation, we speak of *the EM-factorisation* $f = me$.

Axiom (iii) can be equivalently replaced with the following one, based on the orthogonality relation of 2.4.3:

(iii′) (*orthogonality*) for every $e \in E$ and every $m \in M$, $e \perp m$.

(The equivalence is proved in Exercise 2.5.3(a).)

The opposite category C^{op} inherits an *opposite* factorisation system, namely (M^{op}, E^{op}).

We say that the factorisation system (E, M) is *proper*, or *epi-mono*, if $E \subset$ Epi and $M \subset$ Mono. This need not be the case, as we have already seen in 2.5.1(b); moreover, every category admits two trivial factorisation systems: (Iso, Mor) and (Mor, Iso), which are not proper – in general.

2.5.3 Exercises and complements, I

A factorisation system (E, M) is given in C.

(a) Axioms (iii) and (iii′) are equivalent (in the presence of (i) and (ii)).

Hints. Use the left diagram below to prove that (iii) $\Rightarrow$ (iii'), and the right one to prove the converse

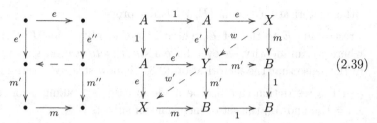

$$(2.39)$$

(b) $E \cap M$ coincides with the set of isomorphisms of C. The factorisation $u = me \colon A \to B$ of an isomorphism u consists of two isomorphisms.

Hints. For the first claim, write $u \in E \cap M$ as $u = 1_B.u = u.1_A$ and apply uniqueness.

(c) Each of the sets E, M determines the other, as an orthogonal class in C (in the sense of 2.4.3)

- $E = \perp M = \{f \in \text{Mor C} \mid \forall m \in M, \ f \perp m\}$,
- $M = E\perp = \{f \in \text{Mor C} \mid \forall e \in E, \ e \perp f\}$.

Hints. It suffices to prove that every morphism $f \in \perp M$ belongs to E. The EM-factorisation $f = me$ gives the left commutative diagram below (because $f \perp m$), and the right square proves that w is an isomorphism

$$
\begin{array}{ccc}
A & \xrightarrow{\ f\ } & B \\
{\scriptstyle e}\downarrow & {\scriptstyle w}\nearrow & \downarrow{\scriptstyle 1} \\
X & \xrightarrow[\ m\ ]{} & B
\end{array}
\qquad\qquad
\begin{array}{ccc}
A & \xrightarrow{\ e\ } & X \\
{\scriptstyle e}\downarrow & {\scriptstyle wm}\ {\scriptstyle 1} & \downarrow{\scriptstyle m} \\
X & \xrightarrow[\ m\ ]{} & B
\end{array}
\qquad (2.40)
$$

(d) For a composite gf in C, we have:

- if $gf \in E$ and $f \in E$, then $g \in E$,
- if $gf \in M$ and $g \in M$, then $f \in M$.

Hints. Assuming that $gf \in E$ and $f \in E$, one can prove that $g \perp m$ (for every $m \in M$) by applying the hypotheses to the commutative diagram

$$
\begin{array}{ccccc}
\bullet & \xrightarrow{\ f\ } & \bullet & \xrightarrow{\ g\ } & \bullet \\
{\scriptstyle uf}\downarrow & & \downarrow{\scriptstyle u} & & \downarrow{\scriptstyle v} \\
\bullet & \xrightarrow[\ 1\ ]{} & \bullet & \xrightarrow[\ m\ ]{} & \bullet
\end{array}
$$

(e) The factorisation systems of C are ordered by the relation expressed by the following equivalent conditions

$$E \subset E', \qquad M' \subset M. \qquad (2.41)$$

In fact, because of the contravariant Galois connection (2.31), $E \subset E'$ implies $M = E\bot \supset E'\bot = M'$, and conversely. The trivial factorisation systems (Iso, Mor) and (Mor, Iso) on C are, respectively, the smallest and the largest in this ordering.

(f) A *strict factorisation system* for C is a pair (E_0, M_0) of subsets of Mor C such that:

(i) E_0 and M_0 are closed under composition and contain all the identities,

(ii) every morphism f has a strictly unique factorisation $f = m_0 e_0$, with $e_0 \in E_0$ and $m_0 \in M_0$.

Note that this is *not* a factorisation system, because E_0 and M_0 do not contain all isomorphisms – generally. But there is a unique factorisation system (E, M) containing the former (or *spanned* by it), where $f = m_0 e_0$ is in E (resp. M) if and only if m_0 (resp. e_0) is an isomorphism. The only non-obvious point – if not a difficult one – is the closure of E and M under composition. (One can see the proof in [G4], Section 2.)

2.5.4 Exercises and complements, II

(a) A category admits the pair (Epi, Mono) as a factorisation system if and only if

(i) all epimorphisms are strong, and all monomorphisms are strong,

(ii) every morphism f has an epi-mono factorisation.

Hints. By 2.4.4 an epi is strong if and only if it is orthogonal to every mono; and dually.

(b) The following categories have such a factorisation system: Set, Set., Ab, R Mod, Gp. For the last case, one should recall that all epimorphisms in Gp are surjective: see 1.3.4(c).

(c) We shall see in 4.2.3 that every regular category has a factorisation system (rEpi, Mono), where rEpi is the set of regular epis (coinciding with the strong ones). The examples of (b) are part of this case.

(d) In Top one can consider two main factorisation systems: (rEpi, Mono) and (Epi, rMono).

One may prefer to use a 'ternary factorisation system', where a map factorises by a regular epi, a bijective continuous mapping (which is mono and epi) and a regular mono.

2.5.5 *Natural weak factorisation systems*

A *functorial weak factorisation system* in C is assigned by a functorial factorisation of its morphisms

$$X \xrightarrow{\lambda(f)} F(f) \xrightarrow{\rho(f)} Y \qquad\qquad f = \rho(f).\lambda(f), \qquad (2.42)$$

where $F: \mathsf{C}^2 \to \mathsf{C}$ is a functor and $\lambda: \mathrm{Dom} \to F$, $\rho: F \to \mathrm{Cod}$ are natural transformations.

Plainly, every factorisation system (E, M) in C produces such a structure, by *choosing* for every morphism $f: X \to Y$, a particular EM-factorisation $X \to F(f) \to Y$.

A *natural weak factorisation system* is a wider notion that covers various situations, as the graph factorisation of Galois connections in 1.7.6 (and of general adjunctions in 3.3.6). It was introduced in [GT]; see also [Ga]. (Interestingly, these 'graph factorisations' are *mono-epi*.)

2.6 Inductive limits and filtered colimits

The inductive limit of a diagram $X = ((X_i), (u_{ij}))$ is a classical notion, studied in many concrete categories of algebraic or topological character. In category theory it is a kind of *colimit*, and has a natural generalisation to a *filtered colimit*.

These colimits have important, peculiar properties: they are 'often' preserved by forgetful functors of concrete categories, and even created by the forgetful functors of equational algebras. Furthermore, they 'often' commute with finite limits.

The dual notion, classically called a *projective limit*, is less important in the usual categories of structured sets, where – generally speaking – *all* limits are already 'well-behaved'.

We keep this section at an elementary level, adequate to the present goals. More complete expositions can be found in [M4, Bo1].

2.6.1 *Inductive limits*

A *filtered* (ordered) *set* I will be a (small) ordered set where every finite subset has an upper bound. Or, equivalently, a non-empty ordered set such that each pair $i, j \in I$ has an upper bound $k \geqslant i, j$ in I. (This is also called a *directed* set.)

An *inductive limit* is the colimit of a functor $X: I \to \mathsf{C}$ defined on such a category. The functor X will be written in index notation (see 1.4.9(a)), in the form

$$X = ((X_i), (u_{ij})), \quad u_{ij} \colon X_i \to X_j, \qquad\qquad \text{for } i \leqslant j \text{ in } I,$$

$$u_{ii} = \mathrm{id}X_i, \qquad\quad u_{jk}.u_{ij} = u_{ik}, \qquad\quad \text{for } i \leqslant j \leqslant k \text{ in } I. \tag{2.43}$$

A cocone $(Y, (f_i))$ of X is thus an object Y with a family of morphisms f_i in C such that

$$f_i \colon X_i \to Y, \qquad f_j u_{ij} = f_i, \qquad\qquad \text{for } i \leqslant j \text{ in } I. \tag{2.44}$$

The colimit is a universal cocone

$$(L, (u_i)) = \mathrm{Colim}((X_i), (u_{ij})), \tag{2.45}$$

which means that, for every cocone $(Y, (f_i))$, there is a unique $f \colon L \to Y$ in C such that $f_i = f u_i \colon X_i \to Y$, for all $i \in I$.

For sure we do not want to include here the case $I = \emptyset$, as it will be evident in 2.6.2(c).

2.6.2 Exercises and complements, I

Generally speaking, inductive limits in a concrete category C are easy to construct, and an important tool. The reader should find it interesting to complete the details of the following exercises. The first two already show that inductive limits can be much simpler than general colimits.

I is always a filtered set, and $U \colon \mathsf{C} \to \mathsf{Set}$ the forgetful functor of C.

(a) The colimit of a functor $X \colon I \to \mathsf{Set}$ has a simple construction: loosely speaking, we put together all the elements of the system (X_i), and identify them when they 'eventually coincide', along I.

(b) The colimit of a functor $X \colon I \to \mathsf{Ab}$ has a similar construction, and is created by the forgetful functor $U \colon \mathsf{Ab} \to \mathsf{Set}$ (cf. 2.2.8).

(c) The previous construction fails for $I = \emptyset$.

(d) The construction in (b) works, quite similarly, in many categories of 'algebras', like Mon, Gp, Rng, R Mod, etc. In fact, it works in every variety of algebras, as we shall see in Theorem 4.4.1.

(e) For a functor $X \colon I \to \mathsf{Top}$, one takes the colimit L of the 'underlying functor' $UX \colon I \to \mathsf{Set}$, and endows L with the finest topology making all mappings $u_i \colon X_i \to L$ continuous. But this is nothing new, because one can do the same for any colimit, as we have observed in 2.2.5(b).

(f) We have a similar situation in the category pOrd of preordered sets (see 2.2.5(c)).

(g) For ordered sets we come back to something of interest: the forgetful functor $U \colon$ Ord $\to$ Set does not preserve general colimits (see 3.2.3(d)), but does preserve all inductive limits, without creating them.

Solutions. (a) We begin by forming the sum $S = \Sigma X_i$, whose elements are written as pairs (x, i), with $i \in I$ and $x \in X_i$ (as in (2.7)).

The reader will prove that the relation $(x, i) \sim (y, j)$ in S defined by

$$\text{there exists } k \geqslant i, j \text{ in } I \text{ such that } u_{ik}(x) = u_{jk}(y), \qquad (2.46)$$

is an equivalence relation, and that the colimit of X can be obtained as the quotient:

$$L = S/\!\sim, \qquad u_i \colon X_i \to L, \quad u_i(x) = [x, i]. \qquad (2.47)$$

(b) For a functor $X \colon I \to$ Ab, one takes the colimit L of the 'underlying functor' $UX \colon I \to$ Set. The reader will check that L has a structure of abelian group, defined as follows

$$[x, i] + [y, j] = [u_{ik}(x) + u_{jk}(y), k], \qquad \text{for } i, j \leqslant k \text{ in } I. \qquad (2.48)$$

Moreover this is the unique structure of abelian group of L that makes all $u_i \colon X_i \to L$ into homomorphisms. Finally $(L, (u_i))$ is the colimit of X in Ab, *created* by $U \colon$ Ab $\to$ Set (as defined in 2.2.8).

(c) The colimit of $X \colon \emptyset \to$ Ab is the trivial group $\{0\}$, and is not created by $U \colon$ Ab $\to$ Set.

(g) For a functor $X \colon I \to$ Ord, one takes the colimit L of the 'underlying functor' $UX \colon I \to$ Set, and endows L with the finest (or 'smallest') order that makes all mappings $u_i \colon X_i \to L$ monotone

$$\xi \leqslant \eta \quad \text{if } \xi = [x, i], \ \eta = [y, i] \ \text{ for some } i \in I \text{ and } x \leqslant y \text{ in } X_i. \quad (2.49)$$

2.6.3 Exercises and complements, II

(a) In Set every object X is the inductive limit of the filtered set $\mathcal{P}_f(X)$ of its finite subsets, ordered by inclusion.

(b) In Ab every object A is the inductive limit of the filtered set $\mathrm{Sub_{fg}}(A)$ of its *finitely generated* subgroups, ordered by inclusion.

(c) Finite subgroups would not give the same, generally.

(d) The reader can easily extend point (b) to various categories of 'algebras', like Mon, Gp, Rng, R Mod, etc. Or to any variety of algebras (after reading Chapter 4, if needed).

*(e) A compactly generated space is the inductive limit of the filtered set of its compact subspaces. (We recall that a topological space X is said

to be *compactly generated* if every subset C that intersects every compact subspace K of X in a set closed in K, is itself closed in X.)

This point is an extension of point (a), viewing a set as a discrete space and a finite subset as a compact subspace.

2.6.4 Filtered colimits

More generally, a *filtered category* S is a (small) *non-empty* category such that (writing $I = \mathrm{Ob}\,\mathsf{S}$):

(i) for each pair $i, j \in I$ there is a pair of arrows $a\colon i \to k$, $b\colon j \to k$ in S,

(ii) for each pair of arrows $a, b\colon i \to j$ there is an arrow $c\colon j \to k$ such that $ca = cb$.

If S is an ordered set, this coincides with the definition of a filtered set, since condition (ii) is trivially satisfied.

A *filtered colimit* is the colimit of a functor $X\colon \mathsf{S} \to \mathsf{C}$ defined on a filtered category. Writing again the functor X in index notation

$$X = ((X_i), (u_a)), \quad u_a\colon X_i \to X_j, \qquad \text{for } i \text{ and } a\colon i \to j \text{ in } \mathsf{S}, \quad (2.50)$$

a cocone $(Y, (f_i))$ of X is an object Y equipped of a family of morphisms in C such that

$$f_i\colon X_i \to Y, \quad f_j u_a = f_i\colon X_i \to Y, \qquad \text{for } i \text{ and } a\colon i \to j \text{ in } \mathsf{S}. \quad (2.51)$$

The colimit is a universal cocone $(L, (u_i))$: for every cocone $(Y, (f_i))$ there exists a unique $f\colon L \to Y$ in C such that $f_i = f u_i\colon X_i \to Y$, for all $i \in I$.

2.6.5 Exercises and complements

Filtered colimits in a concrete category C are constructed by extending what we have seen in 2.6.2.

(a) Let $X\colon \mathsf{S} \to \mathsf{Set}$ be a functor defined on a filtered category. One forms again the sum $S = \Sigma X_i$, with elements (x, i), for $i \in I$ and $x \in X_i$. The reader will prove that the relation $(x, i) \sim (y, j)$ in S defined by

there exist $a\colon i \to k$, $b\colon j \to k$ in S such that $u_a(x) = u_b(y)$, $\quad (2.52)$

is an equivalence relation, and that the colimit of X can be obtained as:

$$L = S/\!\sim, \qquad u_i\colon X_i \to L, \quad u_i(x) = [x, i]. \quad (2.53)$$

(b) The other constructions of 2.6.2 are extended in the same way.

2.6.6 Theorem

Filtered colimits in Set *commute with finite products, in a sense made precise in the proof.*

Note. This statement is a particular instance of a much more general result: see 2.6.7(b).

Proof We begin by considering binary products. Let $X = ((X_i), (u_a))$ and $Y = ((Y_i), (v_a))$ be two functors $S \to$ Set defined on the same filtered category. Denoting by $X \times Y \colon S \to$ Set the functor

$$(X \times Y)_i = X_i \times Y_i, \quad w_a = u_a \times v_a \colon X_i \times Y_i \to X_j \times Y_j, \qquad (2.54)$$

we want to prove that $\mathrm{Colim}(X \times Y) \cong \mathrm{Colim}X \times \mathrm{Colim}Y$, by a canonical bijection.

In fact, there is a canonical mapping $(f, g) \colon \mathrm{Colim}(X \times Y) \to \mathrm{Colim}X \times \mathrm{Colim}Y$, with components f, g produced by the cocones

$$f_i = u_i p_i \colon X_i \times Y_i \to X_i \to \mathrm{Colim}X,$$
$$g_i = v_i q_i \colon X_i \times Y_i \to Y_i \to \mathrm{Colim}Y. \qquad (2.55)$$

(The cocone conditions are plainly satisfied: for an arrow $a \colon i \to j$ in S we have: $f_j w_a = u_j p_j w_a = u_j u_a p_i = u_i p_i = f_i$ and, similarly, $g_j w_a = g_i$.)
The mapping (f, g) acts as follows (for $i \in I$ and $(x, y) \in X_i \times Y_i$):

$$(f, g) \colon \mathrm{Colim}(X \times Y) \to \mathrm{Colim}X \times \mathrm{Colim}Y,$$
$$[x, y, i] \mapsto ([x, i], [y, i]). \qquad (2.56)$$

It is surjective, since an arbitrary pair $([x, i], [y, j])$ can be rewritten as $([u_a x, k], [v_b y, k])$, for any pair of arrows $a \colon i \to k$, $b \colon j \to k$ of S. To prove that it is injective, let us suppose that

$$[x, y, i], \, [x', y', j] \in \mathrm{Colim}(X \times Y),$$

$$[x, i] = [x', j] \text{ in } \mathrm{Colim}X, \qquad [y, i] = [y', j] \text{ in } \mathrm{Colim}Y.$$

Then there exists two pairs of arrows $a \colon i \to k$, $b \colon j \to k$ and $c \colon i \to k'$, $d \colon j \to k'$ such that

$$u_a x = u_b x' \text{ in } X_k, \qquad v_c y = v_d y' \text{ in } Y_{k'}.$$

By Lemma 2.6.8 (below) there exist two arrows $r \colon k \to k''$, $s \colon k' \to k''$ such that

$$ra = sc \colon i \to k'', \qquad rb = sd \colon j \to k''. \qquad (2.57)$$

Then

$$u_{ra}x = u_{rb}x' \text{ in } X_{k''}, \qquad v_{sc}y = v_{sd}y' \text{ in } Y_{k''}, \tag{2.58}$$

which means that $(x,y) \in X_i \times Y_i$ and $(x',y') \in X_j \times Y_j$ coincide in $X_{k''} \times Y_{k''}$, under the arrows $ra = sc: i \to k''$ and $rb = sd: j \to k''$. Therefore $[x,y,i] = [x',y',j]$.

Finally, the 0-ary power of X is the constant functor $X^0: \mathsf{S} \to \mathsf{Set}$, that sends everything to the terminal object $\top = \{*\}$. Its colimit is $\top = (\mathrm{Colim}X)^0$. $\qquad\square$

2.6.7 Exercise and complements

(a) Extend this result to Mon, Gp, Rng, R Mod.

*(b) More generally, filtered colimits commute with finite limits in all varieties of algebras: see [M4, Bo1].

*(c) A category S is filtered if (and only if) every functor $X \to \mathsf{S}$ defined on a finite category has a cocone. The reader can prove this fact, or find a proof in [Bo1], Lemma 2.13.2. Our Lemma below is a particular case, sufficient for the case of finite products, in Theorem 2.6.6.

2.6.8 Lemma

In the filtered category S, *four arrows* a, b, c, d *are given, as in the diagram below. Then there exist two arrows* $r: k \to k''$, $s: k' \to k''$ *making the diagram commutative*

$$ra = sc,$$

$$\tag{2.59}$$

$$rb = sd.$$

Proof First we have two arrows $m: k \to k_1$, $n: k' \to k_1$

$$\tag{2.60}$$

We add two arrows u, v such that $uma = unc$ and $vmb = vnd$. Then we add two arrows x, y making the right square commutative (which can be done in two steps, using 2.6.4(i), (ii)).

Now the maps $r = xum\colon k \to k''$ and $s = yvn\colon k' \to k''$ satisfy our conditions:

$$ra = xuma = xunc = yvnc = sc, \qquad rb = xumb = yvmb = yvnd = sd.$$

$\square$

2.7 Universal arrows, limits and free objects

Universal arrows yield a general, fundamental way of formalising universal properties in a category X, with respect to a functor $U\colon A \to X$ and an object X of X. Obviously, there are two forms, dual to each other.

In particular, limits and colimits, free objects and representable functors can be defined by universal arrows.

2.7.1 Universal arrows

A *universal arrow from the object X to the functor U* is a pair

$$(A, \eta\colon X \to UA),$$

consisting of an object A of A and an arrow η of X which is universal.

This means that every similar pair $(B, f\colon X \to UB)$ factorises uniquely through (A, η): there exists a unique map $g\colon A \to B$ in A such that the following triangle commutes in X

$$
\begin{array}{ccc}
X & \xrightarrow{\ \eta\ } & UA \\
 & \searrow{\scriptstyle f} & \downarrow{\scriptstyle Ug} \\
 & & UB
\end{array}
\qquad Ug.\eta = f. \qquad (2.61)
$$

In other words we are saying that, for every B in A, the following mapping is bijective

$$A(A, B) \to X(X, UB), \qquad g \mapsto Ug.\eta. \qquad (2.62)$$

The pair (A, η) is then determined up to isomorphism: if also (B, f) is a solution, the unique A-morphism $g\colon A \to B$ such that $Ug.\eta = f$ is invertible.

Dually, a *universal arrow from the functor U to the object X* is a pair $(A, \varepsilon\colon UA \to X)$ consisting of an object A of A and an arrow ε of X such that every similar pair $(B, f\colon UB \to X)$ factorises uniquely through

(A, ε): there exists a unique $g \colon B \to A$ in A such that the following triangle commutes in X

$$UA \xrightarrow{\ \varepsilon\ } X$$
$$Ug \uparrow \quad \nearrow f \qquad\qquad \varepsilon.Ug = f. \tag{2.63}$$
$$UB$$

2.7.2 Limits by universal arrows and representability

For a small category S, consider the category C^S of functors $S \to \mathsf{C}$ and their natural transformations, with the diagonal functor (introduced in 1.6.1)

$$D \colon \mathsf{C} \to \mathsf{C}^S, \qquad (DA)_i = A, \quad (DA)_a = \mathrm{id}A, \quad (Df)_i = f. \tag{2.64}$$

Let a functor $X \colon S \to \mathsf{C}$ be given. A cone of X of vertex A is the same as a natural transformation $f \colon DA \to X$.

The limit of X in C is the same as a universal arrow $(L, \varepsilon \colon DL \to X)$ from the functor D to the object X of C^S (see 2.7.1). Dually, the colimit of X in C is the same as a universal arrow $(L', \eta \colon X \to DL')$ from the object X of C^S to the functor D.

As another characterisation, assuming that C has small hom-sets, consider the contravariant functor G *of sets of cones* of X

$$G \colon \mathsf{C}^{\mathrm{op}} \to \mathsf{Set}, \qquad G(A) = \mathsf{C}^S(DA, X). \tag{2.65}$$

Then, by the Yoneda Lemma 1.6.6(b), X has a limit (L, ε) if and only if the functor G is representable, with representative object $X_0 = L$, via the element $x_0 = \varepsilon \in G(L)$.

Dually, X has a colimit (L', η) if and only if the covariant functor $G'(A) = \mathsf{C}^S(X, DA)$ *of cocones* of X is representable, with representative object L', via the element $\eta \in G'(L')$.

2.7.3 Free objects

If $U \colon \mathsf{A} \to \mathsf{Set}$ is a concrete category (see 1.4.8), the *free* A-*object over* the set X is defined as a universal arrow $(A, \eta \colon X \to UA)$ from the set X to the forgetful functor U. (It may exist or not, of course.)

The mapping $\eta \colon X \to UA$ can be called *the insertion of the basis*: in fact, the definition means that every mapping $f \colon X \to U(B)$ with values in the 'underlying set' of any object B in A can be uniquely 'extended' to a morphism $g \colon A \to B$ in A, such that $U(g).\eta = f$.

The examples below can help to make this clear. In particular, the injectivity of η is examined in 2.7.4(a) and 2.7.5.

One can extend this terminology to any functor $U \colon \mathsf{A} \to \mathsf{X}$ which is *viewed*, in a given context, as a 'forgetful functor', even if it is not faithful.

2.7.4 Exercises and complements, I

(a) A reader which is not familiar with these notions might begin by constructing the universal arrow from a set X to the forgetful functor $U \colon R\,\mathsf{Mod} \to \mathsf{Set}$, using the free left module generated by X

$$A = \bigoplus_{x \in X} R. \tag{2.66}$$

This is a direct sum in $R\,\mathsf{Mod}$ of copies of the left module R (see 2.1.6(c)): an element of $F(X)$ is a quasi-null family $(\lambda_x)_{x \in X}$ of scalars $\lambda_x \in R$. The mapping

$$\eta \colon X \to UA, \qquad (\eta x)_x = 1, \quad (\eta x)_y = 0 \qquad (y \neq x), \tag{2.67}$$

sends each element of X to the corresponding Kronecker family in R. Then every mapping $f \colon X \to U(B)$ with values in a left R-module can be uniquely extended to a homomorphism $g \colon A \to B$ in $R\,\mathsf{Mod}$, by linear combinations in B (with quasi-null scalar coefficients)

$$g((\lambda_x)_{x \in X}) = \Sigma_{x \in X}\, \lambda_x . f(x). \tag{2.68}$$

Excluding the trivial case $R = \{0\}$, the mapping η is injective. One often identifies an element $x \in X$ with the family $\eta(x) \in A$, so that the quasi-null family $(\lambda_x) \in A$ can be uniquely rewritten as a *formal linear combination* $\Sigma\, \lambda_x . x$ of elements of the basis, with quasi-null coefficients in R.

If R is the trivial ring, all R-modules are trivial (i.e. singletons): the trivial R-module is free *on each set.*

(b) Similarly we can form the universal arrow from a set X to the forgetful functor $U \colon \mathsf{Mon} \to \mathsf{Set}$ using the free monoid $A = \Sigma_{n \in \mathbb{N}}\, X^n$ of *finite words* on the alphabet X. The reader can easily work out the details (or find them in 3.6.1).

Many other exercises of this kind will be listed and commented in Section 4.1.

(c) In a group G, the *subgroup of commutators* $[G, G]$ is the subgroup generated by all commutators

$$[x, y] = xyx^{-1}y^{-1},$$

for $x, y \in G$. Plainly, G is commutative if and only if this subgroup is trivial.

The reader will prove that this subgroup is necessarily normal in G, and that the canonical projection on the *abelianised* quotient group G^{ab}

$$G^{ab} = G/[G,G], \qquad \eta: G \to G^{ab}, \qquad (2.69)$$

gives the universal arrow from the group G to the embedding $U:$ Ab $\to$ Gp.

(d) A functor $U: A \to$ Set is representable (see 1.6.4) if and only if the singleton $\{*\}$ has a free object A_0 in A, so that

$$U(A) \cong \mathsf{Set}(\{*\}, U(A)) \cong \mathsf{A}(A_0, A). \qquad (2.70)$$

This property often holds in a concrete category, and ensures – by 2.3.5(b) – that U *preserves monos* (and reflects them, if it is faithful).

2.7.5 Exercises and complements, II

Let us come back to a general universal arrow $(A, \eta: X \to UA)$, for a functor $U: A \to X$.

(a) Prove that the arrow η is a monomorphism of X if and only if there exists a monomorphism $f: X \to UB$, for some object B in A.

(b) This condition obviously fails for the embedding $U:$ Ab $\to$ Gp, whenever X is a non-commutative group (see 2.7.4(c)).

(c) The forgetful functor $U: A \to$ Set of a concrete category 'usually' satisfies this condition, so that the insertion of the basis $\eta: X \to UA$ (if it exists) is indeed an *embedding*.

For varieties of algebras it is always the case, with the exception of two 'trivial varieties', as we shall see in Corollary 4.4.3.

2.7.6 Exercises and complements, III

(a) Universal arrows compose: for a composed functor $UV:$ B $\to$ A $\to$ X, given

- a universal arrow $(A, \eta: X \to UA)$ from an object X to U,
- a universal arrow $(B, \zeta: A \to VB)$ from the previous object A to V,

we have a universal arrow from X to UV constructed as follows

$$(B, U\zeta.\eta: X \to UA \to UV(B)). \qquad (2.71)$$

For instance, the free ring on a set can be easily constructed in two steps, along the composed forgetful functor Rng $\to$ Mon $\to$ Set (as will be done in Exercise 4.1.6(b)).

(b) Universal arrows for 'two-dimensional categories' will be considered in 7.1.5.

2.7.7 *Proposition* (Universal arrows and representability)

Let A *and* X *be categories with small hom-sets.*

(a) (Representable functors by universal arrows) *A functor* $G: A \to$ Set *is represented by the object* A_0 *of* A, *via* $x_0 \in G(A_0)$, *if and only if:*

(*) *the pair* $(A_0, x_0: \{*\} \to G(A_0))$ *is a universal arrow from the singleton to the functor* G.

(b) (Universal arrows by representable functors) *Given a functor* $U: A \to X$ *and an object* X *in* X, *we have an associated functor*

$$G: A \to \text{Set}, \qquad G = X(X, U(-)). \tag{2.72}$$

A universal arrow $(A, \eta: X \to UA)$ *from* X *to* U *is the same as representing the functor* G *by the object* A *and the element* $\eta \in G(A) = X(X, UA)$.

Proof (a) Letting $F = A(A_0, -)$, the present condition (*) is equivalent to condition (*) of 1.6.6.

(b) By definition, a universal arrow $(A, \eta: X \to UA)$ gives, for every B in A, a bijection

$$A(A, B) \to G(B) = X(X, UB), \qquad g \mapsto U(g)\eta, \tag{2.73}$$

which means that the object A represents the functor G, via the element $\eta \in GA$.   □

2.8 *Other universals

This miscellaneous section is about three kinds of universal solutions: the splitting of idempotents, Kan extensions and coends. These issues, even though important in category theory, will have a limited use in this book and can be skipped until needed.

2.8.1 *Absolute limits*

The limit $(L, (u_i: L \to X_i))$ of a functor $X: S \to C$ is said to be an *absolute limit* if it is preserved (as a limit) by every functor $C \to D$, with values in an arbitrary category.

Let us begin by noting that there are some trivial limits which always exist and are preserved by any functor, like unary products (with $S = 1$).

More generally, every constant functor $X: S \to C$ (with $S \neq 0$) has an absolute limit.

As another extension, if S has an initial object h any functor $X: S \to C$

has absolute limit X_h, with structural map $u_i \colon X_h \to X_i$ given by the unique arrow $h \to i$ of S. (This also holds for a large category S.)

But there is an interesting, non-trivial case.

2.8.2 Exercises and complements (The splitting of idempotents)

As we have seen in 1.4.9(d), an idempotent endomorphism $e \colon X \to X$ can be equivalently described as a functor $F \colon E \to C$ defined on the idempotent monoid $E = \{1, \underline{e}\}$.

(a) The reader should find it interesting (and easy) to prove that the following conditions on the idempotent $e \colon X \to X$ are equivalent:

(i) the idempotent e *splits* in C: it factorises as $e = mp \colon X \to A \to X$, with $pm = 1_A$ (so that m is a split mono and p is a split epi),

(ii) the idempotent e has an epi-mono factorisation $e = mp$,

(iii) the corresponding functor $F \colon E \to C$ has a limit $m \colon A \to X$,

(iv) the pair $1, e \colon X \to X$ has an equaliser $m \colon A \to X$,

(iii*) the functor F has a colimit $p \colon X \to A$,

(iv*) the pair $1, e \colon X \to X$ has a coequaliser $p \colon X \to A$.

(b) Now, every functor defined on C plainly preserves the existing splittings of idempotents, which are thus absolute limits and colimits.

(c) After biproducts (in 2.1.9, 2.2.6) this is another case where *the limit and colimit objects coincide*; here the coincidence is even necessary, as soon as our (co)limit exists. Again (and of course) the limit-cone $m \colon A \to X$ and the colimit-cocone $p \colon X \to A$ are distinct.

(d) All idempotents split in every category with equalisers (or coequalisers). But we shall see in 6.6.2(g) that this splitting also holds in the category Rel Ab, which lacks general equalisers.

*(e) In Rel Set not all idempotents split. This category lacks equalisers and coequalisers.

2.8.3 The idempotent completion

Every category C has a diagonal embedding (see 1.6.1)

$$D \colon C \to C^E, \qquad C^E = \mathsf{Cat}(E, C), \qquad (2.74)$$

with values in the functor-category C^E, called the *idempotent completion* of C.

In fact, an object $E \to \mathsf{C}$ amounts to an idempotent endomorphism e of C, and will be written thus. A morphism of C^E

$$(e, f, e') \colon e \rightarrowtail e'$$

(where $e \colon X \to X$ and $e' \colon Y \to Y$) is determined by a morphism $f \colon X \to Y$ of C such that $fe = f = e'f$. The composition is induced by that of C

$$(e', g, e'').(e, f, e') = (e, gf, e'') \colon e \rightarrowtail e'', \qquad \mathrm{id}(e) = (e, e, e). \qquad (2.75)$$

The embedding $D \colon \mathsf{C} \to \mathsf{C}^E$ sends X to 1_X and $f \colon X \to Y$ to

$$(1_X, f, 1_Y) \colon 1_X \rightarrowtail 1_Y.$$

Finally, in C^E all *idempotents split*: an idempotent $(e, f, e) \colon e \rightarrowtail e$ comes from a morphism

$$f \colon X \to X, \qquad fe = f = ef, \quad ff = f, \qquad (2.76)$$

and splits as the composite

$$(e, f, e) = (f, f, e).(e, f, f) \colon e \rightarrowtail f \rightarrowtail e, \qquad (e, f, f).(f, f, e) = \mathrm{id}(f).$$

* The *weak universal property* of C^E (a biuniversal property in Cat, see 7.1.5) is slightly more complicated to prove: every functor $F \colon \mathsf{C} \to \mathsf{D}$ with values in a category where all idempotents split can be extended to a functor $G \colon \mathsf{C}^E \to \mathsf{D}$ (so that $GD = F$), and the extension is determined up to functorial isomorphism.*

2.8.4 Kan extensions

Given two functors $F \colon \mathsf{S} \to \mathsf{C}$ and $T \colon \mathsf{S} \to \mathsf{T}$ with the same domain, a *right Kan extension* of F along T is a pair (R, ε) formed of a functor $R \colon \mathsf{T} \to \mathsf{C}$ and a natural transformation $\varepsilon \colon RT \to F$

$$(2.77)$$

which is universal: for every functor $R' \colon \mathsf{T} \to \mathsf{C}$ and natural transformation $\varepsilon' \colon R'T \to F$ there is a unique natural transformation $\rho \colon R' \to R$ such that $\varepsilon' = \varepsilon.\rho T$.

The solution, written as $R = \mathrm{Ran}_T F$, is determined up to a (unique) natural isomorphism. In fact, leaving apart problems of smallness, the property above means that $(R, \varepsilon \colon RT \to F)$ is a universal arrow from the functor $C^T = -.T$ to the object F of the category C^S

$$C^T \colon C^T \to C^S, \qquad R \mapsto RT, \quad \rho \mapsto \rho T. \tag{2.78}$$

Dually (by reversing natural transformations), the *left Kan extension* $L = \mathrm{Lan}_T F \colon T \to C$ is a functor equipped with a natural transformation $\eta \colon F \to LT$, such that every similar pair $(L', \eta' \colon F \to L'T)$

$$\tag{2.79}$$

factorises as $\eta' = \lambda T . \eta$, by a unique natural transformation $\lambda \colon L \to L'$.

2.8.5 Limits and colimits as Kan extensions

In particular, the limit of a functor $X \colon S \to C$ amounts to the right Kan extension $R = \mathrm{Ran}_T X$ along the functor $T \colon S \to \mathbf{1}$ (with values in the terminal category)

$$\tag{2.80}$$

while the colimit of $X \colon S \to C$ amounts to the left Kan extension $L = \mathrm{Lan}_T X \colon \mathbf{1} \to C$.

2.8.6 Coends

Let us be given a functor $F \colon X^{\mathrm{op}} \times X \to C$. Its *coend* is an object C of C equipped with a family of morphisms $u_X \colon F(X, X) \to C$ indexed by the objects of X. This family must be:

(i) *coherent*: for every morphism $f \colon X \to Y$ in X we have

$$u_X . F(f, 1) = u_Y . F(1, f) \colon F(Y, X) \to C,$$

(ii) *universal*: for every coherent family $(v_X \colon F(X, X) \to C')_X$ there is a unique morphism $c \colon C \to C'$ such that $u_X . c = v_X$, for all X.

The coend of F is written as $\int^X F(X, X)$.

If X is small and C is cocomplete, the coend of F exists and can be obtained as the coequalizer of the two morphisms

$$a, b \colon \Sigma_{f \colon X \to Y} F(Y, X) \rightrightarrows \Sigma_X F(X, X),$$

whose co-component on the morphism $f \colon X \to Y$ comes, respectively, from

$$a_f = F(f, 1) \colon F(Y, X) \to F(X, X),$$

$$b_f = F(1, f) \colon F(Y, X) \to F(Y, Y).$$

(Dually, *ends* can be constructed by means of limits.)

A classical example from Algebraic Topology, the geometric realisation of simplicial sets and cubical sets, will be described in 5.3.5 and 5.4.6.

3

Adjunctions and monads

Adjunctions, a crucial step in category theory, were introduced by Kan in 1958 [K2]. They extend Galois connections (defined in Section 1.7) from ordered sets to general categories. The closely related notion of 'monad' is also studied here.

The Adjoint Functor Theorem of Freyd, dealt with in Section 3.5, is an effective tool to prove the existence of an adjoint, which might be difficult to construct.

Many concrete examples can be found in this chapter, then in Chapter 4 (for algebraic structures) and Chapter 5 (for topological structures).

3.1 Adjoint functors

We begin with two well known situations, which help us to introduce the adjunction of functors. Then we give a definition, in four equivalent forms that extend those of Galois connections, in 1.7.1.

3.1.1 Examples

(a) The forgetful functor U: Top $\to$ Set has two well known 'best approximations' to an inverse which does not exist, the functors D, C: Set $\to$ Top, where DX (resp. CX) is the set X with the discrete (resp. indiscrete) topology. In fact they provide the set X with the finest (resp. the coarsest) topology.

For every set X and every space Y we can 'identify' the following hom-sets

$$\mathsf{Top}(D(X), Y) = \mathsf{Set}(X, U(Y)),$$
$$\mathsf{Set}(U(Y), X) = \mathsf{Top}(Y, C(X)),$$
$$(3.1)$$

since every mapping $X \to Y$ becomes continuous if we put on X the discrete

109

topology, and every mapping $Y \to X$ becomes continuous if we put on X the indiscrete one.

These two facts will tell us that $D \dashv U$ and $U \dashv C$, respectively: D is left adjoint to U, and C is right adjoint to the latter. This presentation of the two adjunctions corresponds to form (i) of a Galois connection $f \dashv g$, in 1.7.1, which says that the hom-sets $Y(f(x), y)$ and $X(x, g(y))$ have the same cardinal: either 0 or 1.

(b) In other cases, it may be convenient to follow an approach corresponding to the form (ii) of 1.7.1, and based on universal arrows (defined in Section 2.7).

To wit, let us start from the forgetful functor $U \colon R\,\mathsf{Mod} \to \mathsf{Set}$. We have already seen, in 2.7.4(a), that, for every set X, there is a universal arrow from X to the functor U

$$F(X) = \bigoplus_{x \in X} R, \qquad \eta \colon X \to UF(X), \tag{3.2}$$

consisting of the free R-module $F(X)$ on the set X, with the insertion of the basis η.

The fact that this universal arrow exists *for every X in* Set allows us to construct a 'backward' functor $F \colon \mathsf{Set} \to R\,\mathsf{Mod}$, *left adjoint* to U.

F is already defined on the objects. For a mapping $f \colon X \to Y$ in Set

$$
\begin{array}{ccccc}
X & \xrightarrow{\;\eta X\;} & UF(X) & & F(X) \\
{\scriptstyle f}\big\downarrow & & \big\downarrow{\scriptstyle U(Ff)} & & \big\downarrow{\scriptstyle Ff} \\
Y & \xrightarrow[\;\eta Y\;]{} & UF(Y) & & F(Y)
\end{array}
\tag{3.3}
$$

the universal property of ηX implies that there is precisely one homomorphism $F(f) \colon F(X) \to F(Y)$ such that $U(F(f))$ makes the square above commutative.

(Concretely, $F(f)$ is just the R-linear extension of f; but the general pattern is better perceived if we proceed in a formal way.)

Further applications of the universal property would show that F preserves composition and identities (as we shall see in the general context). Let us note that we have extended the given object-function $F(X)$ in the only way that makes the family (η_X) into a natural transformation $\eta \colon 1 \to UF \colon \mathsf{Set} \to \mathsf{Set}$, called the *unit* of the adjunction.

(c) The relationship between these functors U and F (assuming now that we have defined both) can be equivalently expressed in form (i), by a family of bijective mappings (for X in Set and A in $R\,\mathsf{Mod}$)

$$\varphi_{XA} \colon R\,\mathsf{Mod}(FX, A) \to \mathsf{Set}(X, UA), \tag{3.4}$$

which is natural in X and A, as will be made precise in 3.1.2.

Concretely, $\varphi_{XA}(g\colon FX \to A)$ is the restriction of the homomorphism g to the basis X of $F(X)$; its bijectivity means that every map $f\colon X \to UA$ has a unique extension to a homomorphism $g\colon FX \to A$.

Formally, given the family (ηX), we define

$$\varphi_{XA}(g\colon FX \to A) = Ug.\eta X\colon X \to UA.$$

The other way round, given the family (φ_{XA}), we define

$$\eta X = \varphi_{X,FX}(\mathrm{id}FX)\colon X \to UFX.$$

3.1.2 Definition

An *adjunction* $F \dashv G$, with a functor $F\colon \mathsf{C} \to \mathsf{D}$ *left adjoint* to a functor $G\colon \mathsf{D} \to \mathsf{C}$, can be equivalently presented in four main forms.

(i) We assign two functors $F\colon \mathsf{C} \to \mathsf{D}$ and $G\colon \mathsf{D} \to \mathsf{C}$ together with a family of bijections

$$\varphi_{XY}\colon \mathsf{D}(FX, Y) \to \mathsf{C}(X, GY) \qquad (X \text{ in } \mathsf{C}, Y \text{ in } \mathsf{D}), \tag{3.5}$$

which is natural in X, Y. More formally, the family (φ_{XY}) is a functorial isomorphism

$$\varphi\colon \mathsf{D}(F(-), =) \to \mathsf{C}(-, G(=))\colon \mathsf{C}^{\mathrm{op}} \times \mathsf{D} \to \mathsf{Set}. \tag{3.6}$$

(ii) We assign a functor $G\colon \mathsf{D} \to \mathsf{C}$ and, for every object X in C, a *universal arrow* (see 2.7.1)

$$(F_0X,\ \eta X\colon X \to GF_0X) \qquad \text{from the object } X \text{ to the functor } G.$$

(ii*) We assign a functor $F\colon \mathsf{C} \to \mathsf{D}$ and, for every object Y in D, a *universal arrow*

$$(G_0Y,\ \varepsilon Y\colon FG_0Y \to Y) \qquad \text{from the functor } F \text{ to the object } Y.$$

(iii) We assign two functors $F\colon \mathsf{C} \to \mathsf{D}$ and $G\colon \mathsf{D} \to \mathsf{C}$, together with two natural transformations

$$\eta\colon \mathrm{id}\mathsf{C} \to GF \quad (\text{the } unit), \qquad \varepsilon\colon FG \to \mathrm{id}\mathsf{D} \quad (\text{the } counit),$$

which satisfy the *triangular identities*:

$$\varepsilon F.F\eta = \mathrm{id}F, \qquad\qquad G\varepsilon.\eta G = \mathrm{id}G, \tag{3.7}$$

$$F \xrightarrow{F\eta} FGF \xrightarrow{\varepsilon F} F \qquad\qquad G \xrightarrow{\eta G} GFG \xrightarrow{G\varepsilon} G$$
$$\underset{\mathrm{id}F}{\longrightarrow} \qquad\qquad\qquad\qquad \underset{\mathrm{id}G}{\longrightarrow}$$

The proof of the equivalence will be given in Theorem 3.1.5. Essentially:
- given (i) one defines

$$\eta X = \varphi_{X,FX}(\mathrm{id}FX)\colon X \to GFX,$$

$$\varepsilon Y = (\varphi_{GY,Y})^{-1}(\mathrm{id}GY)\colon FGY \to Y,$$

- given (ii) one defines $F(X) = F_0 X$, the morphism $F(f\colon X \to X')$ by the universal property of ηX and the morphism $\varphi_{XY}(g\colon FX \to Y)$ as $Gg.\eta X\colon X \to GY$,

- given (iii) one defines the mapping $\varphi_{XY}\colon D(F(X),Y) \to C(X,G(Y))$ as above, and a backward mapping $\psi_{XY}(f\colon X \to GY) = \varepsilon Y.Ff$ (which are inverse to each other by the triangular identities).

Finally, the naturality of φ in (3.6) means that, for every pair of morphisms $u\colon X' \to X$ and $v\colon Y \to Y'$ (in C and D, respectively) we have a commutative square in Set

$$Gv.\varphi_{XY}(g).u = \varphi_{X'Y'}(v.g.Fu) \tag{3.8}$$

$$
\begin{array}{ccc}
D(F(X),Y) & \xrightarrow{\;\varphi_{XY}\;} & C(X,G(Y)) \\
{\scriptstyle v.-.Fu}\Big\downarrow & & \Big\downarrow{\scriptstyle Gv.-.u} \\
D(F(X'),Y') & \xrightarrow[\;\varphi_{X'Y'}\;]{} & C(X',G(Y'))
\end{array}
$$

3.1.3 Comments and complements

(a) The previous forms have different features.

Form (i) is the classical definition of an adjunction, and is at the origin of the name, by analogy with adjoint maps of Hilbert spaces.

Form (ii) is used when one starts from a functor G and wants to build its left adjoint, possibly less easy to define. Form (ii*) is used in a dual way.

The 'algebraic' form (iii) is free of quantifiers and adequate to the formal theory of adjunctions (because it makes sense in an abstract 2-category, see 7.1.2).

(b) Let S be a small category. A functor $F\colon C \to D$ can be canonically 'extended' to functor categories on S, composing each functor $X\colon S \to C$ with F

$$F^S\colon C^S \to D^S, \qquad F^S(X) = FX\colon S \to D. \tag{3.9}$$

A natural transformation $\varphi\colon F \to G\colon C \to D$ can be similarly extended

$$\varphi^S\colon F^S \to G^S\colon C^S \to D^S,$$

$$\varphi^S(X) = \varphi X\colon FX \to GX\colon S \to D. \tag{3.10}$$

These extensions preserve all compositions, of functors and natural transformations.

(c) (*Adjunctions and functor categories*) It follows that an adjunction (η, ε): $F \dashv G$ has a canonical extension to an adjunction

$$(\eta^S, \varepsilon^S) \colon F^S \dashv G^S \colon \mathsf{C}^S \to \mathsf{D}^S, \tag{3.11}$$

using the form (iii) of Definition 3.1.2.

*(d) Adjunctions are related to Kan extensions in a complex way. The reader can see [M4], Section X.7.

3.1.4 Examples, exercises and complements

(a) An adjoint equivalence $(F, G, \eta, \varepsilon)$, defined in 1.5.5, amounts to an adjunction where the unit and counit are invertible; then $F \dashv G \dashv F$.

(b) Various examples of Galois connections have been discussed in Section 1.7.

(c) For a 'forgetful functor' $U \colon \mathsf{A} \to \mathsf{X}$ the existence of the left adjoint $F \dashv U$ is equivalent to saying that every object X of X has a free object $(FX, \eta \colon X \to UFX)$ in A (as defined in 2.7.3).

We shall see, in Theorem 4.4.2, that the forgetful functor $U \colon \mathsf{A} \to \mathsf{Set}$ of a variety of algebras has always a left adjoint $F \colon \mathsf{Set} \to \mathsf{A}$, the *free-algebra functor* for A. In many cases there is a simple construction of F: we have seen some of them in 2.7.4 and 3.1.1, and many others will be analysed in Section 4.1.

(d) We have seen that the forgetful functor $U \colon \mathsf{Top} \to \mathsf{Set}$ has both adjoints $D \dashv U \dashv C$, where DX (resp. CX) is the set X with the discrete (resp. indiscrete) topology. Let us note that D has no left adjoint, while C has no right adjoint, as will be proved in 3.2.3(c). (In Section 3.8 we shall see that these adjunctions are idempotent.)

Many other adjoints of interest in Topology will be examined in Chapter 5.

(e) The forgetful functor $U \colon \mathsf{pOrd} \to \mathsf{Set}$ of the category of preordered sets has similar adjoints $D \dashv U \dashv C$, where DX is the set X with the discrete order $x = x'$, while CX is the same set with the indiscrete or chaotic preorder, namely $x, x' \in X$.

Here D has also a left adjoint $\pi_0 \colon \mathsf{pOrd} \to \mathsf{Set}$, where $\pi_0(X)$ is the *set of connected components* of the preordered set X, namely the quotient of the underlying set $|X|$ modulo the equivalence relation generated by the preorder $x \prec x'$ (i.e. the least equivalence relation containing the former).

(f) Prove that the embedding J: pOrd $\to$ Cat of preordered sets into small categories (see 1.2.3) has a left adjoint po: Cat $\to$ pOrd, using the preordered set po(C) defined in 1.4.5. Prove that J does not preserve coequalisers, while po does not preserve equalisers.

(g) Starting from the (non-faithful) functor Ob : Cat $\to$ Set, there is a chain of adjunctions

$$\pi_0 \dashv D \dashv \mathrm{Ob} \dashv C,$$

which extends that of point (e).

Hints. A quick solution can be obtained from the previous points, as shown in Chapter 8.

It may be more interesting to give a direct construction. First, we say that a category C is *connected* if it is not empty, and every pair of objects is linked by a 'zig-zag path of arrows'

$$\bullet \longrightarrow \bullet \longleftarrow \bullet \longrightarrow \bullet \ \ldots \ \bullet \longrightarrow \bullet$$

Now $\pi_0(\mathsf{C})$ can be defined as the set of the connected components of C, that is its maximal connected subcategories. The empty category has none, and a category is connected if and only if $\pi_0(\mathsf{C})$ is the singleton. Every category is the categorical sum of its connected components.

3.1.5 Theorem (Characterisation of adjoints)

The presentations (i)–(iii) of an adjunction in Definition 3.1.2 are equivalent.

Proof (i) $\Rightarrow$ (ii). Given (i), one defines $\eta X = \varphi_{X,FX}(\mathrm{id}FX): X \to GFX$.
The naturality of φ_{XY} on $\mathrm{id}X$ and $g: FX \to Y$ gives

$$\varphi_{XY}(g) = G(g).\varphi_{X,FX}(\mathrm{id}FX) = G(g).\eta X.$$

The bijectivity of φ_{XY} gives the universal property of ηX: for every $f: X \to GY$ in C there exists precisely one $g: FX \to Y$ in D such that $G(g).\eta X = f$.

(ii) $\Rightarrow$ (i). Conversely, given (ii), one defines $F(X) = F_0 X$. Then the morphism $F(f: X \to X')$ is defined by the universal property of ηX, as the unique morphism of D such that

$$
\begin{array}{ccc}
X & \xrightarrow{\ \eta X\ } & GF(X) \\
{\scriptstyle f}\downarrow & & \downarrow{\scriptstyle G(Ff)} \\
X' & \xrightarrow[\ \eta X'\ ]{} & GF(X')
\end{array}
\qquad G(Ff).\eta X = \eta X'.f. \qquad\qquad (3.12)
$$

The reader will verify that G is a functor, using the following diagrams

$$
\begin{array}{ccc}
X & \xrightarrow{\eta X} & GF(X) \\
f\downarrow & & \downarrow G(Ff) \\
X' & \xrightarrow{\eta X'} & GF(X') \\
f'\downarrow & & \downarrow G(Ff') \\
X'' & \xrightarrow[\eta X'']{} & GF(X'')
\end{array}
\qquad
\begin{array}{ccc}
X & \xrightarrow{\eta X} & GF(X) \\
1\downarrow & & \downarrow G(F1_X) \\
X & \xrightarrow[\eta X]{} & GF(X)
\end{array}
\qquad (3.13)
$$

Finally we define $\varphi_{XY}(g\colon FX \to Y) = Gg.\eta X\colon X \to GY$. It is a bijective mapping, because of the universal property of ηX.

The commutativity of the square (3.8) follows easily:

$$
Gv.\varphi_{XY}(g).u = Gv.Gg.\eta X.u = Gv.Gg.GFu.\eta X'
$$
$$
= G(v.g.Fu).\eta X' = \varphi_{X'Y'}(v.g.Fu).
$$

(i) $\Rightarrow$ (iii). Given (i), one defines

$$
\eta X = \varphi_{X,FX}(\mathrm{id}FX)\colon X \to GFX,
$$
$$
\varepsilon Y = (\varphi_{GY,Y})^{-1}(\mathrm{id}GY)\colon FGY \to Y.
$$

The naturality of both families is easily verified. As to the first triangular identity $\varepsilon F.F\eta = \mathrm{id}F$, we have:

$$
\varphi_{X,FX}(\varepsilon FX.F\eta X) = \varphi_{X,FX}(\varepsilon FX).\eta X = \mathrm{id}(GFX).\eta X
$$
$$
= \eta X = \varphi_{X,FX}(\mathrm{id}FX),
$$

whence $\varepsilon FX.F\eta X = \mathrm{id}FX$, for every X in C.

(iii) $\Rightarrow$ (i). Given (iii), one defines the mappings

$$
\varphi_{XY}\colon \mathsf{D}(F(X),Y) \to \mathsf{C}(X,G(Y)), \qquad \varphi_{XY}(g) = Gg.\eta,
$$
$$
\psi_{XY}\colon \mathsf{C}(X,G(Y)) \to \mathsf{D}(F(X),Y), \qquad \psi_{XY}(f) = \varepsilon Y.Ff,
$$

and proves that they are inverse to each other, by the triangular identities.

The naturality of φ is proved as in (ii) $\Rightarrow$ (i), using the commutativity of square (3.12), which here follows from the naturality of η (instead of being a consequence of the definition of $F(f)$). $\qquad \square$

3.1.6 Theorem (Uniqueness)

Given a functor $G\colon \mathsf{D} \to \mathsf{C}$, its left adjoint is uniquely determined up to isomorphism.

More precisely, we suppose that there are two left adjoints $F\colon \mathsf{C} \to \mathsf{D}$ and $F'\colon \mathsf{C} \to \mathsf{D}$, with units $\eta\colon 1 \to GF$ and $\eta'\colon 1 \to GF'$.

Then there is a unique natural transformation $\varphi\colon F \to F'$ such that $G\varphi.\eta = \eta'$, and it is a functorial isomorphism.

Proof Let X be an object of C. By the uniqueness property of universal arrows, in 2.7.1, there is a unique morphism $\varphi X\colon F(X) \to F'(X)$ in D such that the following square commutes

$$
\begin{array}{ccc}
X & \xrightarrow{\ \eta X\ } & GF(X) \\
{\scriptstyle 1}\big\downarrow & & \big\downarrow{\scriptstyle G\varphi X} \\
X & \xrightarrow[\ \eta' X\ ]{} & GF'(X)
\end{array}
\qquad (3.14)
$$

and φX is invertible. To prove that the family of all φX is natural, let us take a morphism $f\colon X \to X'$ in C. In the left diagram below, the left square and the outer rectangle commute, by naturality of η and η', respectively

$$
\begin{array}{ccccc}
X & \xrightarrow{\ \eta X\ } & GF(X) & \xrightarrow{\ G(\varphi X)\ } & GF'(X) \\
{\scriptstyle f}\big\downarrow & {\scriptstyle =} & \big\downarrow{\scriptstyle GFf} & {\scriptstyle ?} & \big\downarrow{\scriptstyle GF'f} \\
X' & \xrightarrow[\ \eta X'\ ]{} & GF(X') & \xrightarrow[\ G(\varphi X')\ ]{} & GF'(X')
\end{array}
\qquad (3.15)
$$

Therefore

$$
G(F'f.\varphi X).\eta X = G(\varphi X').\eta X'.f = G(\varphi X'.Ff).\eta X,
$$

and the naturality of φ follows from the universal property of ηX:

$$
F'f.\varphi X = \varphi X'.Ff.
$$

$\square$

3.2 Properties of adjunctions

This section takes on the general study of adjoint functors.

3.2.1 A category of adjunctions

As a consequence of the composition of universal arrows (in 2.7.6), two consecutive adjunctions

$$
\begin{aligned}
F\colon \mathsf{C} \rightleftarrows \mathsf{D}\colon G, &\qquad \eta\colon 1 \to GF, &\quad \varepsilon\colon FG \to 1, \\
H\colon \mathsf{D} \rightleftarrows \mathsf{E}\colon K, &\qquad \rho\colon 1 \to KH, &\quad \sigma\colon HK \to 1,
\end{aligned}
\qquad (3.16)
$$

give a composed adjunction from the first to the third category

$$HF: \mathsf{C} \rightleftarrows \mathsf{E} : GK,$$

$$G\rho F.\eta: 1 \to GF \to GK.HF, \qquad (3.17)$$

$$\sigma.H\varepsilon K: HF.GK \to HK \to 1.$$

There is thus a category AdjCat of small categories and adjunctions, with morphisms written in one of the following forms, and *conventionally directed as the left adjoint*

$$(F, G, \eta, \varepsilon): \mathsf{C} \rightarrow \mathsf{D}, \qquad F \dashv G: \mathsf{C} \rightarrow \mathsf{D}.$$

We also write $(\eta, \varepsilon): F \dashv G$, but one should not view this as an arrow from F to G in some category: there is no composition of this kind.

Duality of categories interchanges left and right adjoint, unit and counit. AdjCat is thus a category with involution (as defined in 1.8.1)

$$\mathsf{C} \mapsto \mathsf{C}^{\mathrm{op}},$$

$$((F, G, \eta, \varepsilon): \mathsf{C} \rightarrow \mathsf{D}) \mapsto ((G^{\mathrm{op}}, F^{\mathrm{op}}, \varepsilon^{\mathrm{op}}, \eta^{\mathrm{op}}): \mathsf{D}^{\mathrm{op}} \rightarrow \mathsf{C}^{\mathrm{op}}). \qquad (3.18)$$

3.2.2 The preservation of (co)limits

A left adjoint preserves (the existing) colimits, a right adjoint preserves (the existing) limits, as verified below, in Exercise 3.2.3(a). It follows that a left adjoint preserves epimorphisms and a right adjoint preserves monomorphisms (by 2.3.5(b)).

For a covariant Galois connection $f \dashv g$, this amounts to saying that f preserves the existing joins and g the existing meets, as already shown in 1.7.3.

As we know, an equivalence of categories $F: \mathsf{C} \to \mathsf{D}$ can always be completed to an adjoint equivalence, where $F \dashv G \dashv F$; therefore F preserves limits and colimits.

The standard way of proving that a functor cannot have a left (resp. a right) adjoint is to find a limit (resp. a colimit) that it does not preserve.

3.2.3 Exercises and complements

These exercises are important; solutions are given below.

(a) In an adjunction $F \dashv G: \mathsf{C} \rightarrow \mathsf{D}$, the left adjoint F preserves all colimits. The reader can easily give an analytic proof.

*(b) For a more formal alternative proof, one can use the characterisation of colimits as representative objects, in 2.7.2.

(c) We have seen that the forgetful functor U: Top $\to$ Set has both adjoints $D \dashv U \dashv C$. Accordingly U preserves all limits and colimits. Prove that this chain of adjunctions cannot be extended. *Hints:* certain products or sums are not preserved, by D or C.

(d) (*Colimits of ordered sets*) The embedding U: Ord $\to$ pOrd has a left adjoint F, that takes every preordered set X to the quotient set $|X|/\sim$, equipped with the induced order: $[x] \leqslant [y]$ if $x \prec y$ in X (as in 1.2.2). Show how this can be used to compute a colimit in Ord.

Colimits of Hausdorff spaces can be computed in a similar way, as quotients of colimits in Top (see 5.1.4(b)).

(e) The adjunction po: Cat $\rightleftarrows$ pOrd : J in Exercise 3.1.4(f) cannot be extended.

Solutions. (a) For a functor X: S $\to$ C, a colimit $(L, (u_i: X_i \to L))$ is transformed by F into a cocone $(FL, (Fu_i: FX_i \to FL))$ of the composed functor FX: S $\to$ D.

The latter is also universal. In fact, if $(Y, (g_i: FX_i \to Y))$ is a cocone of FX in D, the adjunction gives a cocone $(GY, (f_i: X_i \to GY))$ of X in C; then we have a unique map $f: L \to GY$ such that $fu_i = f_i$ (for all $i \in \mathrm{Ob}\,S$), which corresponds to a unique map $g: FL \to Y$ such that $g.Fu_i = g_i$ (for all i).

*(b) The colimit L of the functor X: S $\to$ C represents the covariant functor of cocones of X

$$H = \mathsf{C}^{\mathsf{S}}(X, D(-)): \mathsf{C} \to \mathsf{Set},$$

via the morphism $u: X \to DL$ in C^{S}. Then FL represents the functor $K = \mathsf{D}^{\mathsf{S}}(FX, D(-)): \mathsf{D} \to \mathsf{Set}$ via the morphism $Fu: FX \to DFL$, by composing three functorial isomorphisms in the variable Y

$$\mathsf{D}(FL, Y) \to \mathsf{C}(L, GY),$$

$$\mathsf{C}(L, GY) \to \mathsf{C}^{\mathsf{S}}(X, DGY) = \mathsf{C}^{\mathsf{S}}(X, G^{\mathsf{S}}(DY)),$$

$$\mathsf{C}^{\mathsf{S}}(X, G^{\mathsf{S}}(DY)) \to \mathsf{D}^{\mathsf{S}}(F^{\mathsf{S}}(X), DY) = K(Y).$$

The first comes from the adjunction $F \dashv G$, the second from L representing the functor H: C $\to$ Set, the third from the extended adjunction $F^{\mathsf{S}} \dashv G^{\mathsf{S}}$ (see 3.1.3(c)).

(c) The functor D: Set $\to$ Top does not preserve infinite products and cannot have a left adjoint, while C: Set $\to$ Top does not preserve (even binary) sums and cannot have a right adjoint.

(d) We use the fact that the left adjoint F: pOrd $\to$ Ord preserves colimits.

Given a functor X: S $\to$ Ord we have already seen, in 2.2.5(c), that the colimit of UX in pOrd is the colimit $(L, (u_i\colon |X_i| \to L))$ in Set, equipped with the finest preorder that makes all mappings u_i monotone. The colimit of X in Ord is the associated ordered set $L/{\sim}$, or more precisely the cocone $(F(L), (Fu_i\colon X_i \to FL))$ produced by the left adjoint $F\colon$ pOrd $\to$ Ord.

(e) We have already seen in 3.1.4(f) that the functor po: Cat $\to$ pOrd does not preserve equalisers, while J does not preserve coequalisers.

3.2.4 Limit functors as right adjoints

The description of the limit of a functor $X\colon$ S $\to$ C as a universal arrow

$$(L(X), \varepsilon X\colon DL(X) \to X)$$

from the diagonal functor $D\colon$ C $\to$ C^{S} to the object X of C^{S} (in 2.7.2) shows that the existence of all S-based limits in C amounts to the existence of a right adjoint $L\colon$ C$^{\mathsf{S}} \to$ C (the *limit functor*) for D, with counit ε.

The limit cone has components $(\varepsilon X)_i\colon L(X) \to X_i$, for $i \in$ Ob S.

Dually, a *colimit functor* $L'\colon$ C$^{\mathsf{S}} \to$ C is left adjoint to D, with a unit $\eta X\colon X \to DL'(X)$ and cocone-components $(\eta X)_i\colon X_i \to L'(X)$.

3.2.5 Theorem (Faithful and full adjoints)

Suppose we have an adjunction $F \dashv G$, *with counit* $\varepsilon\colon FG \to 1$. *Then*

(i) *G is faithful if and only if all the components εY are epi,*

(ii) *G is full if and only if all the components εY are split mono,*

(iii) *G is full and faithful if and only if the counit ε is invertible.*

Proof (i) We are given two maps $g, g'\colon Y \to Y'$ in D and will use, for both implications, the following diagram of two commutative squares

$$
\begin{array}{ccc}
FG(Y) & \xrightarrow{\;\varepsilon Y\;} & Y \\
{\scriptstyle FGg}\big\downarrow\big\downarrow{\scriptstyle FGg'} & & {\scriptstyle g}\big\downarrow\big\downarrow{\scriptstyle g'} \\
FG(Y') & \xrightarrow[\;\varepsilon Y'\;]{} & Y'
\end{array}
\qquad (3.19)
$$

If the components of ε are epi, suppose that $Gg = Gg'$; then $g.\varepsilon Y = g'.\varepsilon Y$ and $g = g'$. Conversely, suppose that G is faithful and $g.\varepsilon Y = g'.\varepsilon Y$; then $\varepsilon Y'.F(Gg) = \varepsilon Y'.F(Gg')$, whence $Gg = Gg'$ and $g = g'$.

(ii) If εY is a split mono, there is some $p\colon Y \to FGY$ in D such that $p.\varepsilon Y = 1_{FGY}$. For a morphism $f\colon GY \to GY'$ in C, we let

$$g = \varepsilon Y'.Ff.p\colon Y \to Y'.$$

Then the two squares of the diagram below commute, and $G(g) = f$ by the universal property of $\varepsilon Y'$

$$
\begin{array}{ccc}
FG(Y) & \xrightarrow{\varepsilon Y} & Y \\
{\scriptstyle FGg}\downarrow\,\,\downarrow{\scriptstyle Ff} & & \downarrow{\scriptstyle g} \\
FG(Y') & \xrightarrow[\varepsilon Y']{} & Y'
\end{array}
\tag{3.20}
$$

Conversely, we suppose that G is full. Then there exists some morphism $p\colon Y \to FGY$ in D such that $G(p) = \eta GY\colon GY \to GFGY$. To prove that $v = p.\varepsilon Y$ coincides with 1_{FGY} we apply the universal property of ηGY in the following diagram

$$
\begin{array}{ccc}
G(Y) & \xrightarrow{\eta GY} & GFG(Y) \\
{\scriptstyle 1}\downarrow & & {\scriptstyle G(1)}\downarrow\,\,\downarrow{\scriptstyle G(v)} \\
G(Y) & \xrightarrow[\eta GY]{} & GFG(Y)
\end{array}
\tag{3.21}
$$

$$
G(v).\eta GY = \eta GY.G\varepsilon Y.\eta GY = \eta GY = G(1_{FGY}).\eta GY.
$$

$\square$

3.2.6 Reflective and coreflective subcategories

A subcategory D $\subset$ C is said to be *reflective* (notice: not 'reflexive') if the inclusion functor $U\colon$ D $\to$ C has a left adjoint, and *coreflective* if U has a right adjoint. For a full reflective subcategory the counit ε is invertible (by Theorem 3.2.5).

For instance Ab is reflective in Gp, with reflector $(-)^{\mathrm{ab}}\colon$ Gp $\to$ Ab sending a group G to the abelianised group $G^{\mathrm{ab}} = G/[G, G]$ (see 2.7.4(c)). The unit-component $\eta G\colon G \to G^{\mathrm{ab}}$ is the canonical projection, and the counit $\varepsilon A\colon A \to A/[A, A]$ is invertible.

If X is an ordered set, an ordered subset $A \subset X$ can be viewed as a (full) subcategory, possibly reflective or coreflective in X.

Exercises and complements. (a) In Ab the full subcategory tAb formed by all torsion abelian groups is coreflective.

(b) In Ab the full subcategory tfAb formed by all torsion-free abelian groups is reflective. *(The reader may know, or easily prove, that a free abelian group is torsion free. But the additive group $\mathbb{Q}$ is also torsion free.)*

(c) The ordered set $\mathbb{Z}$ is reflective and coreflective in the ordered line $\mathbb{R}$ while $\mathbb{Q}$ is neither.

(d) Let A be a lower unbounded, closed subset of the euclidean line $\mathbb{R}$. Prove that A is reflective in the ordered set $\mathbb{R}$.

3.2.7 *Lemma* (Lifting left adjoints)

Let us suppose that we have an adjunction $(\eta, \varepsilon) \colon F \dashv U$

$$
\begin{array}{ccc}
\mathsf{Y} & \underset{V}{\overset{G}{\rightleftarrows}} & \mathsf{A} \\
{\scriptstyle W}\downarrow & & \parallel \\
\mathsf{X} & \underset{U}{\overset{F}{\rightleftarrows}} & \mathsf{A}
\end{array}
\qquad (3.22)
$$

where the right adjoint factorises as $U = WV$, *with* W *full and faithful.*

Then $G = FW \colon \mathsf{Y} \to \mathsf{C}$ *is left adjoint to* V, *with unit the unique natural transformation* $\vartheta \colon 1 \to VG$ *such that* $W\vartheta = \eta W$.

Proof Since W is full and faithful, for every Y in Y there is a unique morphism $\vartheta Y \colon Y \to VG(Y)$ in Y such that

$$
W(\vartheta Y) = \eta(WY) \colon WY \to UFWY = W(VGY).
$$

The pair $(G(Y), \vartheta Y \colon Y \to VG(Y))$ is a universal arrow from the object Y to the functor $V \colon \mathsf{A} \to \mathsf{Y}$. Indeed, for every similar pair $(A, f \colon Y \to VA)$, a morphism $g \colon GY \to A$ satisfies $Vg.\vartheta Y = f$ (as in the left diagram below) if and only if $W(Vg.\vartheta Y) = W(f)$

$$
\begin{array}{ccccc}
Y & \xrightarrow{\vartheta Y} & VGY & \qquad WY & \xrightarrow{\eta WY} & UFWY = W(VGY) \\
& {\scriptstyle f}\searrow & \downarrow{\scriptstyle Vg} & & {\scriptstyle Wf}\searrow & \downarrow{\scriptstyle Ug} \\
& & VA & & & UA = W(VA)
\end{array}
\qquad (3.23)
$$

and there is precisely one solution, because of the universal property of $\eta(WY)$, as in the right diagram above.

Finally we only have to check that the action of the functor G on the morphisms of Y, as determined by the universal property of ϑ, coincides with that of the functor $FW \colon \mathsf{Y} \to \mathsf{C}$.

This amounts to saying that the family $\vartheta Y \colon Y \to VG(Y) = VFW(Y)$ is a natural transformation $1 \to VFW$; but this follows from the naturality of the family $W(\vartheta Y) = (\eta W)Y$, and the faithfulness of W. $\qquad \square$

3.2.8 Exercises and complements (Chains of adjunctions, II)

(a) Chains of Galois connections between ordered sets X, Y have been considered in 1.7.7. Each of them can be transformed into a chain of adjunctions between the categories C^X and C^Y, for any category C.

We now investigate some non-trivial chains of adjunctions of period 2, or even 1, between categories – a situation which does not exist for ordered sets, where we know that a relation $f \dashv g \dashv f$ forces f and g to be inverse to each other (see 1.7.2(b)).

(b) Let $D: C \to C^2$ be the diagonal functor of a category C into $C \times C$. By 3.2.4, D has a left adjoint S and a right adjoint P if (and only if) C has binary sums and products

$$S, P: C^2 \to C, \qquad\qquad S \dashv D \dashv P,$$
$$S(X, Y) = X + Y, \qquad P(X, Y) = X \times Y. \qquad (3.24)$$

When C is Abm, Ab or R Mod, these two functors coincide (as we have seen in 2.1.9): the direct sum (or biproduct) $B(X, Y) = X \oplus Y$ is at the same time

- left adjoint to D, with a unit $\eta: (X, Y) \to (X \oplus Y, X \oplus Y)$ giving the two injections of the sum,

- right adjoint to D, with a counit $\varepsilon: (X \oplus Y, X \oplus Y) \to (X, Y)$ giving the two projections of the product.

We have thus a periodic chain ... $B \dashv D \dashv B \dashv D$ ...

More generally, the same holds in every semiadditive category, see Section 6.4.

(c) In 2.8.2–2.8.3 we have considered the two-element idempotent monoid $E = \{1, \underline{e}\}$ and the diagonal functor $D: C \to C^E$. We have seen that in C all idempotents split if and only C has all E-based limits, if and only if it has all E-based colimits.

Moreover, if this is the case, the limit-object and colimit-object coincide, and (by 3.2.4) we have one functor $L: C^E \to C$ which is:

- left adjoint to D, with a unit $p_e: e \to 1_{L(e)}$ giving the epimorphism of the splitting $e = m_e p_e$,

- right adjoint to D, with a counit $m_e: 1_{L(e)} \to e$ giving the monomorphism of the splitting $e = m_e p_e$.

Here $L(e)$ is (a choice of) the image of e in C. This gives another periodic chain of adjunctions

$$... L \dashv D \dashv L \dashv D ...$$

(d) For a fixed set A there is an obvious endofunctor

$$F \colon \mathsf{Rel\,Set} \to \mathsf{Rel\,Set}, \qquad F(X) = X \times A, \tag{3.25}$$

given by the cartesian product (let us recall that this is *not* a categorical product in Rel Set, see 2.2.6(c)). The reader will prove that $F \dashv F$, which gives a chain of adjunctions of period 1.

3.3 Comma categories

Comma categories are an important construction, with strong relationships with all the main tools of category theory, like universal arrows, limits and adjoints.

They were introduced in Lawvere's thesis, in 1963, reprinted in [Lw2].

3.3.1 *Comma categories and slice categories*

For functors $F \colon \mathsf{X} \to \mathsf{Z}$ and $G \colon \mathsf{Y} \to \mathsf{Z}$ with the same codomain Z, one constructs a *comma category* $(F \downarrow G)$ equipped with functors P, Q and a natural transformation π

$$
\begin{array}{c}
\end{array}
\tag{3.26}
$$

(The original notation was (F, G), whence the name.)

The objects of $(F \downarrow G)$ are the triples $(X, Y, z \colon FX \to GY)$ formed of an object of X, an object of Y and a morphism of Z. A morphism of $(F \downarrow G)$

$$(f, g) \colon (X, Y, z \colon FX \to GY) \to (X', Y', z' \colon FX' \to GY'),$$

comes from a pair of maps $f \colon X \to X'$, $g \colon Y \to Y'$ that form a commutative square in Z, namely $z'.Ff = Gg.z$.

Composition and identities come from those of $\mathsf{X} \times \mathsf{Y}$

$$(f', g').(f, g) = (f'f, g'g), \qquad \mathrm{id}(X, Y, z) = (\mathrm{id}X, \mathrm{id}Y). \tag{3.27}$$

The functors P, Q and the natural transformation $\pi \colon FP \to GQ$ are obvious projections.

There is an obvious universal property, that makes the triple (P, Q, π) a sort of *directed 2-dimensional pullback*: for every category C equipped with similar data

$$P' \colon \mathsf{C} \to \mathsf{X}, \quad Q' \colon \mathsf{C} \to \mathsf{Y}, \quad \pi' \colon FP' \to GQ',$$

there is precisely one functor $W\colon \mathsf{C} \to (F\downarrow G)$ which commutes with the structural data

$$W(C) = (P'C, Q'C, \pi'C), \qquad W(c) = (P'c, Q'c),$$
$$PW = P', \qquad QW = Q', \qquad \pi W = \pi'. \tag{3.28}$$

As a matter of notation, one writes $(F\downarrow \mathrm{id}\mathsf{Z})$ as $(F\downarrow \mathsf{Z})$, and $(\mathrm{id}\mathsf{Z}\downarrow G)$ as $(\mathsf{Z}\downarrow G)$. Moreover an object Z_0 of the category Z can be viewed as a functor $Z_0\colon \mathbf{1} \to \mathsf{Z}$; therefore the comma category $(F\downarrow Z_0)$ has objects $(X, z\colon FX \to Z_0)$, while $(Z_0\downarrow G)$ has objects $(Y, z\colon Z_0 \to GY)$.

In particular we have the *slice categories*

$$(Z_0\downarrow \mathsf{Z}) = \mathsf{Z}\backslash Z_0, \qquad (\mathsf{Z}\downarrow Z_0) = \mathsf{Z}/Z_0, \tag{3.29}$$

of *objects* $(Z, z\colon Z_0 \to Z)$ *below* Z_0 and *objects* $(Z, z\colon Z \to Z_0)$ *above* Z_0, respectively.

3.3.2 Exercises and complements

(a) The categories Set$_\bullet$ and Top$_\bullet$ can be identified with the slice categories Set$\backslash\{*\}$ and Top$\backslash\{*\}$, respectively.

(b) Less trivially the category gRng of 'general rings' (associative, possibly non-unital) is equivalent to the category Rng$/\mathbb{Z}$ of *copointed unital rings*, or rings over the initial object $\mathbb{Z}$.

3.3.3 Theorem (Limits in comma categories)

(a) If the categories X *and* Y *have all limits, preserved by* $F\colon \mathsf{X} \to \mathsf{Z}$ *and* $G\colon \mathsf{Y} \to \mathsf{Z}$, *then the comma category* $\mathsf{W} = (F\downarrow G)$ *has all limits, preserved by the functor* $(P, Q)\colon \mathsf{W} \to \mathsf{X} \times \mathsf{Y}$. *The same is true of colimits.*

(b) If the category X *has all limits, preserved by* $F\colon \mathsf{X} \to \mathsf{Z}$, *and* $Z \in \mathrm{Ob}\,\mathsf{Z}$, *then the slice categories* $(F\downarrow Z)$ *and* $(Z\downarrow F)$ *have all limits, preserved by the forgetful functor with values in* X. *The same is true of colimits.*

Note. If in (a) we also assume that F, G create all limits, we can deduce that $(P, Q)\colon \mathsf{W} \to \mathsf{X} \times \mathsf{Y}$ also creates them. The proof is essentially the same as below.

Proof The argument is standard, and the reader may prefer to work it out. It suffices to prove (a), since (b) is a consequence. Below, it is understood that the index i varies in Ob S.

A functor $W\colon \mathsf{S} \to (F\downarrow G)$ has vertices

$$W_i = (PW_i, QW_i, \pi W_i) = (X_i, Y_i, z_i\colon FX_i \to GY_i).$$

The composite $(X, Y) = (P, Q).W \colon \mathsf{S} \to \mathsf{X} \times \mathsf{Y}$ has a limit (L, M), with cone

$$w_i = (u_i, v_i) \colon (L, M) \to (X_i, Y_i).$$

Since F and G preserve limits, FL and GM are limits of FX and GY in Z, with cones

$$Fu_i \colon FL \to FX_i, \qquad Gv_i \colon GM \to GY_i.$$

The natural transformation $\pi W \colon FX \to GY \colon \mathsf{S} \to \mathsf{Z}$ (of components $\pi W_i = z_i$) determines a unique morphism $z \colon FL \to GM$ in Z consistent with these cones:

$$Gv_i.z = z_i.Fu_i \colon FL \to GY_i \qquad \text{(for all } i \in \mathrm{Ob}\,\mathsf{S}).$$

Finally, the object $(L, M, z \colon FL \to GM)$ of $(F \downarrow G)$ is the limit of W, with cone $(u_i, v_i) \colon (L, M, z) \to (X_i, Y_i, z_i)$. □

3.3.4 Comma categories and universal arrows

A universal arrow from the object X to the functor $U \colon \mathsf{A} \to \mathsf{X}$ (see 2.7.1) is the same as an initial object $(A, \eta \colon X \to UA)$ of the comma category $(X \downarrow U)$.

Dually, a universal arrow from the functor U to the object X is the same as a terminal object $(A, \varepsilon \colon UA \to X)$ of the comma category $(U \downarrow X)$.

3.3.5 Comma categories and limits

We have seen that limits and colimits in the category C, based on a small category S, can be viewed as universal arrows of the diagonal functor $D \colon \mathsf{C} \to \mathsf{C}^{\mathsf{S}}$ (in 2.7.2). Therefore they also amount to terminal or initial objects in comma categories of D.

In fact a cone $(A, (f_i \colon A \to X_i))$ of the functor $X \colon \mathsf{S} \to \mathsf{C}$ is the same as an object of the comma category $(D \downarrow X)$ and the limit of X is the same as the terminal object of $(D \downarrow X)$.

Dually a cocone $(A, (f_i \colon X_i \to A))$ of X is an object of the comma category $(X \downarrow D)$, and the colimit of X is the initial object of $(D \downarrow X)$.

3.3.6 Comma categories and adjunctions

As an extension of the case of Galois connections in 1.7.6, the *graph* of the adjunction $(F, G, \eta, \varepsilon) \colon \mathsf{X} \rightharpoonup \mathsf{Y}$ will be the comma category

$$\mathsf{G}(F, G) = (F \downarrow \mathsf{Y}), \tag{3.30}$$

with objects $(X, Y, c\colon FX \to Y)$ and morphisms

$$(f, g)\colon (X, Y, c\colon FX \to Y) \to (X', Y', c'\colon FX' \to Y'), \qquad c'.Ff = g.c.$$

The adjunction $F \dashv G$ has a *graph factorisation* in AdjCat

$$X \; \underset{G'}{\overset{F'}{\rightleftarrows}} \; \mathsf{G}(F, G) \; \underset{G''}{\overset{F''}{\rightleftarrows}} \; Y \qquad\qquad (3.31)$$

$$F'(X) = (X, FX, 1_{FX}), \quad G'(X, Y, d) = X, \quad \eta'X = 1_X,$$

$$F''(X, Y, c) = Y, \quad G''(Y) = (GY, Y, \varepsilon Y\colon FG(Y) \to Y), \quad \varepsilon''Y = 1_Y.$$

This forms, also here, a *natural weak factorisation system* in AdjCat, in the sense of [GT] (see 2.5.5), which is *mono-epi* (because $G'F'$ and $F''G''$ are identities).

One can replace $\mathsf{G}(F, G)$ with the comma category $\mathsf{G}'(F, G) = (X \downarrow G)$, which is isomorphic to the former because of the adjunction (and coincides with it when X and Y are order categories). Or one can replace (3.31) with a factorisation in three adjunctions, where the central one is a pair of isomorphisms, inverse to each other

$$X \; \rightleftarrows \; \mathsf{G}(F, G) \; \rightleftarrows \; \mathsf{G}'(F, G) \; \rightleftarrows \; Y. \qquad (3.32)$$

3.3.7 A second digression on categorical equivalence

Continuing the argument of 1.5.9, let us note that the comma category $(F \downarrow G)$ was defined in 3.3.1 by a strict construction, and characterised up to isomorphism by a universal property, as a sort of 'directed 2-dimensional pullback'.

Weakening this issue up to categorical equivalence seems to complicate things without any advantage, as examined in [G8], Subsection 1.8.9, and briefly reviewed below. Some notions of homotopy theory are required to follow the argument.

*A reader familiar with homotopy pullbacks [Mat] will recognise a comma category as similar to a *standard homotopy pullback*, defined by its universal property up to homeomorphism; and will know that a *general homotopy pullback* is defined up to homotopy equivalence (the 'topological version' of categorical equivalence).

Both notions are important: the standard form gives a functor, and the weak definition is closed under pasting – while a pasting of standard homotopy pullbacks is only homotopy equivalent to the appropriate standard solution.

Now, the second fact fails in the case of comma categories (essentially because a natural transformation is a kind of 'directed homotopy', generally non-reversible): a pasting of comma squares gives a category which – in general – is not categorically equivalent to the appropriate comma: the latter is a full subcategory of the former, reflective or coreflective according to the direction of comma squares (see [G8], loc. cit.).

(In fact, a more appropriate categorical version of a homotopy pullback would be an 'iso-comma', where we replace the natural transformation of the universal property by a functorial isomorphism; the solution is an obvious full subcategory of $(F \downarrow G)$. In this case the E-notion would indeed be closed under pasting. However, iso-commas are of no interest here.)*

3.4 Monoidal categories and exponentials

Loosely speaking, a monoidal category is a category equipped with a 'tensor product', like $A \otimes B$ in Ab or – more simply – the cartesian product $X \times Y$ in Set. These structures are related to 'internal homs' (or exponentials), via adjunctions.

The examples dealt with in 3.4.7 include the categories Set, Ord, Cat, Set., and $\mathcal{S}$, Ab and R Mod, K Prj, Rel Set, Rng, Ban and Ban$_1$.

Monoidal categories were introduced in 1963 as 'categories with multiplication', in two articles by Bénabou [Be1] and Mac Lane [M3], then renamed by Eilenberg. Other standard references for this topic are Mac Lane [M4], Eilenberg–Kelly [EiK] and Kelly [Kl2].

A reader interested in Computer Science will likely know that the notion of cartesian closed category (exposed in 3.4.5) is 'equivalent' to typed λ-calculus, as shown in Lambek and Scott [LaS], or Barr and Wells [BarW].

3.4.1 Monoidal categories

A *monoidal category* is a category C equipped with a system $(\otimes, E, \kappa, \lambda, \rho)$. It consists of a functor in two variables, often called a *tensor product*

$$\mathsf{C} \times \mathsf{C} \to \mathsf{C}, \qquad (A, B) \mapsto A \otimes B, \qquad (3.33)$$

and an object E, called the *unit*. The operation $- \otimes -$ is assumed to be associative, up to a functorial isomorphism of components

$$\kappa(A, B, C) \colon A \otimes (B \otimes C) \to (A \otimes B) \otimes C \qquad (associator),$$

and the object E is assumed to be an identity, up to functorial isomorphisms

$$\lambda(A) \colon E \otimes A \to A, \qquad \rho(A) \colon A \otimes E \to A \qquad (unitors).$$

All these isomorphisms must satisfy coherence conditions, that will be analysed in the more general framework of weak double categories (see 7.2.3); this allows us to 'forget' them and write

$$A \otimes (B \otimes C) = A \otimes B \otimes C = (A \otimes B) \otimes C, \qquad E \otimes A = A = A \otimes E.$$

The case of a *strict* monoidal category, where the comparison isomorphisms are identities, is less frequent but has important examples: see 3.4.2(g), 7.3.3(d) and 7.3.4.

A monoidal category with small hom-sets has a canonical forgetful functor, represented by its unit

$$U = \mathsf{C}(E, -) \colon \mathsf{C} \to \mathsf{Set}, \tag{3.34}$$

which need not be faithful (as some examples in 3.4.2 will show). For every pair of objects A, B in C there is a canonical mapping

$$\varphi_{AB} \colon U(A) \times U(B) \to U(A \otimes B), \qquad (a, b) \mapsto a \otimes b \tag{3.35}$$

where $a \colon E \to A$, $b \colon E \to B$ and $a \otimes b \colon E = E \otimes E \to A \otimes B$. It is easy to prove that the family (φ_{AB}) is a natural transformation between two functors $\mathsf{C} \times \mathsf{C} \to \mathsf{Set}$.

A *symmetric* monoidal category is further equipped with a symmetry isomorphism, coherent with the other ones

$$s(A, B) \colon A \otimes B \to B \otimes A. \tag{3.36}$$

The latter cannot be omitted: note that $s(A, A) \colon A \otimes A \to A \otimes A$ is not the identity, in general.

3.4.2 Exercises and complements

(a) A category C with finite products has a symmetric monoidal structure given by the categorical product $A \times B$, with unit the terminal object $\top$. The reader will define the comparison isomorphisms, using the universal property of products. This structure is called the *cartesian* (monoidal) structure of C.

It is the important monoidal structure of categories like Set, Ord, Cat, Top, ... The reader will determine the canonical forgetful functor $\mathsf{C}(\top, -)$ of each of them, and show that it is not faithful when C is Cat.

However, if a category C with finite products *is pointed*, its cartesian structure will likely be of little interest, since the associated forgetful functor $U = \mathsf{C}(0, -) \colon \mathsf{C} \to \mathsf{Set}$ is constant at the singleton. (An even stronger shortcoming of cartesian structures of pointed categories will turn out in 3.4.7(b).)

In fact Ab, $\mathsf{Set}_\bullet$, Ban, etc. also have an important symmetric monoidal structure – examined below – which is not cartesian. In many cases such structures arise out of the notion of a 'bi-morphism' $A \times B \to C$, that 'preserves the structure in each variable'.

(b) The important symmetric monoidal structure of the category Ab of abelian groups is the usual tensor product $A \otimes B$, with unit the additive group $\mathbb{Z}$.

More generally the important symmetric monoidal structure of the category R Mod, of modules on a commutative unital ring R, is the tensor product over R, which the reader will likely have already seen in some text dealing with linear algebra (e.g. [Bor1], Chapter II).

$A \otimes_R B$ is characterised up to isomorphism by the fact that this module is equipped with an R-*bilinear* mapping $\eta \colon A \times B \to A \otimes_R B$ (linear in each variable) such that every bilinear mapping $\varphi \colon A \times B \to C$ factorises through η by a unique homomorphism h

$$
\begin{array}{ccc}
A \times B & \xrightarrow{\ \eta\ } & A \otimes_R B \\
& \searrow{\varphi} & \downarrow{h} \\
& & C
\end{array}
\tag{3.37}
$$

Note that η and φ are bilinear mappings while h is linear: this diagram lives and commutes in Set. *An enriched form of the diagram will be given in 7.2.8(c).*

The proof of the existence and the construction of the associator require some work (and can be found in any text on linear algebra or homological algebra), while the unitors and the symmetry $A \otimes_R B \to B \otimes_R A$ follow easily from the universal property. Many other properties of the functors $- \otimes_R A$, like the preservation of colimits, are straightforward consequences of the adjunction of 3.4.6.

The reader will note that the forgetful functor of $(R \text{ Mod}, \otimes, R)$ is the usual one, up to isomorphism. (On the other hand, the forgetful functor of the cartesian structure is constant at the trivial module, as we have already remarked.)

(c) The important symmetric monoidal structure of Set. is given by the *smash product* $(X, x_0) \wedge (Y, y_0)$.

It can be realised as the quotient of the product $(X, x_0) \times (Y, y_0) = (X \times Y, (x_0, y_0))$ that identifies all the points of the sum $(X, x_0) + (Y, y_0)$, computed as a subset of the product (see 2.1.6)

$$
(X, x_0) \wedge (Y, y_0) = (X \times Y)/((X \times \{y_0\}) \cup (\{x_0\} \times Y)).
\tag{3.38}
$$

The base point of the quotient is the equivalence class of the base point of the product

$$[(x_0, y_0)] = (X \times \{y_0\}) \cup (\{x_0\} \times Y).$$

The reader will note that the tensor product of Set. satisfies a universal property similar to that of the tensor product of modules, with respect to 'bipointed mappings' $X \times Y \twoheadrightarrow Z$, separately pointed in each variable.

The identity is the two-point set $\mathbb{S}^0 = \{-1, 1\}$, pointed at 1. The canonical forgetful functor Set. $\to$ Set is the usual one. All this will be better examined in 5.6.2.

(d) The smash product in Top. is deferred to Section 5.6.

(e) The reader will also note that a similar approach for Ord would simply give the cartesian product, since a mapping $X \times Y \to Z$ which preserves the order in each variable is monotone.

(f) Describe the symmetric monoidal structure of S produced by the equivalence with Set. (in 1.8.6). Determine the associated forgetful functor $S \to$ Set.

(g) The category End(C) of endofunctors of a category C and their natural transformations has a strict monoidal structure given by the composition of endofunctors; it is non-symmetric, generally.

3.4.3 The exponential law for sets

For a fixed set A, we show now that the functor $F = - \times A$

$$F: \text{Set} \to \text{Set}, \qquad F(X) = X \times A, \quad F(f) = f \times A = f \times \text{id}A, \qquad (3.39)$$

admits, as a right adjoint, the functor $G = \text{Set}(A, -)$

$$G: \text{Set} \to \text{Set}, \qquad G(Y) = \text{Set}(A, Y), \quad G(g) = \text{Set}(A, g) = g. -. \qquad (3.40)$$

The adjunction $F \dashv G$ (depending on the *parameter* A), is expressed by a natural isomorphism

$$\varphi: \text{Set}(F(-), =) \to \text{Set}(-, G(=)): \text{Set}^{\text{op}} \times \text{Set} \to \text{Set}. \qquad (3.41)$$

In fact we have a family of bijections, for X, Y in Set

$$\varphi_{XY}: \text{Set}(X \times A, Y) \to \text{Set}(X, \text{Set}(A, Y)) \quad (exponential\ law),$$
$$(u: X \times A \to Y) \mapsto (v: X \to \text{Set}(A, Y)), \qquad (3.42)$$

where the mappings u and v are related by the identity $u(x, a) = v(x)(a)$, for all $x \in X$, $a \in A$.

This family is natural: for two mappings $f: X' \to X$, $g: Y \to Y'$, the following diagram commutes

$$
\begin{array}{ccc}
\mathsf{Set}(X \times A, Y) & \xrightarrow{\varphi_{XY}} & \mathsf{Set}(X, \mathsf{Set}(A, Y)) \\
{\scriptstyle g.-.Ff}\Big\downarrow & & \Big\downarrow{\scriptstyle Gg.-.f} \\
\mathsf{Set}(X' \times A, Y') & \xrightarrow[\varphi_{X'Y'}]{} & \mathsf{Set}(X', \mathsf{Set}(A, Y'))
\end{array}
\tag{3.43}
$$

because, for any mapping $u: X \times A \to Y$, the mappings

$$g.u.(f \times 1_A): X' \times A \to Y',$$

$$(\varphi_{X'Y'})^{-1}(Gg.\varphi_{XY}(u).f): X' \times A \to Y',$$

take the same value at any $(x', a) \in X' \times A$, namely $g(u(fx', a))$.

3.4.4 Comments

The hom-sets of the category Set can be written in 'exponential notation' $\mathsf{Set}(X, Y) = Y^X$, because a mapping $y: X \to Y$ is the same as an indexed family $(y_x)_{x \in X}$ of Y^X. (This is consistent with the exponential of small discrete categories, in 1.6.1).

In this way, the previous natural isomorphism can be rewritten as

$$
\mathsf{Set}(X \times A, Y) \cong \mathsf{Set}(X, Y^A), \qquad Y^{X \times A} \cong (Y^A)^X.
\tag{3.44}
$$

The second form explains the name of *exponential law* for these isomorphisms (and their version in other categories).

The existence of the right adjoint $(-)^A$ of the functor $F = - \times A$ will be expressed saying that the object A is 'exponentiable' in Set (as a category equipped with its cartesian product). The fact that all of them are exponentiable will be expressed saying that Set is 'cartesian closed'.

3.4.5 Exponentiable objects and closed structures

More generally, in a *symmetric* monoidal category C an object A is said to be *exponentiable* (with respect to the tensor product that we are considering) if the functor $- \otimes A: \mathsf{C} \to \mathsf{C}$ has a right adjoint. The latter is generally written as $\mathrm{Hom}(A, -): \mathsf{C} \to \mathsf{C}$ or $(-)^A$, and called an *internal hom*, or an *exponential*.

There is thus a family of bijections, natural in the variables X, Y (and depending on the parameter A)

$$
\varphi_{XY}^A: \mathsf{C}(X \otimes A, Y) \to \mathsf{C}(X, \mathrm{Hom}(A, Y)) \qquad (X, Y \text{ in } \mathsf{C}).
\tag{3.45}
$$

Since adjunctions compose, all the tensor powers $A^{\otimes n} = A \otimes \ldots \otimes A$ are also exponentiable, with

$$\mathrm{Hom}(A^{\otimes n}, -) = (\mathrm{Hom}(A, -))^n. \qquad (3.46)$$

A symmetric monoidal category is said to be *closed* if all its objects are exponentiable.

Then one can form a hom-functor $\mathrm{Hom}\colon \mathsf{C}^{\mathrm{op}} \times \mathsf{C} \to \mathsf{C}$ in two variables. Without entering in details, a morphism $f\colon A \to A'$ gives a natural transformation $- \otimes f\colon - \otimes A \to - \otimes A'\colon \mathsf{C} \to \mathsf{C}$, which has a 'mate' (see 7.1.6) $\mathrm{Hom}(f, -)\colon \mathrm{Hom}(A', -) \to \mathrm{Hom}(A, -)$.

A category C with finite products is said to be *cartesian closed* if all its objects are exponentiable for the cartesian monoidal structure.

In the non-symmetric case one should consider a left and a right internal hom-functor, as is the case with cubical sets (see [K1, BroH2, G7]).

Instead of starting from a monoidal category and introducing the internal hom as a right adjoint, one can define a 'closed category' [EiK], by axiomatising an internal hom-functor $\mathrm{Hom}\colon \mathsf{C}^{\mathrm{op}} \times \mathsf{C} \to \mathsf{C}$, and derive the tensor product as a left adjoint. Concretely, this procedure is even easier in many cases considered below; formally, the axioms for Hom are less evident than those of a tensor product.

3.4.6 The exponential law for modules

After sets, modules on a commutative unital ring R have also an exponential law (and family of adjunctions), based on well-known functors of two variables: the tensor product (recalled in 3.4.2(b)) and the internal hom-functor Hom_R

$$- \otimes_R -\colon R\,\mathsf{Mod} \times R\,\mathsf{Mod} \to R\,\mathsf{Mod},$$
$$\mathrm{Hom}_R\colon R\,\mathsf{Mod}^{\mathrm{op}} \times R\,\mathsf{Mod} \to R\,\mathsf{Mod}. \qquad (3.47)$$

Here $\mathrm{Hom}_R(A, B)$ is the set of homomorphisms $R\,\mathsf{Mod}(A, B)$ equipped with the pointwise sum and scalar product:

$$(u + v)(a) = u(a) + v(a), \qquad (\lambda u)(a) = \lambda.u(a).$$

On the morphisms we have the usual action: $\mathrm{Hom}_R(f, g) = g. - .f$.

In fact, for a fixed R-module A, there is again a natural family of isomorphisms indexed by X, Y in $R\,\mathsf{Mod}$

$$\varphi_{XY}\colon \mathrm{Hom}_R(X \otimes_R A, Y) \to \mathrm{Hom}_R(X, \mathrm{Hom}_R(A, Y)),$$
$$\varphi_{XY}(u)(x) = u(x \otimes -)\colon A \to Y, \qquad (3.48)$$

for $u\colon X \otimes_R A \to Y$ and $x \in X$.

The underlying bijections

$$\varphi_{XY}\colon R\,\mathsf{Mod}(X \otimes_R A, Y) \to R\,\mathsf{Mod}(X, \mathrm{Hom}_R(A, Y)),$$

give an adjunction $F \dashv G$ (depending on the parameter A) between the endofunctors:

$$F\colon R\,\mathsf{Mod} \to R\,\mathsf{Mod},$$
$$F(X) = X \otimes_R A, \qquad F(f) = f \otimes \mathrm{id}A,$$
$$G\colon R\,\mathsf{Mod} \to R\,\mathsf{Mod}, \tag{3.49}$$
$$G(Y) = \mathrm{Hom}_R(A, Y), \qquad G(g) = \mathrm{Hom}_R(A, g) = g.\,-\,.$$

With this structure $R\,\mathsf{Mod}$ is a symmetric monoidal closed category.

3.4.7 Exercises and complements

(a) We have seen that Set is cartesian closed, with $\mathrm{Hom}(A, Y) = \mathsf{Set}(A, Y)$. The category Ord is also cartesian closed, with internal hom

$$\mathrm{Hom}(A, Y) = (\mathsf{Ord}(A, Y), \leqslant),$$

equipped with the pointwise order (1.3).

The category Cat of small categories is cartesian closed, with the internal hom $\mathrm{Hom}(\mathsf{S}, \mathsf{C}) = \mathsf{C}^{\mathsf{S}}$ described in 1.6.1. *Every category of presheaves of sets (see 1.6.3) is cartesian closed.*

(b) A pointed cartesian closed category C is equivalent to $\mathbf{1}$.

(c) We have seen that Ab and $R\,\mathsf{Mod}$ are symmetric monoidal closed, with respect to the usual tensor product and Hom functor.

(d) Similarly $\mathsf{Set}_{\bullet}$ is symmetric monoidal closed, with respect to the smash product described in 3.4.2(c): $\mathrm{Hom}(A, Y)$ is the set of pointed mappings $\mathsf{Set}_{\bullet}(A, Y)$, pointed at the zero morphism $A \to Y$.

Describe the associated symmetric monoidal closed structure of $\mathcal{S}$, whose tensor product was the subject of Exercise 3.4.2(f).

(e) The internal hom-functor $\mathrm{Hom}_R\colon R\,\mathsf{Mod}^{\mathrm{op}} \times R\,\mathsf{Mod} \to R\,\mathsf{Mod}$ produces the *dual-module* functor

$$\mathrm{Hom}_R(-, R)\colon R\,\mathsf{Mod}^{\mathrm{op}} \to R\,\mathsf{Mod}, \quad X \mapsto X^* = \mathrm{Hom}_R(X, R). \tag{3.50}$$

Many important 'dual-object' functors arise in this way, from an internal hom-functor, fixing the covariant variable and leaving free the contravariant one.

(f) Top is not cartesian closed; exponentiable spaces are examined in 5.1.1.

Studying 'convenient categories of topological spaces', that are cartesian closed and sufficiently rich, is a domain of research in Categorical Topology. An interested reader can begin with compactly generated spaces, investigated in Kelley's book [Ke].

(g) We shall see that the symmetric monoidal closed structure of the category K Vct of vector spaces on a field K induces a similar structure on the category K Prj of projective spaces on K (in 6.5.6).

(h) Prove that the category Rel Set is symmetric monoidal closed, with tensor product $X \times A$ (the cartesian product of sets) and internal hom $\mathrm{Hom}(A, Y) = A^\sharp \times Y$ (where the involution $(-)^\sharp$ is the identity on objects, but covers the contravariance in the first variable).

*(i) The category Rng of (unital) rings has a symmetric monoidal structure: the tensor product $R \otimes S$ of two rings is the tensor product $R \otimes_{\mathbb{Z}} S$ of the underlying abelian groups, with the $\mathbb{Z}$-bilinear multiplication generated by letting

$$(x \otimes y).(x' \otimes y') = xx' \otimes yy' \qquad (x, x' \in R, \ y, y' \in S). \tag{3.51}$$

The unit is the ring $\mathbb{Z}$ (which is also the initial object of the category).

*(j) The full subcategory CRng of commutative rings inherits a monoidal structure which is co-cartesian: $R_1 \otimes R_2$ is now the categorical sum of the commutative rings R_i, with 'injections'

$$u_i \colon R_i \to R_1 \otimes R_2, \qquad u_1(x) = x \otimes 1, \quad u_2(y) = 1 \otimes y. \tag{3.52}$$

Note that the 'injections' need not be monomorphisms: $R \otimes 0$ is always the trivial ring, as well as $\mathbb{Z}/2 \otimes \mathbb{Z}/3$.

*(k) The categories Ban and Ban_1 are both symmetric monoidal closed, with unit the scalar field K ($\mathbb{R}$ or $\mathbb{C}$). *In both cases*, $\mathrm{Hom}(A, Y)$ is the set $\mathrm{Ban}(A, Y)$ of all continuous linear mappings $A \to Y$, with the usual K-linear structure and norm

$$\|f\| = \sup\{\|f(x)\| \mid x \in A, \|x\| \leqslant 1\}.$$

The tensor product is obtained by a metric completion of the linear tensor product, see [Se].

The canonical forgetful functors have already been considered in (1.37):

$$U = \mathrm{Ban}(K, -) \colon \mathsf{Ban} \to \mathsf{Set}, \qquad U(X) = |X|,$$
$$B_1 = \mathrm{Ban}_1(K, -) \colon \mathsf{Ban}_1 \to \mathsf{Set}, \quad B_1(X) = \{x \in X \mid \|x\| \leqslant 1\}. \tag{3.53}$$

Note that, applying each of them to the Banach space $\mathrm{Hom}(A, Y)$, we get the correct set of morphisms

$$\mathrm{Ban}(A, Y) = |\mathrm{Hom}(A, Y)|, \qquad \mathrm{Ban}_1(A, Y) = B_1(\mathrm{Hom}(A, Y)). \tag{3.54}$$

3.4.8 Internal monoids

Let $C = (C, \otimes, E)$ be a monoidal category; the comparison isomorphisms are left understood.

An *internal monoid* in C is a triple (M, e, m) consisting of an object M and two morphisms $e \colon E \to M$, $m \colon M \otimes M \to M$ of C, called the *unit* and *multiplication*, which make the following diagrams commute

$$
\begin{array}{ccccc}
M & \xrightarrow{e \otimes M} & M^{\otimes 2} & \xleftarrow{M \otimes e} & M \\
 & \searrow{\scriptstyle 1} & \downarrow{\scriptstyle m} & \swarrow{\scriptstyle 1} & \\
 & & M & &
\end{array}
\qquad
\begin{array}{ccc}
M^{\otimes 3} & \xrightarrow{M \otimes m} & M^{\otimes 2} \\
{\scriptstyle m \otimes M}\downarrow & & \downarrow{\scriptstyle m} \\
M^{\otimes 2} & \xrightarrow{m} & M
\end{array}
\qquad (3.55)
$$

These axioms are called *unitarity* and *associativity*.

A monoid in Set (resp. Top or Ord), with respect to the cartesian product, is an ordinary monoid (resp. a topological or ordered monoid). A monoid in Ab (resp. $R\,\mathsf{Mod}$), with respect to the usual tensor product, is a ring (resp. an R-algebra).

One can define a topological group as an 'internal group' in Top, by adding to an internal monoid M an *inversion* morphism $i \colon M \to M$ satisfying suitable axioms. Lie groups can also be defined in a similar way.

The reader is warned that *this approach would not work in* Ord, because the inversion mapping necessarily reverses the order.

3.4.9 Internal comonoids

Dually, an *internal comonoid* in $(C, \otimes, E)$ is a triple (X, e, d) where the morphisms $e \colon X \to E$ (*counit*) and $d \colon X \to X^{\otimes 2}$ (*comultiplication*) satisfy dual axioms, giving commutative diagrams

$$
\begin{array}{ccccc}
 & & X & & \\
 & \swarrow{\scriptstyle 1} & \downarrow{\scriptstyle d} & \searrow{\scriptstyle 1} & \\
X & \xleftarrow{e \otimes X} & X^{\otimes 2} & \xrightarrow{X \otimes e} & X
\end{array}
\qquad
\begin{array}{ccc}
X & \xrightarrow{d} & X^{\otimes 2} \\
{\scriptstyle d}\downarrow & & \downarrow{\scriptstyle X \otimes d} \\
X^{\otimes 2} & \xrightarrow{d \otimes X} & X^{\otimes 3}
\end{array}
\qquad (3.56)
$$

For a cartesian structure $(C, \times, \top)$, every object X has a structure of internal comonoid, where $e \colon X \to \top$ is determined and $d = (1, 1) \colon X \to X^2$ is the diagonal.

In Cat, letting X be the arrow category **2**, we get a structure of interest (see 7.7.2).

3.5 The Adjoint Functor Theorem

Freyd's Adjoint Functor Theorem is one of the main tools to prove the existence of an adjoint, in a case where a (more or less) explicit construction would be difficult, or too complicated to be useful, or unknown (unless by repeating the argument of the theorem itself).

Another existence theorem for adjoints, the Special Adjoint Functor Theorem, by P. Freyd again, can be found in [M4], Section V.8.

3.5.1 *The Initial Object Theorem* (P. Freyd)

Let C *be a complete category with small hom-sets.* C *has an initial object if and only if it satisfies the following condition:*

(Solution Set Condition) *there exists a* solution set, *i.e. a family* W_i *of objects of* C *indexed by a* small *set* I, *such that:*

- *for every object* X *there exists some* $i \in I$ *and some arrow* $W_i \to X$.

Note. One can drop the local-smallness condition on C, and only require that the product W of the family W_i has a small set of endomorphisms $\mathsf{C}(W, W)$. The proof is the same as below.

Proof The necessity of the previous condition is obvious: an initial object W forms, by itself, a solution set (with $\mathsf{C}(W, W)$ a singleton).

Conversely, let us suppose that the above condition is satisfied and let $W = \prod W_i$ with projections $p_i \colon W \to W_i$. The object W is *weakly initial*: in fact, for every object X there exists *some* arrow $W \to W_i \to X$.

We want to prove that the initial object Z can be obtained from the *global equaliser* $h \colon Z \to W$ of the whole set of endomorphisms of W, i.e. the limit of the (small) diagram formed by them. This means that:

(a) for every $f \colon W \to W$, $fh = h$,

(b) every arrow $h' \colon Z' \to W$ which equalises all the endomaps of W factorises as $h' = kh$, by a unique $k \colon Z' \to Z$.

The morphism h is mono and the object Z is weakly initial, because of the arrow $h \colon Z \to W$. To prove that it is the initial object of the category, we suppose to have two maps $f, g \colon Z \to X$, and prove that they coincide.

Let $e \colon Z_0 \to Z$ be their equaliser. Since W is weakly initial, we have a map $k \colon W \to Z_0$ and a diagram

$$
\begin{array}{ccc}
W & \overset{k}{\longrightarrow} & Z_0 \\
 & \underset{h}{\searrow} & \downarrow e \\
 & & Z
\end{array}
\qquad\qquad (3.57)
$$

Now, the maps id W and $hek\colon W \to W$ must be equalised by $h\colon Z \to W$, whence $h = (hek)h = h(ekh)$. Cancelling the monomorphism h we get $ekh = \mathrm{id}\, Z$. Therefore e is (a split) epi and $f = g$. □

3.5.2 The Adjoint Functor Theorem (Freyd [Fr1])

Let $G\colon \mathsf{D} \to \mathsf{C}$ be a functor defined on a category D with small hom-sets and all small limits. Then G has a left adjoint if and only if it satisfies the following conditions:

(i) G preserves all small limits,

(ii) (Solution Set Condition) for every object X in C there exists a family of objects Y_i in D and arrows $w_i\colon X \to G(Y_i)$ in C, indexed by a small set I_X, such that every map $f\colon X \to G(Y)$ (with Y in D) factorises as

$$G(g).w_i\colon X \to G(Y_i) \to G(Y),$$

for some $i \in I_X$ and $g \in \mathsf{D}(Y_i, Y)$.

If C has also small hom-sets, one can simplify the Solution Set Condition:

(iii) for every object X in C there exists a small set $S(X) \subset \mathrm{Ob}\,\mathsf{D}$, such that every map $f\colon X \to G(Y)$ (with Y in D) factorises as

$$G(g).w\colon X \to G(Y_0) \to G(Y),$$

for some $Y_0 \in S(X)$, $w\colon X \to G(Y_0)$ in C and $g \in \mathsf{D}(Y_0, Y)$.

Proof The necessity of (i), (ii) is obvious: a right adjoint preserves all the existing limits, and a universal arrow $(Y_0, \eta\colon X \to G(Y_0))$ from X to G gives a solution set formed by a single object of D and a single arrow of C.

Conversely, assuming (i) and (ii), it is sufficient to prove that, for every object X in C, the comma category $(X \downarrow G)$ has an initial object (by 3.3.4); this follows from the previous Initial Object Theorem.

In fact $(X \downarrow G)$ has small hom-sets, because D has. It is small complete (by Theorem 3.3.3), because C is and G preserves small limits. Finally condition (ii) precisely gives a solution set $(Y_i, w_i\colon X \to G(Y_i))$ for $(X \downarrow G)$, in the sense of Theorem 3.5.1.

The simplified version for C locally small is an obvious consequence. □

3.5.3 Some points on cardinals

We recall some basic points of the theory of cardinal sets, that are useful in order to construct solution sets and prove the existence of an adjoint. Definitions and proofs can be found in [Je].

(a) We write as $\sharp X$ the cardinal of a set X, and $\aleph_0 = \sharp \mathbb{N}$ (read as *aleph zero*). The cardinal sets are totally ordered by the relation $\alpha \leqslant \beta$, meaning that there exists an injective mapping $\alpha \to \beta$.

(b) For cardinals α, β, the symbols $\alpha + \beta$ and $\alpha.\beta$ usually denote the cardinal of their sum or product, respectively. If one at least of α, β is infinite, then $\alpha + \beta = \alpha.\beta = \max(\alpha, \beta)$. One always has $\alpha < 2^\alpha = \sharp(\mathcal{P}\alpha)$.

(c) The cardinal of a small set is small. For a small cardinal α, the set of all cardinals $\leqslant \alpha$ is small.

3.5.4 Exercises and comments

(a) (Solved exercise) The reader will find it useful to apply the Adjoint Functor Theorem to prove the existence of the free group on an arbitrary set X, or equivalently of the left adjoint to the forgetful functor $U \colon \mathsf{Gp} \to \mathsf{Set}$. The argument sets a pattern that can be followed in many other cases.

One can fill-in the following outline. First we know that Gp has small hom-sets and small limits, preserved by U. Let us fix a (small) set X.

Every mapping $f \colon X \to U(G)$ with values in a group factorises through the subgroup $G_0 = \langle f(X) \rangle$ generated by $f(X)$ in G. But $\sharp f(X) \leqslant \sharp X$, whence the cardinal $\alpha = \sharp G_0$ satisfies:

$$\alpha \leqslant \beta_X = \max\{\sharp X, \aleph_0\}.$$

An arbitrary bijection $|G_0| \to \alpha$ allows us to make the cardinal α into a group G_1 isomorphic to G_0, so that f factorises as

$$U(g).f_1 \colon X \to U(G_1) \to U(G),$$

for a monomorphism $g \colon G_1 \to G_0 \subset G$.

The reader will now consider the small set $C(X)$ of all cardinals $\leqslant \beta_X$, and prove that the groups whose underlying set *belongs* to $C(X)$, like the previous G_1, form a *small set* $S(X)$. This gives a solution set for X, in the simplified form 3.5.2(iii) – legitimate because also Set has small hom-sets.

The usual construction of the free group on the set X will be sketched in 4.1.3(c).

(b) A similar procedure can be followed for any variety of algebras, as we shall do in Theorem 4.4.2.

(c) A similar, even easier, procedure can also be followed to prove that the full subcategory $\mathsf{Hsd} \subset \mathsf{Top}$ of Hausdorff spaces is reflective. A 'construction' of the reflector can be found in Proposition 5.1.3.

*(d) Use the Adjoint Functor Theorem to prove that Gp has small sums – the 'free products'.

3.6 Monads and algebras

Monads and their algebras yield a far-reaching formalisation of the 'algebraic character' of a category over another – typically used over the category of sets but also of interest in many other cases.

Given a functor $G\colon A \to X$, this theory answers the question: is A 'algebraic' over X, via G? More precisely, we say that the functor G is 'monadic' if it has a left adjoint $F\colon X \to A$ and the monad $T = FG\colon X \to X$ (i.e. the trace left by the adjunction on the category X) allows us to reconstruct the category A as a category of T-algebras.

We follow the classical terminology and notation of Mac Lane [M4], Chapter VI. Monads are also called 'triples' or 'dual standard constructions' by other authors [BarB, Bc, Ds].

The 'algebraic character' considered here can be related to infinitary operations (see 3.6.6). *Finitary algebraic theories*, or Lawvere theories, are studied in Lawvere's thesis (1963), recently reprinted with comments and a supplement in [Lw2]; see also [Bo2], Chapter 3.

The relationship of monads, Lawvere theories and Universal Algebra with theoretical Computer Science is investigated in Hyland–Power [HyP].

3.6.1 An example

As an informal presentation of this subject, we start from the forgetful functor of monoids, written as $G\colon \mathsf{Mon} \to \mathsf{Set}$. The free monoid on a set X can be constructed as a sum in Set of the cartesian powers X^n

$$F(X) = \Sigma_{n\in\mathbb{N}} X^n,$$
$$(x_1, ..., x_p).(y_1, ..., y_q) = (x_1, ..., x_p, y_1, ..., y_q), \tag{3.58}$$

so that an element of $F(X)$ is a *word* $(x_1, ..., x_p)$ in the alphabet X, and two words are multiplied by concatenation. The empty word $e = (-) \in X^0$ acts as the identity element. We have thus the left adjoint $F \dashv G$, with the following unit and counit (for a set X and a monoid M)

$$\eta_X\colon X \to GF(X), \qquad \eta_X(x) = (x),$$
$$\varepsilon_M\colon FG(M) \to M, \qquad \varepsilon_M(m_1, ..., m_n) = m_1 * ... * m_n. \tag{3.59}$$

Here η_X sends each $x \in X$ to the unary word (x) in the underlying set of $F(X)$, while ε_M sends a word of the free monoid on the set $G(M)$ to the product of its elements in M (writing this multiplication as $*$).

The adjunction leaves, on the domain Set of the left adjoint, a 'trace' formed of an endofunctor T and two natural transformations, called the *unit* and the *multiplication*

$$T = GF \colon \mathsf{Set} \to \mathsf{Set},$$
$$\eta \colon \mathrm{id} \to T, \qquad \mu = G\varepsilon F \colon T^2 \to T.$$

(3.60)

We shall see that these data (T, η, μ) form a *monad* on Set, namely the monad associated to the adjunction, often simply written as T. The important fact is that *the monad allows us to reconstruct the category of monoids*, and the whole adjunction.

Loosely speaking, the monad T has a category of T-*algebras* Set^T, where an object is a pair $(X, a \colon TX \to X)$ consisting of a set X and a mapping $a \colon TX \to X$ satisfying some axioms. We have thus a binary multiplication $x * y = a(x, y)$ on X, and the axioms of T-algebras force this multiplication to be a monoid operation, with $a(x_1, ..., x_n) = x_1 * ... * x_n$.

Finally we conclude that Mon is, up to isomorphism of categories, a category of T-algebras over Set, for the monad associated to the functor $G \colon \mathsf{Mon} \to \mathsf{Set}$, and we express this fact saying that Mon is monadic over Set via G.

Let us note that the free semigroup on the empty set is empty: the category of 'non-empty semigroups', discussed in 1.1.6, lacks such a free object and is not monadic over Set. Similar arguments apply to other 'artificial cases', like the category of 'non-abelian groups'.

We expose now the theory of monads and their links with adjunctions. The exercises are deferred to the next section.

3.6.2 Monads

A *monad* on the category X is a triple (T, η, μ) where $T \colon \mathsf{X} \to \mathsf{X}$ is an endofunctor, while $\eta \colon 1 \to T$ and $\mu \colon T^2 \to T$ are natural transformations (called the *unit* and *multiplication* of the monad) which make the following diagrams commute

(3.61)

These axioms are called *unitarity* and *associativity*. In fact, a monad on the category X is an internal monoid (see 3.4.8) in the category $\mathsf{C} = \mathrm{End}(\mathsf{X})$ of endofunctors of X and their natural transformations, equipped with the strict (non-symmetric) monoidal structure given by the composition of endofunctors (already mentioned in 3.4.2(g)).

For an ordered set X, viewed as a category, a monad $T\colon X \to X$ is just a monotone function which is *inflationary* ($1_X \leqslant T$) and *idempotent* ($T^2 = T$); this is also called a *closure* operator on X. Obvious examples come from the topological closure $\mathcal{P}S \to \mathcal{P}S$ of a topological space S (which, moreover, has to preserve finite unions). Idempotent monads on arbitrary categories will be studied in Section 3.8.

The theory of monads is elegant and simple, and can be approached as a sequence of exercises on the calculus of natural transformations (in 1.5.4). In particular, the self-interchange property of a natural transformation of endofunctors will often be used.

3.6.3 From adjunctions to monads

An adjunction

$$F\colon X \rightleftarrows A \colon G, \qquad \eta\colon 1 \to GF, \quad \varepsilon\colon FG \to 1, \tag{3.62}$$

has an *associated monad* (T, η, μ) on X, which can be viewed as 'the trace of the adjunction on the domain of the left adjoint'.

Namely: $T = GF\colon X \to X$, η is the unit of the adjunction and

$$\mu = G\varepsilon F\colon GF.GF \to GF.$$

Exercises and complements. (a) The triple (T, η, μ) is indeed a monad.

*(b) Typically, a variety of algebras A has an associated monad $T = UF\colon$ Set $\to$ Set produced by the forgetful functor $U\colon A \to$ Set and its left adjoint $F\colon$ Set $\to A$, the free-algebra functor.

This will be seen in Theorem 4.4.2.

3.6.4 Eilenberg–Moore algebras

The other way round, from monads to adjunctions, there are two main constructions; we begin with the more important one.

Given a monad (T, η, μ) on X one defines the category X^T of *T-algebras*, or *Eilenberg–Moore algebras* for T: these are pairs $(X, a\colon TX \to X)$ consisting of an object X of X and a map a (called the *algebraic structure*) satisfying two coherence axioms:

$$a.\eta X = 1_X, \qquad\qquad a.Ta = a.\mu X, \tag{3.63}$$

$$
\begin{array}{ccc}
X \xrightarrow{\ \eta X\ } TX & \qquad & T^2X \xrightarrow{\ Ta\ } TX \\
\ \searrow\ \ \downarrow a & & \mu X \downarrow \qquad\quad \downarrow a \\
1 \quad\ X & & TX \xrightarrow[\ a\ ]{} X
\end{array}
$$

A morphism of T-algebras $f: (X, a) \to (Y, b)$ is a morphism $f: X \to Y$ of X which preserves the algebraic structures, in the sense that $f.a = b.Tf$. They compose as in X.

The category X^T is thus a full subcategory of the comma $(T \downarrow X)$.

One can find this category written as $\mathsf{Alg}(T)$, but we prefer to follow Mac Lane's notation: we shall reserve the symbol $\mathsf{Alg}(T)$ for the category of algebraic objects of an *idempotent* monad, which is isomorphic to X^T (in Section 3.8).

3.6.5 From monads to adjunctions and back

Given a monad (T, η, μ) on X there is an adjunction with the category X^T of T-algebras

$$F^T: X \rightleftarrows X^T : G^T,$$

$$\eta^T = \eta: 1 \to G^T F^T, \qquad \varepsilon^T: F^T G^T \to 1, \tag{3.64}$$

whose associated monad coincides with the given one. We give an outline of the procedure, and the reader is invited to complete the (easy) missing verifications, directly or in the exercises below.

First we have an obvious (faithful) forgetful functor

$$G^T: X^T \to X,$$

$$G^T(X, a) = X, \qquad G^T(f: (X, a) \to (Y, b)) = (f: X \to Y). \tag{3.65}$$

Backwards we have the functor

$$F^T: X \to X^T,$$

$$F^T(X) = (TX, \mu X: T^2 X \to TX), \tag{3.66}$$

$$F^T(f: X \to Y) = Tf: (TX, \mu X) \to (TY, \mu Y).$$

Now $G^T F^T = T: X \to X$. The unit of the adjunction is $\eta^T = \eta: 1 \to T$, while the counit is defined as follows

$$\varepsilon^T: F^T G^T \to 1, \qquad \varepsilon^T(X, a) = a: (TX, \mu X) \to (X, a). \tag{3.67}$$

The triangular identities of (3.7) are satisfied, so that $F^T \dashv G^T$, and the functor F^T gives indeed the *free T-algebra* on an object X of X (with respect to the forgetful functor G^T).

The monad $(G^T F^T, \eta^T, \mu^T)$ associated to this adjunction, with $\mu^T = G^T \varepsilon^T F^T$, coincides with the original one.

Exercises. (a) The functor $F^T: X \to X^T$ is well defined, in (3.66).

(b) Verify the adjunction $F^T \dashv G^T$.

(c) The new multiplication μ^T coincides with the original one.

3.6.6 Monadicity

On the other hand, if we start from an adjunction $(F, G, \eta, \varepsilon): \mathsf{X} \rightharpoonup \mathsf{A}$ as in 3.6.3, form the associated monad (T, η, μ) and then the associated adjunction $(F^T, G^T, \eta, \varepsilon^T): \mathsf{X} \rightharpoonup \mathsf{X}^T$, we get a comparison between the two adjunctions which may be an isomorphism or not; if it is the case, we think of A as a 'category of algebras' over X.

In fact there is a *comparison functor*:

$$K: \mathsf{A} \to \mathsf{X}^T, \qquad K(A) = (GA, G\varepsilon A: GFGA \to GA),$$
$$K(h: A \to B) = Gh: (GA, G\varepsilon A) \to (GB, G\varepsilon B). \tag{3.68}$$

(To verify that $K(A)$ is a T-algebra we use again the self-interchange property of ε. Note that, if the functor G is faithful, K is also.)

K links the two adjunctions (which share the unit), in the sense that – as one can easily verify

$$
\begin{array}{ccc}
\mathsf{X} \underset{G}{\overset{F}{\rightleftarrows}} \mathsf{A} & KF = F^T, & G^T K = G, \\
1 \big\downarrow \quad \big\downarrow K & & \\
\mathsf{X} \underset{G^T}{\overset{F^T}{\rightleftarrows}} \mathsf{X}^T & \varepsilon^T K = K\varepsilon & (\eta^T = \eta).
\end{array}
\tag{3.69}
$$

Now a functor $G: \mathsf{A} \to \mathsf{X}$ is said to be *monadic* (or *algebraic*), or to make A *monadic over* X, if it has a left adjoint $F: \mathsf{X} \to \mathsf{A}$ and moreover the comparison functor $K: \mathsf{A} \to \mathsf{X}^T$ defined above is an *isomorphism* of categories. (This point is commented in the Remarks below.)

We shall see, in 4.4.6, that the forgetful functor $U: \mathsf{A} \to \mathsf{Set}$ of any variety of algebras is monadic. But the present *formalisation of the algebraic character of a category* is much wider: for instance the category of compact Hausdorff spaces is monadic over Set (see 5.1.9), and this monadicity depends on the closure operator of such spaces, a sort of 'infinitary operation'.

There are various 'monadicity theorems' (also called 'tripleability theorems') that give sufficient (or necessary and sufficient) conditions for a functor to be monadic: the interested reader can see [M4], Chapter VI, [Bo2], Section 4.4, and [Bc, Ds].

Remarks. (a) We are following Mac Lane ([M4], Section VI.3) in defining monadicity in a strong sense, *up to isomorphism of categories*, because concrete examples, based on structured sets, generally fall in this case.

On the other hand [Bo2], in Section 4.4, and other authors only ask that the comparison K be an equivalence of categories – another aspect of the opposition that we have considered in 1.5.9 and 3.3.7.

(b) The difference can be appreciated at the light of this example. The category Set. of pointed sets is a variety of algebras, produced by a single 0-ary operation under no axioms, and is thus monadic over Set, via its forgetful functor U. The equivalent category S of sets and partial mappings (see 1.8.6) is equipped with the composed functor

$$US \colon S \to \mathsf{Set}_{\bullet} \to \mathsf{Set},$$

and the associated comparison functor is now an equivalence of categories.

We prefer to view S as 'weakly algebraic' over Set, rather than as 'algebraic'.

3.6.7 *Theorem* (Limits of T-algebras)

Let S be a small category. Given a monad (T, η, μ) on the category X, we suppose that X has the limits of all functors $S \to X$.

Then the category X^T has all the S-based limits, created by the forgetful functor $G^T \colon X^T \to X$ (in the sense of 2.2.8).

Proof The argument is standard; writing it down makes a useful exercise.

For a functor $X \colon S \to X$ we write $(LX, (u_i \colon LX \to X_i))$ the limit cone (for $i \in S = \mathrm{Ob}\, S$). For a functor $Y \colon S \to X^T$, we let

$$X = G^T Y \colon S \to X, \qquad Y_i = (X_i, a_i \colon TX_i \to X_i).$$

There is an obvious structure of algebra on LX, defined by the universal property of the limit

$$a \colon T(LX) \to LX, \qquad u_i a = a_i . T(u_i) \colon T(LX) \to X_i, \tag{3.70}$$

because the maps $a_i . T(u_i)$ form a cone of X. In fact, if $s \colon i \to j$ is in S, we have a commutative diagram

$$
\begin{array}{ccccc}
T(LX) & \xrightarrow{\;Tu_i\;} & TX_i & \xrightarrow{\;a_i\;} & X_i \\
\| & & \downarrow{\scriptstyle TX(s)} & & \downarrow{\scriptstyle X(s)} \\
T(LX) & \xrightarrow[\;Tu_j\;]{} & TX_j & \xrightarrow[\;a_j\;]{} & X_j
\end{array}
\tag{3.71}
$$

The axioms for the structure a are 'detected' by the jointly monic morphisms u_i of the limit cone

$$u_i(a.\eta_{LX}) = a_i . T(u_i) . \eta_{LX} = a_i . \eta_{X_i} . u_i = u_i,$$

$$u_i(a.\mu_{LX}) = a_i . T(u_i) . \mu_{LX} = a_i . \mu_{X_i} . T^2 u_i = a_i . T a_i . T^2 u_i = u_i(a_i . T a_i).$$

The cone-morphisms give a family of morphisms of T-algebras

$$u_i \colon (LX, a) \to (X_i, a_i),$$

because of (3.70), and a is the only structure on LX that makes all u_i morphisms of X^T. This family is a cone of Y, because the faithful functor G^T reflects commutative diagrams.

Finally the universal property of this cone follows easily: if

$$f_i \colon (X', a') \to (X_i, a_i)$$

is a cone of Y, there is precisely one map $f \colon X' \to LX$ such that $u_i f = f_i$ (for all i), and this f lifts to the unique $f \colon (X', a') \to (LX, a)$ that solves the same condition in X^T. $\qquad\square$

3.6.8 Kleisli algebras and their adjunction

The second construction of an adjunction from a monad (T, η, μ) on X is based on *Kleisli algebras* for T and their category X_T. All proofs are deferred to the exercises below.

An object of X_T is an object of X, *viewed as the basis of a free T-algebra*. A morphism $f^\sharp \colon X \nrightarrow Y$ of X_T is 'represented' by an arbitrary morphism $f \colon X \to TY$ of X (which tells us how $f^\sharp$ acts on the basis).

By definition, the composite $g^\sharp f^\sharp$ of $f^\sharp$ with $g^\sharp \colon Y \nrightarrow Z$ is represented by the map

$$\mu Z.Tg.f \colon X \to TY \to T^2 Z \to TZ, \tag{3.72}$$

while the identity $\mathrm{id}X \colon X \nrightarrow X$ is represented by $\eta X \colon X \to TX$.

The forgetful functor G_T of Kleisli algebras and its left adjoint F_T are now:

$$G_T \colon \mathsf{X}_T \to \mathsf{X}, \qquad G_T(X) = TX,$$
$$G_T(f^\sharp \colon X \nrightarrow Y) = \mu Y.Tf \colon TX \to TY,$$
$$F_T \colon \mathsf{X} \to \mathsf{X}_T, \qquad F_T(X) = X, \tag{3.73}$$
$$F_T(f \colon X \to Y) = (\eta Y.f)^\sharp \colon X \nrightarrow Y.$$

Again $G_T F_T = T$. The unit and counit of the adjunction $F_T \dashv G_T$ are

$$\eta_T = \eta \colon 1 \to T = G_T F_T,$$
$$\varepsilon_T(X) = (1_{TX})^\sharp \colon F_T G_T(X) = TX \nrightarrow X. \tag{3.74}$$

Also here the monad associated to this adjunction is the original one, since the new multiplication coincides with the previous one:

$$(G_T.\varepsilon_T.F_T)(X) = G_T(1_{TX}) = \mu X.$$

If we start from an adjunction $(F, G, \eta, \varepsilon)\colon \mathsf{X} \rightarrowtail \mathsf{A}$, as in 3.6.3, form the associated monad (T, η, μ) and then the Kleisli adjunction

$$(F_T, G_T, \eta, \varepsilon_T)\colon \mathsf{X} \rightarrowtail \mathsf{X}_T,$$

there is again a comparison functor which links the two adjunctions

$$H\colon \mathsf{X}_T \to \mathsf{A}, \qquad H(X) = FX,$$

$$H(f^\sharp\colon X \rightarrowtail Y) = \varepsilon FY.Ff\colon FX \to FY, \tag{3.75}$$

$$HF_T = F, \quad GH = G_T, \quad H\varepsilon_T = \varepsilon H.$$

Exercises. (a) The category X_T is well defined, in (3.72).

(b) Verify the adjunction $F_T \dashv G_T$.

(c) The graph-morphism $H\colon \mathsf{X}_T \to \mathsf{A}$ is indeed a functor, full and faithful. It gives an equivalence of categories $H'\colon \mathsf{X}_T \to \mathsf{B}$ with values in the full subcategory B of A formed by the objects isomorphic to some 'free object' $F(X)$, for X in X.

(d) Verify the coherence relations (3.75).

3.6.9 Comonads and coalgebras

Dually, a *comonad* on the category A is a triple (S, ε, δ) formed of an endofunctor $S\colon \mathsf{A} \to \mathsf{A}$ with natural transformations $\varepsilon\colon S \to 1$ (*counit*) and $\delta\colon S \to S^2$ (*comultiplication*) which make the following diagrams commute

$$\tag{3.76}$$

A comonad on the category A is an internal comonoid (see 3.4.9) in the monoidal category $\mathrm{End}(\mathsf{A})$.

The category $^S\mathsf{A}$ of S-*coalgebras* $(A, c\colon A \to SA)$ is also defined by duality.

An adjunction $(F, G, \eta, \varepsilon)\colon \mathsf{X} \rightarrowtail \mathsf{A}$, as in 3.6.3, gives a comonad (S, ε, δ) on A (the domain of the right adjoint G), where $S = FG\colon \mathsf{A} \to \mathsf{A}$, the natural transformation ε is the counit of the adjunction and

$$\delta = F\eta G\colon FG \to FG.FG.$$

3.7 Exercises and complements on monads

This sections contains exercises on monads, their (Eilenberg–Moore) algebras and Kleisli algebras.

3.7.1 Exercises and complements

(a) The forgetful functor $U:$ Mon $\to$ Set is monadic. The reader can prove directly this fact, following the outline of 3.6.1, or take it as granted by a general result on varieties of algebras (Theorem 4.4.6).

(b) Similarly, the forgetful functor $U:$ Ab $\to$ Set is monadic.

We are also interested in the Kleisli category Set$_T$ of the associated monad: an object is a set and a map $f^\sharp: X \nrightarrow Y$ can be described as a $\mathbb{Z}$-*weighted mapping between sets*, that associates to each $x \in X$ some formal linear combination

$$f(x) = \Sigma_{y \in Y} \, f_y(x) \, y$$

of elements of Y, with quasi-null integral coefficients $f_y(x) \in \mathbb{Z}$.

The reader can now easily compute the Kleisli composition with another weighted mapping $g^\sharp: Y \nrightarrow Z$.

(c) The Kleisli category Set$_T$ is isomorphic to the category of free abelian groups with an assigned basis, and arbitrary homomorphisms between them; it is equivalent to the full subcategory of Ab formed by the free abelian groups.

(d) One can extend all this to R-*weighted mappings*, for a unital ring R. Or replace Ab and R Mod with other variety of algebras.

These facts are part of a general result, in 3.6.8.

3.7.2 Exercises (The monad of upper-complete lattices)

(a) Define a suitable monad $T:$ Set $\to$ Set, where $T(X) = |\mathcal{P}(X)|$ is the *set* of subsets of X, so that the category Set^T of T-algebras is isomorphic to the *category* VSlt *of upper-complete semilattices*, i.e. complete lattices with mappings that preserve arbitrary joins.

(b) It may be easier to start from the forgetful functor $U:$ VSlt $\to$ Set, define its left adjoint and deduce the structure of the associated monad T.

(c) Prove that the Kleisli category Set$_T$ is isomorphic to Rel Set.

(d) Define a similar monad $T':$ Set $\to$ Set, where $T'(X) = |\mathcal{P}_f(X)|$ is the set of finite subsets of X. Determine its categories of algebras and Kleisli algebras.

3.7.3 *Exercises* (Categories as algebras)

These exercises are left to the reader.

(a) We write as Gph the *category of small graphs* and their morphisms (in the sense of 1.1.2). Prove that the forgetful functor $U \colon \mathsf{Cat} \to \mathsf{Gph}$ has a left adjoint F, where the free category $F(X)$ on the graph X has the same objects, and a morphism $\xi = (\xi_1, ..., \xi_n) \colon x \to x'$ is a finite paths of consecutive arrows of X

$$x = x_0 \to x_1 \to ... \to x_n = x'. \qquad (3.77)$$

This includes the empty path $\mathrm{id}(x)$ at any vertex x of X.

(b) Prove that Cat is monadic over Gph.

(c) In the category Gph_T of Kleisli algebras (of the associated monad), a morphism $f \colon X \nrightarrow Y$ between graphs sends an arrow $\xi \colon x \to x'$ of X to a formal composite $f(\xi) = (\eta_1, ..., \eta_n) \colon f(x) \to f(x')$ of arrows of Y.

3.8 *Idempotent monads and idempotent adjunctions

Loosely speaking, being an algebra for an idempotent monad $T \colon \mathsf{X} \to \mathsf{X}$ reduces to a *property* of the objects of X. For instance, an algebra for the abelianisation monad Gp $\to$ Gp (examined in 3.8.5(c)) is 'the same' as a commutative group. Thus, the category X^T of T-algebras can be replaced with an isomorphic category $\mathsf{Alg}(T)$, which is the full subcategory of X formed by its 'algebraic objects'.

The theory of idempotent monads and idempotent adjunctions is well known in the domain of category theory, but it may be difficult to find it in a single text. (Some of the main facts are exposed in [Bo2] Section 4.2, [LaS] Lemma 4.3 and [AT] Section 6.)

We give here a rather detailed treatment of this topic, even though this part will be marginally used in the rest of the book. A reader might be satisfied with browsing definitions and statements, and omitting the proofs; alternatively, one might view each statement as a list of exercises, and write down the proofs.

An adjunction is always 'equipped' with the associated monad, defined in 3.6.3.

3.8.1 *Idempotent monads*

A monad $T = (T, \eta, \mu)$ on the category X is said to be *idempotent* if the multiplication $\mu \colon T^2 \to T$ is invertible. Then the inverse of μ is the transformation $T\eta = \eta T \colon T \to T^2$.

Equivalently one can define an idempotent monad on X as a pair (T, η) formed of a functor $T\colon X \to X$ and a natural transformation $\eta\colon 1 \to T$ such that $T\eta = \eta T\colon T \to T^2$ is invertible.

In this approach one defines $\mu = (T\eta)^{-1}\colon T^2 \to T$. The first axiom $\mu.\eta T = 1 = \mu.T\eta$ obviously holds; for the second it suffices to cancel $T^2\eta = T\eta T$ from the identity

$$(\mu.T\mu).T^2\eta = \mu = (\mu.\mu T).T\eta T.$$

Dually we define *idempotent comonads*.

3.8.2 Theorem and Definition (Idempotent adjunctions)

Let $F \dashv G$ be a pair of adjoint functors (or more generally an adjunction in a 2-category, see 7.1.2)

$$F\colon X \rightleftarrows A \colon G, \qquad \eta\colon 1_X \to GF, \quad \varepsilon\colon FG \to 1_A,$$
$$\varepsilon F.F\eta = 1_F, \qquad G\varepsilon.\eta G = 1_G. \tag{3.78}$$

The following conditions are equivalent:

(i) one of the four natural transformations which appear in the triangular identities (namely $F\eta$, εF, ηG and $G\varepsilon$) is invertible,

(ii) all of them are invertible, i.e. $F\eta.\varepsilon F = 1_{FGF}$ and $\eta G.G\varepsilon = 1_{GFG}$,

(iii) the associated monad $T = GF\colon X \to X$ is idempotent, i.e. $\mu = G\varepsilon F\colon T^2 \to T$ is invertible (and then $T\eta = \eta T$ is its inverse),

(iv) $T\eta \, (= GF\eta)$ is invertible,

(v) $\eta T \, (= \eta GF)$ is invertible,

(iii*) the associated comonad $S = FG\colon A \to A$ is idempotent, i.e. $\delta = F\eta G\colon S \to S^2$ is invertible (and then $S\varepsilon = \varepsilon S$ is its inverse),

(iv*) $\varepsilon S \, (= \varepsilon FG)$ is invertible,

(v*) $S\varepsilon \, (= FG\varepsilon)$ is invertible.

When these conditions hold we say that the adjunction is idempotent. *Then the same is true of the opposite adjunction $G^{\mathrm{op}} \dashv F^{\mathrm{op}}$.*

Proof The triangle identities show that $F\eta$ is invertible if and only if εF is; the same holds for ηG and $G\varepsilon$. Moreover (iii) $\Leftrightarrow$ (iv) $\Leftrightarrow$ (v).

The main point is proving that (v) implies that $F\eta$ and εF are invertible, i.e. $F\eta.\varepsilon F = 1_{FGF}$.

The natural transformation $F\eta.\varepsilon F$ forms the upper row of the following commutative diagram

$$
\begin{array}{ccccc}
FGF & \xrightarrow{\ \varepsilon F\ } & F & \xrightarrow{\ F\eta\ } & FGF \\
{\scriptstyle FGF\eta}\big\downarrow & & \big\downarrow{\scriptstyle F\eta} & & \big\downarrow{\scriptstyle FGF\eta} \\
FGFGF & \xrightarrow[\ \varepsilon FGF\ ]{} & FGF & \xrightarrow[\ F\eta GF\ ]{} & FGFGF
\end{array}
\qquad (3.79)
$$

Now $F\eta GF$ is invertible by (v) and $(\varepsilon F.F\eta)GF = 1_{FGF}$, by (3.78). It follows that the lower row of the diagram $(F\eta.\varepsilon F)GF$ is the identity. Therefore $FGF\eta.(F\eta.\varepsilon F) = FGF\eta$ and we get our claim cancelling the transformation $FGF\eta$, which is invertible by (iv) (equivalent to (v)).

Adding an obvious implication we have:

- (ηG and $G\varepsilon$ are invertible) $\Rightarrow$ (v) $\Rightarrow$ ($F\eta$ and εF are invertible).

By duality all the properties of the statement are equivalent. $\qquad\square$

3.8.3 Proposition and Definition (Algebraic objects, I)

Let (T, η) be an idempotent monad on X, as defined in 3.8.1: the transformation $T\eta = \eta T \colon T \to T^2$ is invertible and $\mu = (T\eta)^{-1} \colon T^2 \to T$.

The following conditions on an object X of X are equivalent:

(i) ηX is invertible,

(ii) there exists a morphism $h \colon TX \to X$ such that $h.\eta X = 1_X$,

(iii) there exists an (Eilenberg–Moore) T-algebra (X, h) over X,

(iv) ηX is invertible and $\hat{X} = (X, (\eta X)^{-1})$ is the only T-algebra on X,

(v) X is isomorphic to TX in X.

When they hold we say that X is an algebraic object, *for T. $\mathsf{Alg}(T) \subset \mathsf{X}$ will denote the full subcategory of these objects.*

In this case the algebra $\hat{X} = (X, (\eta X)^{-1})$ is isomorphic to the free T-algebra $(TX, \mu X)$ on X.

Proof (i) $\Rightarrow$ (ii) Obvious.

(ii) $\Rightarrow$ (iii) If $h.\eta X = 1_X$ then (X, h) is a T-algebra because the second axiom is here a consequence:

$$ h.Th.T\eta X = h = h.\mu X.T\eta X. $$

(iii) $\Rightarrow$ (iv) The naturality of η on the morphism h and the property $T\eta = \eta T$ give

$$\eta X.h = Th.\eta TX = Th.T\eta X = T(h.\eta X) = 1_{TX},$$

so that ηX and h are inverses.

(iv) $\Rightarrow$ (v) Obvious.

(v) $\Rightarrow$ (i) The existence of an isomorphism $i\colon X \to TX$ gives $Ti.\eta X = \eta TX.i$, whence ηX is invertible.

The last claim follows from the isomorphism

$$\eta X\colon (X, (\eta X)^{-1}) \to (TX, \mu X).$$

$\square$

3.8.4 Proposition (Algebraic objects, II)

(T, η) is again an idempotent monad on the category X.

(a) An arrow of T-algebras $f\colon \hat{X} \to \hat{Y}$ is an ordinary morphism $f\colon X \to Y$ in X, between algebraic objects (as defined in 3.8.3).

The forgetful functor $G^T\colon \mathsf{X}^T \to \mathsf{X}$ of the category of T-algebras is thus a full embedding

$$G^T(\hat{X}) = X, \qquad G^T(f\colon \hat{X} \to \hat{Y}) = (f\colon X \to Y), \qquad (3.80)$$

and makes X^T isomorphic to the full reflective subcategory $\mathsf{Alg}(T) \subset \mathsf{X}$ of algebraic objects. The counit ε^T is invertible.

(b) Using this isomorphism the adjunction $F^T \dashv G^T$ associated to T can be rewritten in the following equivalent form, where G' is a full embedding with reflector F'

$$F'\colon \mathsf{X} \rightleftarrows \mathsf{Alg}(T) \colon G', \qquad \eta'\colon 1 \to G'F' = T, \quad \varepsilon'\colon F'G' \to 1,$$

$$F'(X) = TX, \qquad F'(f\colon X \to Y) = Tf\colon TX \to TY, \qquad (3.81)$$

$$\eta' = \eta, \qquad \varepsilon'(X) = (\eta X)^{-1}\colon F'G'(X) = TX \to X.$$

If T is the monad of an adjunction $F \dashv G\colon \mathsf{X} \twoheadrightarrow \mathsf{A}$, the comparison functor $K\colon \mathsf{A} \to \mathsf{X}^T$ can be rewritten as $K'\colon \mathsf{A} \to \mathsf{Alg}(T)$, a codomain-restriction of $G\colon \mathsf{A} \to \mathsf{X}$, so that $G = G'K'$. If G is full, then K' is also (and K as well).

(c) The composed functor from X^T to the category X_T of Kleisli algebras

$$F_T G^T\colon \mathsf{X}^T \to \mathsf{X} \to \mathsf{X}_T, \qquad F_T G^T(X, (\eta X)^{-1}) = X, \qquad (3.82)$$

is an equivalence of categories (not an isomorphism, generally).

Proof (a) and (b). It is an obvious consequence of Proposition 3.8.3, applying the naturality of η.

(c) As in 3.6.8 we write as $f^\sharp\colon X \twoheadrightarrow Y$ the X_T-arrow represented by the X-morphism $f\colon X \to TY$; recall that $\mathrm{id}\, X = (\eta X)^\sharp$ and the composite of $f^\sharp$ with $g^\sharp\colon Y \twoheadrightarrow Z$ is represented by $\mu Z.Tg.f\colon X \to TZ$.

The functor $F_T G^T\colon \mathsf{X}^T \to \mathsf{X}_T$ sends the arrow $f\colon \hat X \to \hat Y$ to

$$(\eta Y.f)^\sharp\colon X \twoheadrightarrow Y.$$

Since ηY is invertible, this mapping $\mathsf{X}^T(\hat X, \hat Y) \to \mathsf{X}_T(X, Y)$ is bijective. It is now sufficient to prove that $F_T G^T$ is essentially surjective on the objects.

Let us take an object X of X_T, i.e. an arbitrary object of X, and consider the T-algebra $(TX, \mu X)$; for an arbitrary monad this is the free T-algebra on X, with $F_T G^T(TX, \mu X) = TX$. Here X and TX are *isomorphic objects in* X_T (not in X, generally), by the inverse morphisms

$$(T\eta X.\eta X)^\sharp\colon X \twoheadrightarrow TX, \qquad (1_X)^\sharp\colon TX \twoheadrightarrow X,$$

since

$$(\mu X.T1_X.T\eta X.\eta X)^\sharp = (\eta X)^\sharp = \mathrm{id}(X),$$

$$(\mu TX.T^2\eta X.T\eta X.1_X)^\sharp = (\mu TX.T\eta TX.T\eta X)^\sharp = (\eta TX)^\sharp = \mathrm{id}(TX).$$

$\square$

3.8.5 Examples and complements

(a) We know, from 3.6.2, that a monad T on an ordered set X is always idempotent and amounts to a closure operator; an algebraic object for T is the same as a *closed* element $x = Tx$. Therefore every adjunction

$$(F, G, \eta, \varepsilon)\colon X \twoheadrightarrow \mathsf{A}$$

from the ordered set X to an arbitrary category is idempotent; by duality this also holds for every adjunction $\mathsf{A} \twoheadrightarrow X$.

(b) Let A be a *full reflective subcategory* of X which is *replete* in X: any isomorphism $i\colon A \to X$ of X that involves an object of A belongs to A.

Then the embedding $G\colon \mathsf{A} \subset \mathsf{X}$ has a left adjoint $F \dashv G$, and the counit $\varepsilon\colon FG \to 1$ is invertible (by 3.2.5), which implies that the adjunction is idempotent. Moreover A coincides with $\mathsf{Alg}(GF)$ and G is monadic.

(In fact, every $A \in \mathrm{Ob}\,\mathsf{A}$ is an algebraic object of X, because $\eta(GA)$ is invertible, and conversely an algebraic object $X \cong GF(X)$ belongs to A.)

(c) In particular we have seen (in 3.2.6) that Ab is full reflective in Gp, and obviously replete. The groups which are algebraic for the associated *abelianisation* monad

$$T\colon \mathsf{Gp} \to \mathsf{Gp}, \qquad \eta G\colon G \to T(G) = G/[G,G], \qquad (3.83)$$

are precisely the abelian ones. Similar facts hold for many embeddings of algebraic varieties, like $\mathsf{Gp} \subset \mathsf{Mon}$ or $\mathsf{CRng} \subset \mathsf{Rng}$. But also for $\mathsf{Ord} \subset \mathsf{pOrd}$ (see 3.2.3(d)) and $\mathsf{Hsd} \subset \mathsf{Top}$ (see 5.1.3).

(d) On the other hand the embedding $\mathsf{Rng} \subset \mathsf{gRng}$ (see 3.3.2(b)) is not full, since a homomorphism in gRng need not preserve units, when they exist; thus its left adjoint $(-)^+\colon \mathsf{gRng} \to \mathsf{Rng}$ (that universally adds a unit, as described in 3.3.2) does not yield an idempotent monad: the ring R^{++} will have a new unit added with respect to R^+.

The embedding of monoids into semigroups behaves in a similar way.

(e) An interested reader can prove the following extension of (b). We have an adjunction $F \dashv G$ where the functor G is full and faithful (whence $\varepsilon\colon FG \to 1$ is invertible and the adjunction is idempotent). It follows that the comparison $K\colon \mathsf{A} \to \mathsf{X}^T$ is an equivalence of categories, with weak inverse $FG^T\colon \mathsf{X}^T \to \mathsf{A}$, and G is *monadic in the weak sense* mentioned in Remark 3.6.6(b).

3.8.6 Idempotent comonads and coalgebras

An idempotent comonad (S, ε) on the category A has a dual characterisation, as an endofunctor $S\colon \mathsf{A} \to \mathsf{A}$ with a natural transformation $\varepsilon\colon S \to 1$ such that $S\varepsilon = \varepsilon S\colon S^2 \to S$ is invertible; the comultiplication $\delta\colon S \to S^2$ is the inverse of the latter.

Now an S-coalgebra amounts to a *coalgebraic object* A of A, characterised by the following equivalent conditions:

(i) εA is invertible,

(ii) there exists a morphism $k\colon A \to SA$ such that $\varepsilon A.k = 1_A$,

(iii) there exists an S-coalgebra $(A, k\colon A \to SA)$ on A,

(iv) εA is invertible and $\hat{A} = (A, (\varepsilon A)^{-1})$ is the unique S-coalgebra on A,

(v) A is isomorphic to SA in A.

In this case $\hat{A} = (A, (\varepsilon A)^{-1})$ is isomorphic to the cofree S-coalgebra $(SA, \delta A)$ on A.

The category of S-coalgebras $^S\mathsf{A}$ is linked to the full coreflective subcategory $\mathsf{Coalg}(S) \subset \mathsf{A}$ of coalgebraic objects, by the isomorphism

$$U\colon {}^S\mathsf{A} \to \mathsf{Coalg}(S), \qquad U(A, k) = A,$$
$$U(f\colon (A, k) \to (B, k')) = f\colon A \to B. \qquad (3.84)$$

The adjunction associated to the comonad S

$$^SF:\ ^S\mathsf{A} \rightleftarrows \mathsf{A} :\,^SG, \qquad ^S\eta: 1 \to {}^SG\,^SF, \qquad ^S\varepsilon = \varepsilon:\,^SF\,^SG \to 1, \qquad (3.85)$$

is computed as follows:

$$^SF(A,k) = A, \qquad ^SF(f: (A,k) \to (B,k')) = f: A \to B,$$
$$^SG(A) = (SA, \delta: SA \to S^2 A),$$
$$^SG(f: A \to B) = Sf: {}^SG(A) \to {}^SG(B),$$
$$^S\eta(A,k) = k: (A,k) \to (SA, \delta: SA \to S^2 A).$$

It can be rewritten in the following equivalent form, where F'' is a full embedding with coreflector G''

$$F'': \mathsf{Coalg}(S) \rightleftarrows \mathsf{A} :G'',$$

$$\eta'': 1 \to G''F'', \qquad \varepsilon'' = \varepsilon: F''G'' \to 1,$$

$$G''(A) = SA, \qquad G''(f: A \to B) = Sf: SA \to SB, \qquad (3.86)$$

$$\eta''(A) = (\varepsilon A)^{-1}: A \to SA, \qquad \varepsilon''(A) = \varepsilon A: SA \to A.$$

3.8.7 Examples

Dualising 3.8.5(b), the inclusion functor $U: \mathsf{A} \to \mathsf{X}$ of a full coreflective subcategory has a left adjoint $G: \mathsf{X} \to \mathsf{A}$ with invertible unit η.

We have now an idempotent comonad $S = GU: \mathsf{A} \to \mathsf{A}$; the category of coalgebras $^S\mathsf{A}$ is isomorphic to the full reflective subcategory $\mathsf{Coalg}(S) \subset \mathsf{A}$ of coalgebraic objects, namely the objects A of A such that εA is invertible.

In particular the category $t\mathsf{Ab}$ of torsion abelian groups is full coreflective in Ab: the counit $\varepsilon A: tA \to A$ is the embedding of the torsion subgroup of A (see 3.2.6); the abelian groups which are coalgebraic for the associated comonad $t: \mathsf{Ab} \to \mathsf{Ab}$ are precisely the torsion abelian groups.

3.8.8 Theorem (Factorising idempotent adjunctions)

An idempotent adjunction $(F, G, \eta, \varepsilon): \mathsf{X} \to \mathsf{A}$ *can be factorised as follows, up to isomorphism*

$$\mathsf{X} \underset{G'}{\overset{F'}{\rightleftarrows}} \mathsf{Alg}(T) \underset{G^\sharp}{\overset{F^\sharp}{\rightleftarrows}} \mathsf{Coalg}(S) \underset{G''}{\overset{F''}{\rightleftarrows}} \mathsf{A} \qquad (3.87)$$

where:

- $\mathsf{Alg}(T) \subset \mathsf{X}$ *is the full reflective subcategory of algebraic objects of the idempotent monad* $T = GF: \mathsf{X} \to \mathsf{X}$ *(see 3.8.3),*

- $\mathsf{Coalg}(S) \subset \mathsf{A}$ *is the full coreflective subcategory of coalgebraic objects of the idempotent comonad* $S = FG: \mathsf{A} \to \mathsf{A}$ *(see 3.8.6).*

Proof The reflective adjunction $(\eta', \varepsilon')\colon F' \dashv G'$ is described in (3.81). G' is the full embedding of algebraic objects, with reflector $F'(X) = TX$ and invertible counit $\varepsilon'(X) = (\eta X)^{-1}$. The comparison functor K is now computed as G

$$K\colon A \to \mathsf{Alg}(T), \qquad K(A) = GA, \quad K(f\colon A \to B) = Gf,$$
$$KF = F', \quad G'K = G, \quad K\varepsilon = \varepsilon'K. \tag{3.88}$$

The coreflective adjunction $(\eta'', \varepsilon'')\colon F'' \dashv G''$ is described in (3.86). F'' is the full embedding of coalgebraic objects, with coreflector $G''(A) = SA$ and invertible unit $\eta''(A) = (\varepsilon A)^{-1}$. The comparison functor H is computed as F

$$H\colon X \to \mathsf{Coalg}(S), \qquad H(X) = FX, \quad H(f\colon X \to Y) = Ff,$$
$$HG = G'', \quad F''H = F, \quad H\eta = \eta'H. \tag{3.89}$$

The given adjunction restricts to an adjunction between our full subcategories

$$(\eta^\sharp, \varepsilon^\sharp)\colon F^\sharp \dashv G^\sharp\colon \mathsf{Alg}(T) \rightarrowtail \mathsf{Coalg}(S),$$
$$F^\sharp X = FX, \quad G^\sharp A = GA, \tag{3.90}$$

because, for every X in X, FX is a coalgebraic object of A (with structure $F\eta X\colon FX \to SFX$), and dually. This is actually an *adjoint equivalence*, because $\eta^\sharp X = \eta X$ is invertible for every algebraic object X, and dually. It is even an isomorphism if T is the identity over all algebraic objects and S is the identity over all the coalgebraic ones.

Finally, when we compose the three adjunctions above we get $F''F^\sharp F' = FGF$, isomorphic to F (by $F\eta = (\varepsilon F)^{-1}$) and $G'G^\sharp G'' = GFG$, isomorphic to G (by $\eta G = (G\varepsilon)^{-1}$). $\qquad\square$

4

Applications in Algebra

The second part of this book deals with applications of category theory in algebra and topology. The reader should have acquired some practice with categories, and the solution of easy exercises will be more frequently left out.

We begin by studying, in Section 4.1, the free objects in various categories of algebraic kind. Regular and Barr-exact categories are introduced in Section 4.2; their categories of relations in Section 4.5.

Varieties of algebras are briefly studied in Sections 4.3 and 4.4, proving that these categories $\mathsf{Alg}(\Omega, \Xi)$ are complete and Barr-exact. The fact that such algebras can be viewed as Eilenberg–Moore algebras for the monad associated to the forgetful functor $\mathsf{Alg}(\Omega, \Xi) \to \mathsf{Set}$ is stated in Theorem 4.4.6; the proof is only referred to [M4].

Other notions of 'exact category' will be presented in 6.3.1.

4.1 Free algebraic structures

This section examines many explicit constructions of free algebraic structures, over weaker structures or sets, given by left adjoints of forgetful functors. The reader should find it useful to consider these exercises, completing the missing details. General results on the existence of such adjoints for varieties of algebras will be given in Section 4.4.

4.1.1 Algebraic structures and forgetful functors

Various categories of sets equipped with a kind of algebraic structure (and their homomorphisms) have been listed in 1.1.1, like the categories Set (of sets), Ab (of abelian groups), Gp (of groups), Mon (of monoids), Abm (of abelian monoids), Rng (of unital rings), CRng (of commutative unital rings), R Mod (of left modules on a fixed unital ring), RAlg (of unital algebras on a fixed commutative unital ring), Set. (of pointed sets).

This list includes Set and Set., where the algebraic structure of the object X is void or reduced to a 0-ary operation $X^0 = \{*\} \to X$, giving the base-point 0_X.

Each of these categories A is made concrete by an obvious (faithful) forgetful functor

$$U: \mathsf{A} \to \mathsf{Set}. \tag{4.1}$$

But there are also 'intermediate' forgetful functors $U: \mathsf{A} \to \mathsf{B}$ that forget part of the structure or part of its properties, like

$$
\begin{array}{ll}
\mathsf{Ab} \to \mathsf{Gp} \to \mathsf{Mon}, & \mathsf{Ab} \to \mathsf{Abm} \to \mathsf{Mon}, \\
R\mathsf{Alg} \to R\,\mathsf{Mod} \to \mathsf{Ab}, & R\mathsf{Alg} \to \mathsf{Rng} \to \mathsf{Mon}.
\end{array}
\tag{4.2}
$$

All these functors have left adjoints, giving free algebraic structures over sets or – more generally – over 'weaker' algebraic structures; many of these constructions are reviewed below.

4.1.2 Free objects

As we have already seen in 2.7.3, given a functor $U: \mathsf{A} \to \mathsf{Set}$, the free A-object over a set X is an object $F(X)$ (or $F_\mathsf{A}(X)$) equipped with a mapping $\eta: X \to U(F(X))$ satisfying the usual universal property. The existence of all free objects amounts to the existence of the left adjoint $F \dashv U$.

These definitions can be extended to any functor $U: \mathsf{A} \to \mathsf{B}$ between arbitrary categories.

In particular, the left adjoint F of a functor $U: \mathsf{A} \to \mathsf{Set}$, has a peculiar property deriving from the fact that every set X is isomorphic to a categorical sum $\Sigma_{x \in X} \{*\}$ (in Set) of copies of the singleton. The functor F, as a left adjoint, preserves sums, and $F(X)$ must be isomorphic to the categorical sum $\Sigma_{x \in X} F(\{*\})$ (in A) of copies of the free A-object on the singleton (independently of the existence of all sums in A). A stronger result in this sense will be given in 4.1.3(e).

In fact, many free 'algebras' are concretely built, below, as categorical sums of the free 'algebra' on one generator.

4.1.3 Exercises and complements, I

(a) Let us recall once more (after 2.7.4 and 3.1.1) that the free left R-module $F(X) = RX$ can be constructed as a direct sum in R Mod of copies of the left module R

$$F(X) = \bigoplus_{x \in X} R. \tag{4.3}$$

In particular, the free abelian group $F(X) = \bigoplus_{x \in X} \mathbb{Z}$ is constructed as a direct sum in Ab of copies of the additive group $\mathbb{Z}$.

(b) The free abelian monoid $F(X) = \mathbb{N}X$ has a similar construction, as a direct sum $\bigoplus_{x \in X} \mathbb{N}$ in Abm of copies of the additive monoid $\mathbb{N}$ (see 2.1.6(c)).

*This can be generalised to left modules over any unital *semiring* R (where the existence of opposites is not assumed).*

(c) The reader may know that the free group $F(X)$ can be constructed as a 'free product', i.e. a categorical sum in Gp, of copies of the additive group $\mathbb{Z}$

$$F(X) = \Sigma_{x \in X} \mathbb{Z}. \tag{4.4}$$

The elements of $F(X)$ can be described as equivalence classes $[w] \in W/{\sim}$ of 'words'

$$w = x_1^{k_1} x_2^{k_2} \dots x_p^{k_p} \qquad (x_i \in X,\ k_i \in \mathbb{Z}), \tag{4.5}$$

in the alphabet X, with integral exponents. Each word $w \in W$ has a 'reduced form' $R(w)$, obtained by adding up the exponents of consecutive terms with the same basis, and cancelling all occurrences of type x^0. The result does not depend on the sequence of reductions; this seemingly obvious point requires some care, but the rest is easy.

We say that $w \sim w'$ when $R(w) = R(w')$. Words are multiplied by concatenation, and the quotient set $W/{\sim}$ has a well defined product: $[w].[w'] = [w.w']$; indeed the concatenation $w.w'$ is equivalent to $R(w).R(w')$, because $R(w.w')$ can be obtained as $R(R(w).R(w'))$. Then we prove that $F(X)$ is a group, with basis the family of the equivalence classes $\eta(x) = [(x)]$.

(d) The existence of the free group $F(X)$ was proved in 3.5.4, in a non-constructive way, by the Adjoint Functor Theorem. The reader may note that, once we know that the universal arrow $\eta X \colon X \to UF(X)$ exists, it is straightforward to show that each element of $F(X)$ can be written in form (4.5); it is also obvious that such a form can be reduced as above, but less obvious that no further identification is needed.

(e) As an overview of the previous cases, let a functor $U \colon \mathsf{A} \to \mathsf{Set}$ be given. We assume that:

(i) the singleton $\{*\}$ has a free object $(A, \eta_0 \colon \{*\} \to UA)$ in A,

(ii) for every (small) set X, the sum $F(X) = \Sigma_{x \in X} A$ exists in A, with injections $u_x \colon A \to F(X)$.

Then one can prove that $F(X)$ is the free A-object over X, with universal map

$$\eta \colon X \to UF(X), \qquad \eta(x) = (Uu_x)(\eta_0(*)). \qquad (4.6)$$

(f) Apply the previous exercise to the forgetful functors $U \colon \mathsf{Top} \to \mathsf{Set}$ and $\mathrm{Ob} \colon \mathsf{Cat} \to \mathsf{Set}$. Note that the second is not faithful.

4.1.4 Exercises and complements, II

The procedure we have followed above is effective when sums in A have a simple construction, but there may be simpler direct constructions.

(a) We have seen in 3.6.1 that the free monoid on the set X has a simple construction as a sum $F(X) = \sum_{n \in \mathbb{N}} X^n$ *of sets*, with multiplication by concatenation.

Compare this form with the construction derived from Exercise 4.1.3(e), as a sum in Mon of free monoids on the singleton.

(b) The free commutative unital ring $F(X)$ can be realised as the polynomial ring $\mathbb{Z}[x \,|\, x \in X]$ with integral coefficients, and indeterminates belonging to the set X (possibly infinite). The basis X is embedded as a set of monomials, by $\eta(x) = x$.

$F(X)$ is thus the sum $\sum_{x \in X} \mathbb{Z}[x]$ in CRng, of polynomial rings in one indeterminate. (Finite sums of commutative rings are described in 3.4.7(j).)

(c) More generally, for a commutative unital ring R, the free commutative unital R-algebra can be realised as the polynomial ring $R[x \,|\, x \in X]$, with indeterminates in X.

Free rings are examined below, in 4.1.6(b).

4.1.5 Exercises and complements, III

Many interesting constructions in algebra arise as a left adjoint to a forgetful functor $U \colon \mathsf{A} \to \mathsf{B}$, between categories of algebraic structures of which the second is 'weaker' than the first – in the sense of having less structure, or less properties, or both.

We mention some cases but the reader can compute many more. (A general result will be given in Theorem 4.4.5.)

(a) As we have already seen (in 3.2.6 and 3.8.5) the embedding $\mathsf{Ab} \subset \mathsf{Gp}$ has a left adjoint, given by the abelianised group $G^{\mathrm{ab}} = G/[G, G]$.

(b) Describe the left adjoint to the forgetful functor $U \colon \mathsf{Rng} \to \mathsf{Mon}$ that drops the additive structure. *Hints:* for a monoid M, use the free abelian group $\mathbb{Z}M$ generated by the *set* M.

This construction is more often considered for a group G, and known as the *group ring* $\mathbb{Z}G$, or $\mathbb{Z}[G]$.

(c) More generally, for a commutative unital ring R, the forgetful functor U: RAlg → Mon has a left adjoint. Again this construction is more often considered for a group G, and known as the *group algebra* RG, or $R[G]$.

(d) The embedding U: Ab → Abm has a left adjoint F, which can be built extending the well-known construction of $\mathbb{Z}$ as a quotient of $\mathbb{N} \times \mathbb{N}$. *Hints.* The solution is easier for abelian monoids that satisfy the cancellation law ($x + z = y + z$ implies $x = y$). The reader may begin with this case, and extend it. (The general solution can be found in Chapter 8.)

(e) For the previous exercise, find an abelian monoid M such that the elements of $F(M)$ are not covered by the elements of M and their inverses. *Hints:* use ordinary fractions.

*(f) The reader will likely know the construction of the field of fractions on an integral ring: this gives the left adjoint to the embedding of the category of fields in the category of integral rings. The topic can be further extended to rings of fractions with respect to a multiplicative part.

4.1.6 Exercises and complements, IV

(a) Composing the forgetful functors Ab → Gp → Set we get the forgetful functor Ab → Set. Since adjoints compose, contravariantly, we deduce that the abelianised group of the free group $\Sigma_{x \in X} \mathbb{Z}$ is the free abelian group $\bigoplus_{x \in X} \mathbb{Z}$, actually a rather obvious fact.

(b) It is more interesting to examine the composed forgetful functor

$$\text{Rng} \to \text{Mon} \to \text{Set}.$$

The free ring on the set X can thus be constructed in two steps: first the free monoid M on X (by finite words on this alphabet, or 'non-commutative monomials', where $xyx \neq x^2y$, see 3.6.1); then the free ring $\mathbb{Z}M$ on this monoid (by $\mathbb{Z}$-linear combinations of these monomials, see 4.1.5(b)).

The result is the ring of 'non-commutative polynomials' with indeterminates in X and integral coefficients, where the 'polynomial' $2 + xyx - 3x^2y$ is not the same as $2 - x^2y$.

(c) The free general ring on a set is similarly obtained replacing monoids with semigroups: it is the ideal of the previous ring generated by the indeterminates, and formed by the 'polynomials' whose constant term is 0.

*(d) A *magma* is a set equipped with a binary operation, under no axioms. The free magma $M(X)$ on a set X has elements like $x((yx)z)$, expressed by (dichotomically) branched strings of generators. The free 'possibly non-associative, possibly

non-unital ring' is built on the free abelian group $\mathbb{Z}M(X)$, extending the product of $M(X)$.

A *magmoid* is a 'possibly non-associative, possibly non-unital category', used in categorical coherence topics: see [G11], Section 2.6.

4.2 Regular and Barr-exact categories

This is a brief introduction to the theory of regular and Barr-exact categories. The reader can find further information in Barr [Bar] and Borceux [Bo2], Chapter 2.

4.2.1 Definition

A *regular* category is a category with finite limits, where the kernel pair $R \rightrightarrows A$ of every map $f: A \to B$ (see 2.3.4) has a coequaliser; moreover, the pullback of any regular epi (see 2.4.4) along any arrow is a regular epi.

(The texts [Bar, Bo2] give a more general definition, where the only limits assumed to exist are the kernel pairs and the pullback of any regular epi along any arrow; but this extension is presented in [Bo2] as "essentially a matter of personal taste".)

4.2.2 Theorem and Definition (Canonical factorisation)

In a regular category C *every map* f *has a* canonical factorisation $f = mp$, *unique up to isomorphism, formed by a regular epimorphism* p *and a monomorphism* m.

In this factorisation, p *is a coequaliser of the kernel pair of* f.

Proof Given f, we form a diagram

$$
\begin{array}{ccc}
R \underset{v}{\overset{u}{\rightrightarrows}} A & \overset{f}{\longrightarrow} & B \\
{\scriptstyle q}\downarrow \quad\quad {\scriptstyle p}\downarrow & {\scriptstyle m}\nearrow & \\
S \underset{s}{\overset{r}{\rightrightarrows}} P & &
\end{array}
\tag{4.7}
$$

where (u, v) is the kernel pair of f and p is the coequaliser of (u, v), a regular epimorphism; then $fu = fv$, and there is a unique factorisation $f = mp$ through p. To prove that m is mono, we let (r, s) be its kernel pair, and it will be sufficient to prove that $r = s$ (by 2.3.5(a)).

From $m(pu) = m(pv)$ it follows that there is precisely one map $q: R \to S$ such that $rq = pu$ and $sq = pv$. Therefore $rq = sq$, and it will be sufficient to prove that q is epi.

We form a commutative diagram where the four horizontal squares are pullbacks and $v = v''v'$

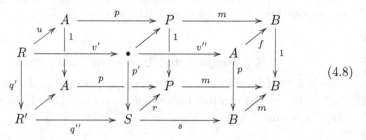

$$(4.8)$$

Trivially, the back squares are also pullbacks. Then the two front squares are pullbacks as well, by Lemma 2.3.7. But a pullback of the regular epi p is a regular epi, whence also p', q' and q'' are, and $q''q' \colon R \to S$ is an epimorphism; it coincides with the morphism q of the first diagram because $r(q''q') = pu = rq$ and $s(q''q') = pv = sq$.

Finally the essential uniqueness of our factorisation is easily proved. If $f = mp$, where p is a regular epi and m is mono, then the kernel pair (u, v) of f is also the kernel pair of p, and the latter is a coequaliser of (u, v): our factorisation is thus determined up to isomorphism.   □

4.2.3 Corollary

In a regular category, regular epimorphisms coincide with strong epimorphisms and are closed under composition.

As a consequence, regular epimorphisms and general monomorphisms form a factorisation system (rEpi, Mono) *(as defined in Section 2.5).*

Proof It is an easy consequence of the previous theorem and Section 2.4.

In fact, every regular epimorphism is strong. Conversely, if $f = mp$ is the canonical factorisation of a strong epimorphism, the monomorphism m is a strong epi, and therefore invertible, while f is a regular epi.   □

4.2.4 Definition (Equivalence relations)

In a category C with finite limits, an (internal) *equivalence relation* in the object A will be a jointly monic pair $u, v \colon R \to A$ (or equivalently a monomorphism $(u, v) \colon R \rightarrowtail A \times A$) that satisfies the following conditions:

(i) (*reflexivity*) the diagonal monomorphism $d_A = (1, 1) \colon A \rightarrowtail A \times A$ is smaller than (u, v): there is a (unique) morphism $h \colon A \to R$ such that $uh = 1_A = vh$,

(ii) (*symmetry*) the monomorphism $(u, v): R \rightarrowtail A \times A$ is equivalent to the monomorphism (v, u): there is a (unique) morphism $h: R \to R$ such that $uh = v$ and $vh = u$ (and then h is an involutive isomorphism: $hh = 1_R$),

(iii) (*transitivity*) in the following diagram, where (P, r, s) is a pullback

$$A \xleftarrow{u} R \xrightarrow{v} A \xleftarrow{u} R \xrightarrow{v} A \tag{4.9}$$

$$r \searrow \qquad \nearrow s$$

$$P$$

there is a (unique) morphism $h: P \to R$ such that $uh = ur$ and $vh = vs$.

4.2.5 Exercises

(a) Verify that, in Set, the conditions (i)–(iii) above amount to the usual ones.

(b) Prove that, in a category C with finite limits, a kernel pair (u, v) is always an equivalence relation.

4.2.6 Definition

A *Barr-exact* category [Bar] is a regular category C where every equivalence relation is *effective*, i.e. a kernel pair of a morphism.

Equivalently, one can require that every equivalence relation be the kernel pair of its coequaliser. This is plainly true in Set, but we shall see that every variety of algebras is Barr-exact, in Theorem 4.4.4.

4.3 Varieties of algebras

Universal Algebra is studied in well-known books, like Grätzer [Gr1] and Cohn [Coh].

As we have already recalled in 1.1.6, and shall make precise below, a 'variety of algebras' includes all algebraic structures of a given signature, satisfying a given set of equational axioms (or universally quantified identities).

We also recall that, as in [Coh], we do not assume that the underlying set of an algebra should be non-empty: a convention which would destroy all the main results below, from the existence of all limits and all free algebras, to the monadicity of the forgetful functor with values in Set.

Universal Algebra has a wide role in theoretical Computer Science: see [We, HyP].

4.3.1 Signatures and algebras

An (algebraic) *signature* (Ω, ar) is a set Ω equipped with a mapping ar: $\Omega \to \mathbb{N}$. Each $\omega \in \Omega$ with $\mathrm{ar}(\omega) = n$ is called an *operator* of *arity n*, or an *n-ary* operator; their set is denoted as $\Omega(n)$. The signature will be written as Ω, unless this leads to ambiguity.

An *Ω-algebra* is a set A equipped with a family of *operations* (ω_A) indexed by Ω

$$\omega_A \in \mathsf{Set}(A^n, A), \qquad \text{for } n \geqslant 0 \text{ and } \omega \in \Omega(n), \qquad (4.10)$$

and the n-ary operation $\omega_A \colon A^n \to A$ is the *realisation* of the operator ω in A. The underlying set of A will be written as $|A|$ when useful.

A *homomorphism* $f \colon A \to B$ is a mapping $|A| \to |B|$ that preserves all the operations

$$f.\omega_A = \omega_B.f^n \colon A^n \to B, \qquad \text{for } \omega \in \Omega(n). \qquad (4.11)$$

We have thus the category $\mathsf{Alg}(\Omega)$ of Ω-algebras and their homomorphisms, made concrete by the underlying-set functor, which is faithful

$$U \colon \mathsf{Alg}(\Omega) \to \mathsf{Set}, \qquad U(A) = |A|. \qquad (4.12)$$

A singleton $\{x\}$ has a unique structure of Ω-algebra, and gives the terminal object $\top$ of $\mathsf{Alg}(\Omega)$.

A 0-ary operator $\omega \in \Omega(0)$ is also called a *constant*, and provides every Ω-algebra A with an element $\omega_A \in A$, so that $|A|$ cannot be empty. If Ω does not contain any 0-ary operator, the empty set $\emptyset$ has a unique structure of Ω-algebra and gives the initial object $\bot$ of $\mathsf{Alg}(\Omega)$.

As an example, the signature Ω of semigroups consists of a single binary operator μ, called *multiplication*. An Ω-algebra is *any* set A equipped with a binary operation $\mu_A(x, y) = xy$; it is a semigroup when the associativity axiom is satisfied.

This is expressed by a pair of 'derived operators', the ternary multiplications

$$\xi' = \mu(p_1^3, \mu(p_2^3, p_3^3)), \qquad \xi'' = \mu(\mu(p_1^3, p_2^3), p_3^3), \qquad (4.13)$$

which must give the same 'derived operations' on A

$$\begin{aligned}
\xi'_A \colon A^3 \to A, \qquad & \xi'_A(x, y, z) = x(yz), \\
\xi''_A \colon A^3 \to A, \qquad & \xi''_A(x, y, z) = (xy)z.
\end{aligned} \qquad (4.14)$$

All this will be formalised defining the clone $\overline{\Omega}$ of derived operators of a signature Ω.

It is important to note that the axioms which one can express in this form are 'universally quantified equations' like, in the previous case:

$$\text{for all } x, y, z \text{ in } A: \qquad x(yz) = (xy)z. \tag{4.15}$$

The derived operators will be built making use of new formal n-ary operators p_i^n (for $n > 0$ and $i = 0, 1, ..., n$) called *projectors*. They operate on every set A as cartesian projections (for $i > 0$), or the terminal map (for $i = 0$)

$$
\begin{aligned}
&p_i^n: A^n \to A, \qquad &&p_i^n(x_1, ..., x_n) = x_i, \qquad \text{for } i = 1, ..., n, \\
&p_0^n: A^n \to \{*\}, \qquad &&p_0^n(x_1, ..., x_n) = *.
\end{aligned}
\tag{4.16}
$$

In particular we write p_1^1 as id. The reader will note that the projector p_0^1 plays a role in expressing the axiom $xx^{-1} = 1$ of groups.

4.3.2 Definition (Varieties of algebras)

A signature Ω has a *clone of derived operators* $\overline{\Omega}$, inductively constructed as follows.

We start from the signature (Ω_0, ar_0), formed by adding to (Ω, ar) all the projectors p_i^n ($0 \leqslant i \leqslant n$), with arity $n \geqslant 0$. Then $(\Omega_{k+1}, \text{ar}_{k+1})$ is obtained by adding to (Ω_k, ar_k), in a disjoint way, new operators of the following form

$$\xi = \omega(\xi_1, ..., \xi_p)$$
$$\text{for } \omega, \xi_1, ..., \xi_p \in \Omega_k, \quad \text{ar}_k(\omega) = p, \quad \text{ar}_k(\xi_i) = n, \tag{4.17}$$

and extending ar_k by letting $\text{ar}_{k+1}(\xi) = n$.

Finally $\overline{\Omega}$ is the union of the increasing sequence of sets (Ω_k), with the extended arity $\text{ar}: \overline{\Omega} \to \mathbb{N}$. The elements of $\overline{\Omega}$ are called the *derived operators* of Ω.

Every Ω-algebra A can be viewed as an $\overline{\Omega}$-algebra, with the following *clone of derived operations*:

- the projector p_i^n is realised on the set A as specified above, in (4.16),

- having defined the realisation in A of all the operators of Ω_k, we realise the operator ξ of formula (4.17) as:

$$\xi_A: A^n \to A,$$
$$\xi_A(x_1, ..., x_n) = \omega_A(\xi_{1A}(x_1, ..., x_n), ..., \xi_{pA}(x_1, ..., x_n)). \tag{4.18}$$

An *axiom* for the signature Ω is a pair (ξ', ξ'') of derived operators with the same arity. An *equational signature* (Ω, Ξ) is a signature equipped with a set Ξ of axioms.

We write as $\mathsf{Alg}(\Omega, \Xi)$ the full subcategory of $\mathsf{Alg}(\Omega)$ of all (Ω, Ξ)-*algebras*, i.e. all Ω-algebras A that satisfy the axioms of Ξ. This plainly means that $\xi'_A = \xi''_A$, for all the axioms $(\xi', \xi'') \in \Xi$.

The category $\mathsf{Alg}(\Omega, \Xi)$ is called a *variety of algebras*. It is made concrete by the restriction $\mathsf{Alg}(\Omega, \Xi) \to \mathsf{Set}$ of the forgetful functor of $\mathsf{Alg}(\Omega)$; the latter coincides with $\mathsf{Alg}(\Omega, \emptyset)$.

Let us note that the Ω-algebra on a singleton $\{x\}$ satisfies any axiom, and gives the terminal object $\top$ of any variety $\mathsf{Alg}(\Omega, \Xi)$.

4.3.3 Exercises and complements

(a) The forgetful functor $\mathsf{Alg}(\Omega, \Xi) \to \mathsf{Set}$ reflects the isomorphisms; in other words a homomorphism $f \colon A \to B$ is invertible if and only if the underlying mapping $|f|$ is bijective. It also reflects the identities.

(b) Set is the category of algebras for $\Omega = \Xi = \emptyset$.

(c) Describe $\mathsf{Set}_\bullet$ as a category of algebras (up to isomorphism of categories, of course).

(d) Describe Ab as a variety $\mathsf{Alg}(\Omega, \Xi)$, where the signature Ω consists of three operators: the *unit* ζ, the *opposite* ι and the *addition* σ (of arity 0, 1 and 2, respectively).

(e) Describe $R\,\mathsf{Mod}$ as a variety $\mathsf{Alg}(\Omega, \Xi)$, where the signature Ω consists of the previous operators, with the addition of a unary operator λ for every scalar $\lambda \in R$.

(f) Describe the category Lth of (bounded) lattices and homomorphisms as a variety $\mathsf{Alg}(\Omega, \Xi)$, where the signature Ω has two constants $0, 1$ and two binary operations $\vee, \wedge$.

(g) Describe the category of boolean algebras as a variety $\mathsf{Alg}(\Omega, \Xi)$.

(h) The category Fld of (commutative) fields has poor categorical properties. Let us begin by observing that a homomorphism $f \colon F \to F'$ is necessarily injective (its kernel must be the null ideal) and induces an isomorphism between the minimal subfields of F and F'; the latter have thus the same characteristic p, which is 0 or a prime number.

Fld is thus the categorical sum of its full subcategories Fld_p, of the fields of characteristic p.

The reader will prove that Fld and each Fld_p lack a terminal object. Therefore, by a remark in 4.3.1, these categories cannot be isomorphic, or even equivalent, to any variety $\mathsf{Alg}(\Omega, \Xi)$.

4.3.4 Trivial varieties

There are two *trivial* varieties of algebras.

The first is formed by all the sets of cardinal 0 or 1, and can be obtained as $\mathsf{Alg}(\emptyset, \Xi)$, where Ξ contains a single axiom, namely (p_1^2, p_2^2) (or $x = y$, informally). In fact, this axiom gives the full subcategory of Set of all the sets X whose projections $X^2 \to X$ coincide.

The second is formed by all the sets of cardinal 1. The reader can describe it as a variety $\mathsf{Alg}(\Omega, \Xi)$, where Ω contains one 0-ary operator. The trivial ring R gives also such a variety $R\,\mathsf{Mod}$, as we have seen in 2.7.4.

We shall see in 4.4.3 that (up to isomorphism) there are no other varieties having finitely many isomorphism-types of algebras, or equivalently a finite skeleton.

> Marginally, we note that the category **1**, consisting of an object and its identity, is not – strictly speaking – a variety of algebras, because we have already remarked that all the singletons belong to any variety. But of course **1** is equivalent to the trivial variety of all singletons.

4.3.5 Subalgebras

For an (Ω, Ξ)-algebra A, a *subalgebra* is a subset $C \subset |A|$ closed under the operations of A:

- for every n-ary operator $\omega \in \Omega$ and every n-tuple $(x_1, ..., x_n) \in C^n$, $\omega_A(x_1, ..., x_n) \in C$.

The set C is viewed as an (Ω, Ξ)-algebra, by means of the restricted operations ω_C

$$\omega_C(x_1, ..., x_n) = \omega_A(x_1, ..., x_n), \qquad \text{for } (x_1, ..., x_n) \in C^n, \qquad (4.19)$$

which automatically satisfy the axioms of Ξ. This is the unique structure of C that makes the inclusion $C \subset A$ into a homomorphism, and therefore a monomorphism of $\mathsf{Alg}(\Omega, \Xi)$.

The set-theoretical intersection of any family (A_i) of subalgebras of an algebra A is a subalgebra $\bigcap A_i$. Each subset $S \subset |A|$ *generates* a subalgebra $\langle S \rangle$, namely the intersection of all subalgebras containing it. $\langle S \rangle$ can be obtained as the union of all subsets $S_k \subset A$, where $S_0 = S$ and

$$S_{k+1} = \{\omega_A(x_1, ..., x_n) \mid n \geqslant 0,\ \omega \in \Omega(n),\ x_1, ..., x_n \in S_k\}.$$

This proves that

$$\sharp\langle S \rangle \leqslant \max\{\sharp S, \sharp\Omega, \aleph_0\}, \qquad (4.20)$$

where $\sharp$ denotes the cardinal of a set (see 3.5.3).

A homomorphism $f: A \to B$ of (Ω, Ξ)-algebras has a (set-theoretical) image $f(A)$ which is a subalgebra of B, since

$$\omega_B(fx_1, ..., fx_n) = f(\omega_A(x_1, ..., x_n)) \qquad \text{for } x_i \in A.$$

4.3.6 Congruences

A *congruence* R in an (Ω, Ξ)-algebra A is an equivalence relation in the set $|A|$ which is consistent with the operations of A:

- for every n-ary operator $\omega \in \Omega$ and every pair $(\underline{x}, \underline{y}) \in R^n$, we have $(\omega_A(\underline{x}), \omega_A(\underline{y})) \in R$,

where (by an abuse of notation) $R^n \subset (|A| \times |A|)^n = |A|^n \times |A|^n$ is the associated equivalence relation of the set $|A|^n$. (The correspondence of this notion with an 'internal' equivalence relation, in the sense of 4.2.4, will be established in Theorem 4.4.4.)

The quotient set $C = |A|/R$ becomes an (Ω, Ξ)-algebra, by means of the induced operations ω_C

$$\omega_C([x_1], ..., [x_n]) = [\omega_A(x_1, ..., x_n)], \qquad \text{for } (x_1, ..., x_n) \in A^n, \qquad (4.21)$$

which automatically satisfy the axioms Ξ. This is the unique structure of C that makes the projection $p: |A| \to |A|/R$ into a homomorphism $p: A \to A/R$, and therefore an epimorphism of $\mathsf{Alg}(\Omega, \Xi)$.

We prove below, in Theorem 4.4.4, that $p: A \to C$ is a regular epimorphism of $\mathsf{Alg}(\Omega, \Xi)$.

For any family (R_i) of congruences of an algebra A, their set-theoretical intersection $\cap R_i$ is again a congruence. Therefore each relation R in the set $|A|$ *generates* a congruence $\overline{R}$, namely the intersection of all congruences containing R.

A homomorphism $f: A \to B$ of (Ω, Ξ)-algebra has a (set-theoretical) equivalence relation R in $|A|$, which is a congruence of A: if $f^n(\underline{x}) = f^n(\underline{y})$ then

$$f\omega_A(\underline{x}) = \omega_B(f^n\underline{x}) = \omega_B(f^n\underline{y}) = f\omega_A(\underline{y}).$$

4.3.7 Canonical factorisation

Putting together the previous results, every homomorphism $f: A \to B$ of (Ω, Ξ)-algebras has a canonical factorisation (derived from the canonical factorisation of the underlying mapping $|f|$), where

$$f = mgp: A \to A/R \to f(A) \to B, \qquad (4.22)$$

- $p: A \to A/R$ is the projection modulo the congruence associated to f,

- $m\colon f(A) \to B$ is the inclusion of the subalgebra $f(A) \subset B$,
- g is the homomorphism produced by the bijection $[x] \to f(x)$, which preserves all operations:

$$g(\omega_C([x_1], ..., [x_n])) = g[\omega_A(x_1, ..., x_n)] = f(\omega_A(x_1, ..., x_n))$$

$$= \omega_B(fx_1, ..., fx_n) = \omega_{f(A)}(g[x_1], ..., g[x_n]),$$

and is an isomorphism of (Ω, Ξ)-algebras, by 4.3.3(a).

*4.3.8 *Complements*

A famous Birkhoff Theorem proves that a full subcategory C of $\mathsf{Alg}(\Omega)$ is a variety of (Ω, Ξ)-algebras (for some set Ξ of axioms) if and only if C is closed in $\mathsf{Alg}(\Omega)$ under *subalgebras, homomorphic images and cartesian products.*

The interested reader can see the proof in [Gr1], Section 2.6, Theorem 3 or [Coh], Chapter 4, Theorem 3.5.

4.4 Limits and free algebras

This section takes on the study of the categories $\mathsf{Alg}(\Omega, \Xi)$, showing that they are complete and Barr-exact. Moreover all of them are monadic over Set, and cocomplete.

4.4.1 Theorem (Limits in a variety)

(a) *The category* $\mathsf{Alg}(\Omega, \Xi)$ *has all limits and coequalisers of kernel pairs, created by the forgetful functor* $U\colon \mathsf{Alg}(\Omega, \Xi) \to \mathsf{Set}$ *(in the sense of 2.2.8).*

(b) *The forgetful functor* U *also creates all filtered colimits.*

Proof (a) Let S be a small category and $X\colon \mathsf{S} \to \mathsf{Alg}(\Omega, \Xi)$ a functor, written in index notation, for $i \in S = \mathrm{Ob}\,\mathsf{S}$ and $a\colon i \to j$ in S:

$$X\colon \mathsf{S} \to \mathsf{C}, \qquad i \mapsto X_i, \quad a \mapsto (u_a\colon X_i \to X_j). \qquad (4.23)$$

Let $(L, (u_i))$ be the limit-cone of UX in Set, with structural mappings $u_i\colon L \to |X_i|$ $(i \in S)$. For every $n \in \mathbb{N}$, L^n is the limit of the functor $(UX)^n$, with the mappings $(u_i)^n\colon L^n \to |X_i|^n$. For every n-ary operator $\omega \in \Omega$ we have a natural transformation

$$\omega\colon (UX)^n \to UX\colon \mathsf{S} \to \mathsf{Set}, \qquad \omega_i = \omega X_i\colon |X_i|^n \to |X_i|, \qquad (4.24)$$

because each homomorphism $u_a\colon X_i \to X_j$ must be consistent with ω.

There is thus unique mapping ω_L (in Set) consistent with the cones

$$\omega_L \colon L^n \to L,$$
$$u_i.\omega_L = \omega X_i.(u_i)^n \colon L^n \to |X_i| \qquad (i \in S), \tag{4.25}$$

and a unique structure $(\omega_L)_{\omega \in \Omega}$ of Ω-algebra on the set L which makes all the mappings $u_i \colon L \to X_i$ into homomorphisms of Ω-algebras. It is easy to see that the Ω-algebra L is actually in $\mathsf{Alg}(\Omega, \Xi)$, and therefore a limit in the latter.

Finally, every homomorphism $f \colon A \to B$ has a kernel pair $R \rightrightarrows A$, where R is the set-theoretical congruence of $|f|$, with the structure of subalgebra of the product $A \times A$. We have seen in 4.3.6 that the coequaliser $|A|/R$ in Set has a unique algebraic structure that makes the projection $p \colon A \to A/R$ a homomorphism; it is easy to deduce that p is the algebraic coequaliser, which is thus created by U.

(b) If S is a filtered category, one can repeat the previous argument, taking into account that filtered colimits in Set commute with finite products (by Theorem 2.6.6). $\qquad \square$

4.4.2 Theorem (Free algebras)

The forgetful functor $U \colon \mathsf{Alg}(\Omega, \Xi) \to \mathsf{Set}$ has a left adjoint: for every set X there exists a free (Ω, Ξ)-algebra $F(X)$, equipped with a universal arrow $\eta X \colon X \to UF(X)$.

Proof We use the Adjoint Functor Theorem, extending the argument of Exercise 3.5.4(a) for the free group.

We know that $\mathsf{Alg}(\Omega, \Xi)$ has small hom-sets and all small limits, preserved by U. Moreover, every set X has a solution set $S(X)$ (in the simplified form 3.5.2(iii)), consisting of the small set of all (Ω, Ξ)-algebras whose underlying set is (precisely) a cardinal $\alpha \leqslant \beta_X = \max\{\sharp X, \sharp \Omega, \aleph_0\}$.

The details are left to the reader. $\qquad \square$

4.4.3 Corollary (The embedding of the basis)

The following conditions on a variety $\mathsf{Alg}(\Omega, \Xi)$ are equivalent:

(i) there exists an algebra with more than one element,

(ii) for each set X the insertion of the basis $\eta X \colon X \to UF(X)$ is injective,

(iii) $\mathsf{Alg}(\Omega, \Xi)$ has infinitely many types of isomorphism,

(iv) $\mathsf{Alg}(\Omega, \Xi)$ is none of the two 'trivial varieties' of 4.3.4.

Proof Let us assume (i). The existence of small products of algebras preserved by U proves that there are algebras of a cardinal higher than any given small cardinal. Therefore their cardinals (and their types of isomorphism) cannot form a finite set. Moreover any set X can be embedded in the underlying set of some algebra, and ηX is injective (by 2.7.5(a)).

Conversely, if no algebra has more than one element, then $\mathsf{Alg}(\Omega, \Xi)$ can only contain the empty set and the singletons, with their unique algebraic structure; since all the singletons must be there, we fall in one of the trivial varieties of 4.3.4. Then conditions (ii) and (iii) fail as well. $\quad\square$

4.4.4 Theorem (Varieties and Barr-exactness)

A variety $\mathsf{Alg}(\Omega, \Xi)$ *is a complete Barr-exact category.*

More particularly, we have the following points, for every algebra A.

(a) The forgetful functor

$$U \colon \mathsf{Alg}(\Omega, \Xi) \to \mathsf{Set}$$

gives a bijective correspondence between equivalence relations $R \rightarrowtail A^2$ *(as defined in 4.2.4) and congruences of* A *(as defined in 4.3.6).*

Every quotient $p \colon A \to A/R$ *modulo a congruence is a regular epimorphism, whence a strong one. Every strong (or regular) epimorphism defined on* A *is of this kind, up to equivalence of epimorphisms.*

(b) A monomorphism $i \colon C \to A$ *is the same as an injective homomorphism, and is equivalent to the embedding of a subalgebra. The subobjects of* A *can (and will) be identified with its subalgebras, and form a complete lattice* $\mathrm{Sub}(A)$.

(c) The factorisation $f = m.(ip)$ *of 4.3.7 is the canonical factorisation regular epi - mono of a regular category. It is strictly determined, if we require that* m *be the inclusion of a subalgebra.*

Proof We already know, by Theorem 4.4.1, that $\mathsf{Alg}(\Omega, \Xi)$ has all limits and coequalisers of kernel pairs, created by the forgetful functor with values in Set. It follows that regular epimorphisms are closed under pullback, as in Set.

Let $f = (u, v) \colon R \rightarrowtail A^2$ be an equivalence relation, as in 4.2.4. Then $|f| \colon |R| \rightarrowtail |A|^2$ is an equivalence relation of $|A|$, and we can think of $|f|$ as an inclusion. Moreover, $|R|$ is a congruence of algebras, because for every n-ary operator ω and $(\underline{x}, \underline{y}) \in |R|^n \subset |A|^n \times |A|^n$

$$(\omega_A(\underline{x}), \omega_A(\underline{y})) = \omega_{A^2}(f^n(\underline{x}, \underline{y})) = f(\omega_R(\underline{x}, \underline{y})) \in |R|.$$

The quotient $p\colon A \to A/R$ (described in 4.3.6) is turned by U into the coequaliser of the kernel pair $|f|\colon |R| \rightarrowtail |A|^2$ of Up; whence $(u,v)\colon R \rightarrowtail A^2$ is the kernel pair of p, which is the coequaliser of (u,v).

We have proved that $\mathsf{Alg}(\Omega, \Xi)$ is complete and Barr-exact. We have also proved point (a), taking into account that in a regular category strong and regular epimorphisms coincide.

Point (b) follows from Theorem 4.4.2: the free (Ω, Ξ)-algebra on the singleton exists, therefore the (faithful) forgetful functor $U\colon \mathsf{Alg}(\Omega, \Xi) \to \mathsf{Set}$ is representable, and preserves and reflects monomorphisms (by 2.3.5(b)). Moreover, we have already seen in 4.3.5 that the subalgebras of A are closed under arbitrary intersections.

Point (c) is a consequence. $\square$

4.4.5 Theorem (Comparing algebras)

Let us be given two equational signatures (Ω, Ξ), (Ω', Ξ'), *with* $\Omega' \subset \Omega$ *and* $\Xi' \subset \Xi$. *Then the obvious forgetful functor*

$$W\colon \mathsf{Alg}(\Omega, \Xi) \to \mathsf{Alg}(\Omega', \Xi')$$

has a left adjoint. W creates limits and coequalisers of kernel pairs.

Proof This is an extension of Theorems 4.4.2 and 4.4.1, as $\mathsf{Set} = \mathsf{Alg}(\emptyset, \emptyset)$.

The proof of the existence of the left adjoint of W is similar to that of Theorem 4.4.2. For the second part of the statement, we consider the forgetful functors

$$\mathsf{Alg}(\Omega, \Xi) \xrightarrow{\ W\ } \mathsf{Alg}(\Omega', \Xi') \xrightarrow{\ V\ } \mathsf{Set} \qquad U = VW. \qquad (4.26)$$

We know that U and V create all limits. Let S be a small category with $S = \mathsf{Ob}\,\mathsf{S}$; all families below have indices $i \in S$.

For a functor $A\colon \mathsf{S} \to \mathsf{Alg}(\Omega, \Xi)$, consider the composites WA and VWA. The latter has a limit $(L'', (r_i\colon X \to VWA_i))$. The limit of A in $\mathsf{Alg}(\Omega, \Xi)$ is the unique cone $(L, (p_i\colon X \to WA_i))$ taken by $U = VW$ to the limit cone $(L'', (r_i))$. Similarly, the limit of WA in $\mathsf{Alg}(\Omega', \Xi')$ is the unique cone $(L', (q_i\colon X \to WA_i))$ taken by V to the cone $(L'', (r_i))$.

The functor W takes $(L, (p_i))$ to a cone of WA, and there is a unique arrow $g\colon WL \to L'$ such that $q_i g = W(p_i)$, for all $i \in S$. But $Vp_i = r_i = V(Wp_i)$, for all i, whence the underlying mapping $V(g)$ is the identity and the homomorphism $g\colon WL \to L'$ is also.

The same argument holds for coequalisers of kernel pairs, which are also created by U and V. $\square$

4.4.6 *Theorem* (Varieties and monadicity)

The forgetful functor $U\colon \mathsf{Alg}(\Omega, \Xi) \to \mathsf{Set}$ *is monadic.*

Proof This important result proves that each variety of algebras is a category of Eilenberg–Moore algebras, for the monad associated to the forgetful functor.

The proof can be found in Mac Lane [M4], Section VI.8. The argument is based on Beck's Monadicity Theorem (or Tripleability Theorem), in [M4], Section VI.7. $\qquad\qquad\square$

4.4.7 *Exercises and complements* (Colimits)

The interested reader can prove that every variety $\mathsf{Alg}(\Omega, \Xi)$ is also cocomplete.

(a) The existence of arbitrary coequalisers follows easily from 4.3.6.

(b) The existence of small sums can be proved with the Adjoint Functor Theorem, using the completeness of the lattices of subalgebras and the upper bound (4.20) for the cardinal of a subalgebra generated by a subset.

(c) In particular $\mathsf{Alg}(\Omega, \Xi)$ always has an initial object, the free algebra on the empty set. Its underlying set can take different forms: for instance in Set, Mon, Rng, RAlg, etc.

4.5 Relations for regular categories

We begin by re-examining the ordered category Rel Set, viewed as a *locally ordered 2-category*. This construction is extended to a regular category C, forming the ordered category of relations Rel (C), where a relation $X \nrightarrow Y$ is a subobject of the product $X \times Y$ (as in [Grt, Bo1]).

Marginally, we briefly sketch a second approach to relations (followed for instance in [Me]), where a relation $X \nrightarrow Y$ is *any* monomorphism with values in $X \times Y$. In this way one gets a laxer structure, called a *locally preordered bicategory*, which is written here as Rel $'$(C).

This matter can also be viewed as an introduction to general 2-categories and bicategories, that will be dealt with in Section 7.1.

4.5.1 *Cartesian relations and locally ordered 2-categories*

Let us recall that $\mathsf{R} = \mathsf{Rel}\,\mathsf{Set}$ has been constructed in 1.8.3, as an involutive ordered category.

A relation $r\colon X \nrightarrow Y$ is a subset $r \subset X \times Y$, and will be called a *cartesian relation* when we want to distinguish it from the similar item of 4.5.5 (a monomorphism $r \colon \bullet \rightarrowtail X \times Y$). Their composition has been recalled in (1.70). Given two parallel relations $r, s \colon X \rightarrow\!\!\!\!\rightarrow Y$, the ordering $r \leqslant s$ means that $r \subset s$ as subsets of $X \times Y$; it agrees with composition.

In this way, each hom-set $\mathsf{R}(X, Y)$ is the ordered set $\mathcal{P}(X \times Y)$, and the relation $r \colon X \rightarrow\!\!\!\!\rightarrow Y$ can be represented as in the diagram below, where the vertical map is the inclusion of a subset

$$X \xleftarrow{\ p'\ } X \times Y \xrightarrow{\ p''\ } Y \qquad\qquad (4.27)$$
$$\uparrow r$$

An ordered category can be viewed as an elementary form of a 2-category, and an introduction to this higher dimensional notion that will be presented in Section 7.1 – in the same way as an ordered set is an elementary form of a category. In this perspective, we think of the relationship $r \leqslant s$ in $\mathsf{R}(X, Y)$ as a 2-*cell* $r \to s$ (determined by r, s), i.e. a higher arrow between parallel morphisms.

These higher arrows have operations of vertical composition and whisker composition as in 1.5.3 (for natural transformations), but much simpler – since here the result is determined by the 'boundaries'

$$X \quad \overset{\overset{r}{\longrightarrow}}{\underset{\underset{t}{\longrightarrow}}{\Downarrow}} \quad Y \qquad\qquad X' \xrightarrow{\ u\ } X \quad \overset{\overset{r}{\longrightarrow}}{\underset{\underset{s}{\longrightarrow}}{\Downarrow}} \quad Y \xrightarrow{\ v\ } Y' \qquad (4.28)$$

Namely, the vertical composition of $r \to s$ with $s \to t$ is the (unique) 2-cell $r \to t$, and the whisker composition above, at the right, is the (unique) 2-cell $vru \to vsu$.

This 'elementary form' is called a *locally ordered 2-category*, where 'locally' refers to the hom-sets. (The definition of a general 2-category will be seen in 7.1.2; here most of the axioms are automatically satisfied, because of the uniqueness of a 2-cell between two given arrows.)

4.5.2 The basic choices

Let C be a regular category. Extending the previous point, we want to construct an involutive ordered category $\mathsf{R} = \mathrm{Rel}\,(\mathsf{C})$ of relations on C, in the 'cartesian form'.

For this goal, we assume that C is equipped with a choice of binary products and subobjects. (In many concrete cases, these selections can be made without any use of the axiom of choice, by cartesian products and subsets in Set, via a forgetful functor $U \colon \mathsf{C} \to \mathsf{Set}$.)

As a consequence, every morphism $f\colon X \to Y$ has *precisely one* factorisation $f = mp$ where p is a regular epi and m is a subobject of Y, and we shall write

$$f = mp, \qquad m = \mathrm{im}\,(f)\colon \mathrm{Im}\,(f) \to Y. \tag{4.29}$$

(We are using a *strict* factorisation system in C, see 2.5.3(f).)

4.5.3 Constructing the category of relations

For every pair of objects X, Y we let $\mathsf{R}(X, Y)$ be the ordered set of subobjects of the product $X \times Y$.

Given two consecutive relations $r = (r', r'')\colon R \rightarrowtail X \times Y$ and $s = (s', s'')\colon S \rightarrowtail Y \times Z$, their composite $s.r = w$ is a subobject of $X \times Z$, determined by a construction based on a pullback P of (r'', s'), and the strict factorisation (4.29), where p is a regular epi and w is a subobject

$$P \xrightarrow{\;p\;} W \xrightarrow{\;w\;} X \times Z$$
$$\underset{(r'u',\,s''u'')}{\longrightarrow} \tag{4.30}$$

$$s.r = \mathrm{im}\,(r'u', s''u'')\colon W \rightarrowtail X \times Z. \tag{4.31}$$

The reader will note that any choice of the pullback gives the same subobject of $X \times Z$. The identity 1_X is the subobject associated to the diagonal monomorphism

$$1_X = \mathrm{im}\,((1,1)\colon X \rightarrowtail X \times X) = (u, u)\colon \Delta_X \rightarrowtail X \times X, \tag{4.32}$$

where u is an isomorphism. The involution sends a subobject $r\colon R \rightarrowtail X \times Y$ to the 'opposite' relation

$$r^\sharp = \mathrm{im}\,(ir\colon R \rightarrowtail X \times Y \to Y \times X), \tag{4.33}$$

where $i = (p_2, p_1)\colon X \times Y \to Y \times X$ is the canonical symmetry of the cartesian product.

For two parallel relations $r, r'\colon X \nrightarrow Y$, the order relation

$$r \leqslant r' \tag{4.34}$$

has already been defined, as the canonical order of subobjects of $X \times Y$.

4.5.4 Theorem and Definition

These definitions produce an involutive ordered category $\mathsf{R} = \mathrm{Rel}\,(\mathsf{C})$, *called the category of relations over the regular category* C.

The embedding $\mathsf{C} \to \mathrm{Rel}\,(\mathsf{C})$ *is the identity on the objects and sends the morphism* $f\colon X \to Y$ *to its graph, namely the subobject associated to the monomorphism* $(1, f)\colon X \to X \times Y$.

The category R *has small hom-sets if and only if every object of* C *has a small set of subobjects.*

Proof First, the associativity of the composition law, for three relations $r\colon X \twoheadrightarrow Y$, $s\colon Y \twoheadrightarrow Z$, $t\colon Z \twoheadrightarrow U$, is proved by the following diagram

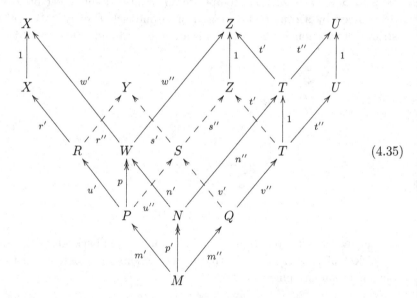

$$(4.35)$$

At the lower level we have a symmetric construction, based on three pullbacks (P, u', u''), (Q, v', v'') and (M, m', m''). At the upper level we first compute

$$sr = \mathrm{im}\,(r'u', s''u'') = (w', w'')\colon W \rightarrowtail X \times Z,$$

then the pullback (N, n', n'') of (w'', t'), so that $t(sr) = \mathrm{im}\,(N \to X \times U)$.

Now, the square $pm' = n'p'$ is a pullback, by Lemma 2.3.7, and p is a regular epi, whence p' is also. But then $\mathrm{im}\,(N \to X \times U)$ coincides with $\mathrm{im}\,(M \to X \times U)$, and this subobject of $X \times U$ is determined by the *symmetric* construction of the lower level.

By symmetry, $(ts)r$ too coincides with this subobject.

Second, the identity $1_X \colon \Delta_X \rightarrowtail X \times X$ defined above gives $r1_X = r$

$$\tag{4.36}$$

because u is an isomorphism and we can take R itself as the pullback of (u, r'); therefore the image of $(uv, r''.1) = (r', r'') \colon R \rightarrowtail X \times Y$ is precisely the relation r. Symmetrically, $1_Y s = s$.

We leave to the interested reader the verification of the remaining points:

- the involution defined in (4.33) is consistent with the composition (in a contravariant way),
- the order defined in (4.34) is consistent with composition and involution,
- the embedding $\mathsf{C} \to \mathrm{Rel}\,(\mathsf{C})$ is a functor.

The last statement, about smallness, is obvious. $\qquad\square$

4.5.5 *Jointly monic relations

We end this chapter by sketching a less elementary construction, that will not be used below. It gives a preordered structure, which we shall write as $\mathsf{R}' = \mathrm{Rel}\,'\mathsf{C}$. It is not a category, but a *locally preordered bicategory*; again, it can serve as an introduction to bicategories, that will be briefly analysed in Section 7.1.

The essential differences are:

- a hom-set $\mathsf{R}'(X, Y)$ is now a *pre*ordered set (and generally a large one),
- the composition of arrows is 'weakly associative' and has 'weak identities', *up* to the equivalence relation associated to the preorder.

A *jointly monic relation* $r = (r', r'') \colon X \nrightarrow Y$ is a jointly monic pair of morphisms of C, as in the left diagram below, with an arbitrary domain R (it is also called a jointly monic *span*, from X to Y)

$$\tag{4.37}$$

$\mathsf{R}'(X, Y)$ is the preordered set of such spans, with a 2-cell $r \prec s$ if there exists a (unique) mapping u that makes the right diagram above commute.

Equivalently, $\mathsf{R}'(X, Y)$ is the set of all monomorphisms $r \colon \bullet \rightarrowtail X \times Y$ with values in $X \times Y$, with the canonical preordering $r \prec s$ of monomorphisms with the same codomain (see 2.4.1). Let us note that the previous $\mathsf{R}(X, Y)$ can be identified with the associated ordered set $\mathsf{R}'(X, Y)/\sim$.

For a consecutive jointly monic relation $s: Y \twoheadrightarrow Z$, the composite $w = sr$ is obtained by:

- *choosing* a pullback (P, u', u'') of the pair (r'', s'),

- then replacing the mapping $(r'u', s''u''): P \to X \times Z$ with an associated jointly monic span, i.e. *any* monomorphism $w: W \rightarrowtail X \times Z$ of the factorisation regular epi - mono $(r'u', s''u'') = wp$, in C

$$
\begin{array}{ccccccc}
X & & & Y & & & Z \\
& \overset{r'}{\searrow} & \overset{r''}{\nearrow} & & \overset{s'}{\nwarrow} & \overset{s''}{\nearrow} & \\
& R & & & & S & \\
& & \overset{u'}{\searrow} & & \overset{u''}{\nearrow} & & \\
& & & P & & &
\end{array}
\tag{4.38}
$$

$$
P \xrightarrow{\ p\ } W \overset{w}{\rightarrowtail} X \times Z
$$
$$
\underset{(r'u',\,s''u'')}{\underbrace{\hspace{3cm}}}
$$

Then one defines the vertical composition and the whisker composition of 2-cells. We only remark that the vertical composition is associated to a preorder, and is therefore strictly associative, with strict identities; on the other hand, the composition of morphisms is weakly associative and has weak identities, up to the equivalence relation associated to the preorder (as follows readily from the first approach).

In this way one gets a *locally preordered bicategory* $\mathsf{R}' = \mathrm{Rel}\,'\mathsf{C}$ (see 7.1.3).

The latter gives back the locally ordered 2-category R, by identifying its arrows up to the equivalence relation associated to the preorder. Moreover, *defining* R *as this quotient gives an ordered category that is independent of any choice.*

We also note that an equivalence relation *in* the object A has been defined in 4.2.4 following this approach.

5

Applications in Topology and Algebraic Topology

Various categorical topics have interesting applications in Topology and Algebraic Topology. One of the first interactions of this kind can be found in Eilenberg–Steenrod's foundations of (topological) homology theories [EiS], whose axioms deal with categories of 'pairs' of topological spaces, homology functors and natural transformations. (Something about these categories will be said in Section 6.7.)

Here we begin by studying various adjunctions between categories of topological spaces. Then, in Section 5.2, the cylinder-cocylinder adjunction is used to define homotopies in Top, and set up their structure. Simplicial sets and cubical sets are dealt with in Sections 5.3 and 5.4, as combinatorial data that represent elementary topological spaces.

Two new sections deal with subjects scantily present in the first edition: Section 5.5 is an introduction to singular homology; Section 5.6 deals with categorical aspects of the smash product of pointed spaces.

The chapter ends with some hints at Directed Algebraic Topology, in Section 5.7.

For a reader who would like to explore the fascinating domain of Algebraic Topology there are excellent texts, like the classical Hilton–Wylie [HiW] or the recent book by Hatcher [Ha], which is freely downloadable. For an easier approach, one could begin with an elementary textbook on singular homology, like Vick [Vi] or Massey [Mas] (for the simplicial or cubical construction, respectively).

As to our notation, a functor will be defined by its action on objects whenever its extension to morphisms is evident (more frequently than above). A component $\varphi X \colon FX \to GX$ of a natural transformation will often be written as $\varphi \colon FX \to GX$.

The intervals of the real line are always denoted by square brackets, like $[a, b]$, $[a, b[$, $]a, b[$, etc. The standard euclidean interval is written as $\mathbb{I} = [0, 1]$, the standard euclidean n-sphere as $\mathbb{S}^n$.

The two-valued index α (or β) varies in the set $\{0,1\}$, also written as $\{-,+\}$.

5.1 Adjoints and limits in Topology

We already know that the forgetful functor $U\colon \mathsf{Top} \to \mathsf{Set}$ has both adjoints $D \dashv U \dashv C$, where DX (resp. CX) is the set X with the discrete (resp. indiscrete) topology (see 3.1.1(a)).

Here we examine various adjunctions of interest in Topology, starting from exponentiation. Some of them are used to compute colimits. Many other similar facts are studied under the heading of 'Categorical Topology' (see [AHS, Ke]).

5.1.1 Exponentiable spaces

We have seen in 3.4.3 that every set A is exponentiable in Set: this means that the endofunctor $F(X) = X \times A$ has a right adjoint $G(Y) = Y^A = \mathsf{Set}(A,Y)$, linked to the former by the exponential law.

In Top the situation is more complex: not every space A is exponentiable, but we can prove that this holds if A is *locally compact*, in the sense that every point has a fundamental system of compact neighbourhoods.

Indeed, let A be such a space and consider the functor

$$F\colon \mathsf{Top} \to \mathsf{Top},$$
$$F(X) = X \times A, \qquad F(f) = f \times A = f \times \mathrm{id}A. \tag{5.1}$$

F has a right adjoint

$$G\colon \mathsf{Top} \to \mathsf{Top},$$
$$G(Y) = Y^A, \quad G(g) = \mathsf{Top}(A,g)\colon h \mapsto gh \qquad (\text{for } g\colon Y \to Y'), \tag{5.2}$$

where the space Y^A is the set $\mathsf{Top}(A,Y)$ endowed with the *compact-open topology*, also called the *topology of uniform convergence on compact subspaces*.

The open subsets of Y^A are generated by the sets

$$W(K,V) = \{h \in \mathsf{Top}(A,Y) \mid h(K) \subset V\}, \tag{5.3}$$

where K is any compact subspace of X and V is open in Y. A general open subset is a union of finite intersections of 'distinguished open sets', of the previous form. (The latter form a subbase of the topology of Y^A.)

The proof is left to the reader, who can follow the outline given in 5.1.2. In particular the standard euclidean interval $[0,1]$ is exponentiable in

Top, with all its cartesian powers. This is a crucial fact in homotopy theory, as we shall see in Section 5.2.

Complements. (a) In other contexts, 'locally compact' can simply mean that every point has a compact neighbourhood. For a Hausdorff space, these (and other) definitions are equivalent.

(b) As a partial converse to the previous result, every exponentiable space which is Hausdorff must be locally compact, as proved in [He], Section 7.3.18. In particular, the rational line $\mathbb{Q}$ is not exponentiable.

(c) As a general characterisation, a topological space is exponentiable if and only if it satisfies a property called *core compactness*, or *quasi local compactness*: every neighbourhood V of a point contains a smaller one W, which is *relatively compact under* V; this relation, denoted by $W \ll V$, means that any open cover of V contains a finite subcover of W.

This characterisation began with an article by Day and Kelly in 1970 [DaK], and was completed by various contributions. One can find the story in [LoR], together with the topology of the exponential space and many related results for other 'topological categories'.

Exponentiable spaces can also be characterised in terms of convergence of ultrafilters: see [Pi].

5.1.2 Exercises and complements

This is an outline of the proof of the previous claim: a locally compact space A is exponentiable in Top. Some points are left as exercises (and solved in Chapter 8).

We have defined the space $G(Y) = Y^A$. We let ε_Y be the evaluation mapping

$$\varepsilon_Y \colon FG(Y) = Y^A \times A \to Y, \qquad \varepsilon_Y(h, a) = h(a). \tag{5.4}$$

We want to prove that (GY, ε_Y) is a universal arrow from the functor F to the object Y. (Note that the action of G on the morphisms is then a consequence: we do not have to prove the continuity of $G(g)$ in (5.2).)

The reader will prove the following points.

(a) For every space Y, the mapping ε_Y is continuous at any point $(h_0, a_0) \in Y^A \times A$.

(b) For every map $f \colon X \times A \to Y$, the mapping

$$g \colon X \to Y^A, \qquad g(x) = f(x, -) \colon A \to Y, \tag{5.5}$$

is well defined, i.e. $f(x, -)$ is continuous on A, for every $x \in X$.

(c) The mapping g is continuous on X, and is determined by the condition $f = \varepsilon_Y.(g \times \mathrm{id}A)$. (This is the only point requiring some care.)

5.1.3 Proposition (The reflector of Hausdorff spaces)

The full subcategory Hsd $\subset$ Top *of all Hausdorff spaces is reflective in* Top, *with a unit whose components* $p_X \colon X \to X/R$ *are projections on a quotient.*

Proof Let X be a space and R the equivalence relation in X where xRx' means that every map $f \colon X \to Y$ with values in a Hausdorff space has $f(x) = f(x')$.

Let $H(X) = X/R$ be the quotient space and $p \colon X \to X/R$ the projection. We prove that the latter is the universal arrow from the space X to the inclusion $U \colon$ Hsd $\subset$ Top.

Plainly, every map $f \colon X \to Y$ with values in a Hausdorff space factorises as $gp \colon X \to X/R \to Y$, by a unique map g. It is thus sufficient to prove that the quotient space X/R is Hausdorff.

For two equivalence classes $[x] \neq [x']$ there exist some map $f \colon X \to Y$, with values in a Hausdorff space, such that $f(x) \neq f(x')$. If V, V' are disjoint open neighbourhoods of these two points in Y, their preimages U, U' are disjoint open neighbourhoods of x, x' in X. Moreover the sets U, U' are saturated for the equivalence relation R_f associated to f, whence they also are for the (finer) equivalence relation R. It follows that $p(U)$ and $p(U')$ are disjoint open neighbourhoods of $[x], [x']$ in X/R. $\square$

5.1.4 Exercises and complements (Separation axioms)

(a) The 'constructive' character of the previous definition $H(X) = X/R$ is limited, because it may be difficult to determine the equivalence relation R of a given space.

However, let us consider – for a given space X – the equivalence relation R' generated by all pairs of points x, x' that do not have a pair of disjoint neighbourhoods in X. Then $R' \subset R$. If *it happens that* the quotient X/R' is Hausdorff, these equivalence relations coincide and our goal is reached. On the contrary, one can reiterate the procedure on the space $X' = X/R'$, and so on, hoping of getting a Hausdorff space.

(b) The functor $U \colon$ Hsd $\subset$ Top creates all limits and all sums, which thus exist in Hsd. It does not create general colimits, but it is easy to prove that they exist: for a functor S $\to$ Hsd, one can compute the colimit in Top and apply the left adjoint $H \colon$ Top $\to$ Hsd (following the same pattern used for colimits of ordered sets, in 3.2.3(d)).

(c) These results can be easily adapted to T_0-spaces (and many other separation axioms):

- the full embedding U: $T_0\mathsf{Top} \subset \mathsf{Top}$ of T_0-spaces has a left adjoint T_0: $\mathsf{Top} \to T_0\mathsf{Top}$,

- this functor U creates all limits and all sums. To compute the colimit of a functor $\mathsf{S} \to T_0\mathsf{Top}$, one can compute the colimit in Top and then apply the reflector T_0.

5.1.5 Topological spaces and preorder relations

The similarity of the theories of preordered sets and topological spaces can easily be made precise, by a full embedding of pOrd in Top.

We start from an idempotent adjunction

$$A: \mathsf{pOrd} \rightleftarrows \mathsf{Top} : W, \qquad A \dashv W. \qquad (5.6)$$

For a space X, the preordered set $W(X)$ has the same underlying set and the *specialisation preorder* $x \prec y$, defined by the following equivalent conditions (where $\overline{y}$ is the closure of $\{y\}$ in X)

$$x \in \overline{y}, \qquad \overline{x} \subset \overline{y}. \qquad (5.7)$$

Plainly, a continuous mapping $f: X \to Y$ becomes a monotone mapping $f: W(X) \to W(Y)$.

The other way round, for a preordered set X, the space $A(X)$ has the same underlying set and the topology whose closed subsets $C \subset X$ are the downward closed ones

$$\text{if } x \prec y \text{ in } X \text{ and } y \in C, \text{ then } x \in C. \qquad (5.8)$$

One verifies easily that:

- this defines an *Alexandrov topology* on X (which means that the family of closed subsets is closed under arbitrary intersections *and* unions),

- a monotone mapping $f: X \to Y$ becomes a continuous mapping $f: A(X) \to A(Y)$.

For a preordered set X, the unit $\eta_X: X \to WA(X)$ is the identity (because the closure of $x \in X$ in the space $A(X)$ is the subset $\downarrow x$). For a space X, the counit is

$$\varepsilon_X: AW(X) \to X, \qquad |\varepsilon_X| = 1_X, \qquad (5.9)$$

because $AW(X)$ has a finer topology than X: every closed subset C of X is downward closed for the specialisation preorder. The triangular identities are trivially satisfied.

Therefore A embeds pOrd in Top, as the full coreflective subcategory of spaces with Alexandrov topology, namely the coalgebraic objects of the adjunction (see 3.8.6–3.8.8).

5.1.6 Exercises and complements (Topology and order)

(a) The previous adjunction can be restricted to an adjunction

$$A \colon \mathsf{Ord} \rightleftarrows \mathsf{T_0Top} \colon W, \qquad A \dashv W, \qquad (5.10)$$

between ordered sets and T_0-spaces. Now A embeds Ord in $\mathsf{T_0Top}$, as the full coreflective subcategory of T_0-spaces with Alexandrov topology.

(b) This adjunction and the previous one (in 5.1.5) can be inserted in a commutative square of adjunctions (i.e. a commutative square in AdjCat, leaving apart questions of size)

$$
\begin{array}{ccc}
\mathsf{pOrd} & \xrightleftarrows[\;W\;]{\;A\;} & \mathsf{Top} \\[2pt]
{\scriptstyle F}\big\downarrow{\scriptstyle\,\vert\,U} & & {\scriptstyle T_0}\big\downarrow{\scriptstyle\,\vert\,U} \\[2pt]
\mathsf{Ord} & \xrightleftarrows[\;W\;]{\;A\;} & \mathsf{T_0Top}
\end{array}
\qquad (5.11)
$$

where the right adjoints are dashed. The adjunction $F \dashv U$ between ordered and preordered sets can be found in 3.2.3(d), while the adjunction $T_0 \dashv U$ between T_0-spaces and spaces is in 5.1.4(c).

(c) Consider the specialisation order of a T_1-space.

(d) Describe the topology of the space $A(\mathbb{R}, \leqslant)$ associated to the ordered real line.

(e) The Sierpinski two-point space, in 1.6.7, is the space $A(\mathbf{2})$ associated to the ordinal $\mathbf{2}$.

5.1.7 Exercises and complements (Metric spaces)

(a) Let us recall that a metric space is a set X equipped with a mapping d (or d_X), its distance (or metric), satisfying the following conditions, for all x, y, z in X

(i) $d \colon X \times X \to [0, +\infty[$,

(ii) $d(x, y) = 0 \;\Leftrightarrow\; x = y$ (*separation*),

(iii) $d(x, y) + d(y, z) \geqslant d(x, z)$ (*triangular inequality*),

(iv) $d(x, y) = d(y, x)$ (*symmetry*).

We have already remarked, in 1.1.5(a), that continuous mappings between metric spaces give a category which does not distinguish important properties of the latter. A mapping $f \colon X \to Y$ is said to be *Lipschitz* if it admits a Lipschitz constant $L \in \mathbb{R}$, such that

$$d_Y(f(x), f(x')) \leqslant L.d_X(x, x'), \qquad \text{for } x, x' \in X, \qquad (5.12)$$

and a *weak contraction* if it admits 1 as a Lipschitz constant, so that

$$d_Y(f(x), f(x')) \leqslant d_X(x, x'), \qquad \text{for } x, x' \in X.$$

A Lipschitz mapping $f \colon X \to Y$ is always continuous; actually, it is *uniformly continuous*:

- for every $\varepsilon > 0$ there is some $\delta > 0$ such that, if $d(x, x') < \delta$ in X, then $d(f(x), f(x')) < \varepsilon$ in Y.

We have thus the categories

- Mtr *of metric spaces and Lipschitz mappings,*

- Mtr$_1$ *of metric spaces and weak contractions,*

with obvious forgetful functors

$$\text{Mtr}_1 \subset \text{Mtr} \to \text{Hsd} \subset \text{Top},$$
$$\text{Ban} \to \text{Mtr}, \qquad \text{Ban}_1 \to \text{Mtr}_1. \tag{5.13}$$

(b) A metric space is *complete* if every Cauchy sequence converges in it. The full subcategory CplMtr $\subset$ Mtr of complete metric spaces is reflective, with reflector given by the completion functor.

The same holds for the full subcategory CplMtr$_1 \subset$ Mtr$_1$.

(c) The classical notion recalled in (a) has been generalised in various way. For a a *pseudometric* (as in Bourbaki [Bor3], Chapter IX) the conditions (i), (ii) are replaced by the following more general ones (while (iii) and (iv) are kept as above)

(i') $d \colon X \times X \to [0, +\infty]$,

(ii') $d(x, x) = 0$.

We shall see that the resulting categories

- psMtr *of pseudometric spaces and Lipschitz mappings,*

- psMtr$_1$ *of pseudometric spaces and weak contractions,*

have better categorical properties than the classical ones.

(Dropping also the symmetry axiom gives a further, important extension, which will be considered in 7.3.4.)

(d) Prove that the forgetful functors

$$U \colon \text{psMtr} \to \text{Set}, \qquad U \colon \text{psMtr}_1 \to \text{Set}, \tag{5.14}$$

have both adjoints $D \dashv U \dashv C$, which give to any set the *discrete pseudometric* (the largest) or the *indiscrete pseudometric* (the smallest). Therefore these functors U preserve (the existing) limits and colimits.

(e) Prove that psMtr$_1$ has all limits and colimits. The pseudometric of a coequaliser is not obvious, and interesting.

(f) For an *extended metric* we only replace (i) with (i'). Prove that the embedding U: eMtr$_1$ $\subset$ psMtr$_1$, defined on the category of extended metric spaces and weak contractions, has a left adjoint $F(X) = X/\sim$, by means of the obvious equivalence relation that turns a pseudometric into an extended metric.

(g) Use this left adjoint to construct all colimits in eMtr$_1$.

*(h) Uniform continuity is based on nearness of pairs of points, rather than nearness of a point to a fixed one. It can be defined on metric spaces, as above, but can also be defined for topological abelian groups, where – for every neighbourhood V of 0 – we can say that two points x, y are *near of order V* if $x - y \in V$.

The 'natural domain' of uniform continuity should extend the previous situations and be independent of auxiliary structures like the real field. This leads to the beautiful theory of 'uniform spaces'; a good reference is Bourbaki [Bor2], Chapter II.

5.1.8 Collapsing a pointed subspace

The quotient X/A of a pointed topological space modulo a pointed subspace (with the same base point) is often used in topology, and will repeatedly turn up in Sections 5.2 and 5.6.

The quotient X/A denotes the topological quotient of X modulo the equivalence relation that collapses the points of A

$$x \sim x' \quad \Leftrightarrow \quad (x, x' \in A \text{ or } x = x' \text{ in } X), \tag{5.15}$$

pointed at the class $[0_X] = A$. It comes equipped with a canonical projection

$$p \colon X \to X/A, \tag{5.16}$$

which is the coequaliser of the inclusion $i \colon A \to X$ and the zero morphism $0 \colon A \to X$. *(This is the same as the cokernel of i, see 5.6.1.)*

Exercises and complements. The solution is straightforward and left to the reader.

(a) In Top$_\bullet$, the space X/A can also be obtained as the pushout of the inclusion i and the (unique) map $A \to \{*\}$

$$
\begin{array}{ccc}
A & \xrightarrow{\ i\ } & X \\
\downarrow & & \downarrow{\scriptstyle u} \\
\{*\} & \xrightarrow[\ v\ ]{} & X/A
\end{array}
\tag{5.17}
$$

(b) In Top the meaning of X/A can (slightly) depend on the context, and this notation will not be used elsewhere, in this book.

If $A \neq \emptyset$, there is no doubt: X/A is defined as the quotient $X/\!\!\sim$ with respect to the equivalence relation (5.15), and *can* be pointed at the class A. Otherwise, according to the context, we may want to form a mere space and take $X/\emptyset = X$, or we may want to form a pointed space and take $X/\emptyset = X + \{*\}$, using the pushout (5.17), in Top.

*(c) For relative homology theories one should use the option $X/\emptyset = X$: see Section 6.7.

*(d) For the cone of a topological space one should use the other option, given by the pushout (5.17): see 5.2.5.

The euclidean spheres also 'agree' with this interpretation of a quotient.

In fact, the sphere $\mathbb{S}^n$ $(n > 0)$ can be viewed as the topological quotient $\mathbb{I}^n/\partial\mathbb{I}^n$ of the standard cube $\mathbb{I}^n$ modulo its boundary in $\mathbb{R}^n$. For $n = 0$ we have $\mathbb{I}^0 = \mathbb{R}^0 = \{*\}$ and $\partial\mathbb{I}^0 = \emptyset$; the discrete two-point space $\mathbb{S}^0 = \{*\} + \{*\}$ still amounts to the quotient $\mathbb{I}^0/\partial\mathbb{I}^0 = \{*\}/\emptyset$, as defined by the pushout (5.17).

(e) Forgetting topologies, the same 'quotients' are defined for pointed sets, and for sets.

5.1.9 *Complements* (Compact Hausdorff spaces as algebras)

The full subcategory CmpHsd $\subset$ Top of *compact Hausdorff spaces* has peculiar properties among the categories of topological spaces.

To begin with, CmpHsd is balanced (see 1.3.1), as a consequence of a classical theorem: any map $f \colon X \to Y$ from a compact space to a Hausdorff space is closed, and any bijective map of this kind is a homeomorphism. Moreover, an epimorphism has a dense image (by the same argument used for Hsd, in 2.3.6(b)), and must be surjective.

We now review some important properties of CmpHsd, without proofs (referred to [M4]).

The embedding $V \colon$ CmpHsd $\subset$ Top has a left adjoint

$$
\beta \colon \text{Top} \to \text{CmpHsd},
\tag{5.18}
$$

called the *Stone-Čech compactification* (see [M4], Section V.8).

The unit $\eta X \colon X \to V\beta(X)$ is a topological embedding if and only if X is a completely regular space. (Classically a 'compactification' was meant to be an embedding, and this procedure was only considered for completely regular spaces.)

The composed adjunction (where D equips a set with the discrete topology)

$$\mathsf{Set} \underset{W}{\overset{D}{\rightleftarrows}} \mathsf{Top} \underset{V}{\overset{\beta}{\rightleftarrows}} \mathsf{CmpHsd} \qquad\qquad \beta D \dashv WV, \qquad (5.19)$$

gives the left adjoint $\beta D \colon \mathsf{Set} \to \mathsf{CmpHsd}$ of the forgetful functor $U = WV \colon \mathsf{CmpHsd} \to \mathsf{Set}$.

The latter is monadic, so that a compact Hausdorff space can be viewed as an algebra $(X, a \colon TX \to X)$, where $T = U\beta D \colon \mathsf{Set} \to \mathsf{Set}$ is the monad associated to the adjunction (5.19) (see [M4], Section VI.9).

The existence of the left adjoint β can be proved by the Special Adjoint Functor Theorem (as in [M4]) or by some 'constructive' procedure which can be found in many books on General Topology (for instance in [Dg, Ke]).

Here we only sketch, without proofs, one of the most frequently used – based on the compact interval $[0, 1]$ of the euclidean line.

We consider the set $C = \mathsf{Top}(X, [0, 1])$ and the set

$$[0, 1]^C = \mathsf{Set}(C, [0, 1]).$$

The latter is given the product topology, which is Hausdorff compact, by Tychonoff's theorem. Its projections are written as $p_f \colon [0, 1]^C \to [0, 1]$, for $f \in C$.

Then one takes the mapping of *evaluation*

$$\mathrm{ev}_X \colon X \to [0, 1]^C,$$
$$\mathrm{ev}_X(x) \colon C \to [0, 1], \qquad\qquad \mathrm{ev}_X(x)(f) = f(x), \qquad (5.20)$$

which is continuous, because all its components are:

$$p_f.\mathrm{ev}_X = f \colon X \to [0, 1].$$

Now we let βX be the (compact) closure of the subset $\mathrm{ev}_X(X)$ in the space $[0, 1]^C$, and take the codomain-restriction of ev_X

$$\eta_X \colon X \to V\beta X, \qquad\qquad \eta_X(x) = \mathrm{ev}_X(x). \qquad (5.21)$$

The delicate point is proving the universal property of this arrow, from the space X to the functor V.

The interval $[0, 1]$ works in this construction because it is a *cogenerator* of the category CmpHsd: this means that if A and B are compact Hausdorff spaces, and $f, g \colon A \to B$ are distinct maps, there is a map $h \colon B \to [0,1]$ that distinguishes them: $hf \neq hg$.

5.2 Cylinder, cocylinder and homotopies

In Top the standard interval $\mathbb{I} = [0,1]$ gives the cylinder endofunctor $I = - \times \mathbb{I}$ and its right adjoint, the cocylinder or path endofunctor $P = (-)^{\mathbb{I}}$.

Homotopies between maps $X \to Y$ are equivalently defined as maps $IX \to Y$ or $X \to PY$. Their structure is produced by a rich structure on the standard interval, covariantly extended to the cylinder functor and contravariantly extended to the path functor.

Similar constructions can be made in the category Top. of pointed spaces.

(Many other situations where homotopies can be considered, classical or non-classical, have a similar framework, organised by a cylinder or cocylinder endofunctor, or both. Many of them are studied in [G8]; this includes cases where we should rather speak of 'directed homotopies', as Cat itself, or the category of preordered topological spaces that will be sketched in Section 5.7.)

In Top and Top. we also have the cone and suspension endofunctors; their right adjoints, the cocone and loop endofunctors, only exist in the pointed case.

We recall that the two-valued index α (or β) varies in the set $\{0,1\}$, also written as $\{-,+\}$.

5.2.1 The structure of the euclidean interval

A path in the space X is a map $a \colon \mathbb{I} \to X$ defined on the standard interval $\mathbb{I} = [0,1]$, with euclidean topology.

The basic, 'first degree' structure of $\mathbb{I}$ consists of four maps, linking $\mathbb{I}$ to its 0-th cartesian power, the singleton $\mathbb{I}^0 = \{*\}$

$$\partial^{\alpha} \colon \{*\} \to \mathbb{I}, \qquad \partial^-(*) = 0, \ \partial^+(*) = 1 \qquad (faces),$$

$$e \colon \mathbb{I} \to \{*\}, \qquad e(t) = * \qquad (degeneracy), \quad (5.22)$$

$$r \colon \mathbb{I} \to \mathbb{I}, \qquad r(t) = 1 - t \qquad (reversion).$$

We shall say that $(\mathbb{I}, \partial^{\alpha}, e)$ is an *internal interval* in Top, and $(\mathbb{I}, \partial^{\alpha}, e, r)$ an *internal interval with reversion*. (We shall use the same terminology replacing Top with any monoidal category, provided that $e\partial^{\alpha} = \mathrm{id}$ and moreover, in the reversible case: $rr = \mathrm{id}$, $er = e$, $\partial^- r = \partial^+$.)

Identifying a point x of the space X with the corresponding mapping $x \colon \{*\} \to X$, this basic structure determines:

(a) the endpoints of a path $a\colon \mathbb{I} \to X$, $a\partial^- = a(0)$ and $a\partial^+ = a(1)$,

(b) the trivial path at the point x, which will be written as $0_x = xe$,

(c) the reversed path of a, written as $-a = ar$.

Two consecutive paths $a, b\colon \mathbb{I} \to X$ (with $a\partial^+ = b\partial^-$, i.e. $a(1) = b(0)$) have a *concatenated path* $a + b$

$$(a + b)(t) = \begin{cases} a(2t), & \text{for } 0 \leqslant t \leqslant 1/2, \\ b(2t - 1), & \text{for } 1/2 \leqslant t \leqslant 1. \end{cases} \tag{5.23}$$

(This operation is not associative, and only works well up to homotopy with fixed endpoints, see 5.2.9.) Formally, we start from the (standard) *concatenation pushout*: pasting two copies of the interval, one after the other, is homeomorphic to $\mathbb{I}$ and can be realised as $\mathbb{I}$ itself

$$\begin{array}{ccc} \{*\} & \xrightarrow{\ \partial^+\ } & \mathbb{I} \\ {\scriptstyle \partial^-} \downarrow & & \downarrow {\scriptstyle c^-} \\ \mathbb{I} & \xrightarrow[\ c^+\ ]{} & \mathbb{I} \end{array} \qquad \begin{array}{l} c^-(t) = t/2, \\[2mm] c^+(t) = (t + 1)/2. \end{array} \tag{5.24}$$

Now, the concatenated path $a + b\colon \mathbb{I} \to X$ comes from the universal property of the pushout, and is characterised by the conditions

$$(a + b).c^- = a, \qquad (a + b).c^+ = b. \tag{5.25}$$

Finally, there is a 'second degree' structure which involves the standard square $\mathbb{I}^2 = [0, 1] \times [0, 1]$ and is used to construct homotopies of paths

$$
\begin{aligned}
g^-&\colon \mathbb{I}^2 \to \mathbb{I}, & g^-(t, t') &= \max(t, t') & &(\textit{lower connection}), \\
g^+&\colon \mathbb{I}^2 \to \mathbb{I}, & g^+(t, t') &= \min(t, t') & &(\textit{upper connection}), \\
s&\colon \mathbb{I}^2 \to \mathbb{I}^2, & s(t, t') &= (t', t) & &(\textit{transposition}).
\end{aligned}
\tag{5.26}
$$

Together with (5.22), these maps complete the structure of $\mathbb{I}$ as an *involutive lattice* in Top. The choice of the superscripts of g^-, g^+ comes from the fact that the unit of g^α is $\partial^\alpha(*)$. Within homotopy theory, the importance of these binary operations has been highlighted by R. Brown and P.J. Higgins [BroH1, BroH2], which introduced the term of *connection*, or *higher degeneracy*.

The reader will recall that a space X is said to be *path-connected* if any two points are linked by a path; this implies that X is connected. The *path-connected components* of a space X are the equivalence classes of the relation $x \sim x'$ given by the existence of a path in X linking these two points.

5.2.2 *Cylinder and homotopies*

Given two maps $f, g\colon X \to Y$ (in Top), a homotopy $\varphi\colon f \to g$ 'is' a map $\varphi\colon X \times \mathbb{I} \to Y$ defined on the cylinder $I(X) = X \times \mathbb{I}$, which coincides with f on the lower basis and with g on the upper one

$$\varphi(x, 0) = f(x), \quad \varphi(x, 1) = g(x) \qquad \text{for } x \in X. \qquad (5.27)$$

This map will be written as $\hat{\varphi}\colon X \times \mathbb{I} \to Y$ when we want to distinguish it from the homotopy $\varphi\colon f \to g$ which it represents.

All this is based on the *cylinder* endofunctor:

$$I\colon \mathsf{Top} \to \mathsf{Top}, \qquad I(X) = X \times \mathbb{I}, \qquad (5.28)$$

with a basic structure of four natural transformations, inherited from the structural maps of the standard interval and written as the latter

$$\begin{aligned}
\partial^\alpha &\colon X \to IX, & \partial^\alpha(x) &= (x, \alpha), & (faces), & \\
e &\colon IX \to X, & e(x, t) &= x & (degeneracy), & \qquad (5.29) \\
r &\colon IX \to IX, & r(x, t) &= (x, 1 - t) & (reversion). &
\end{aligned}$$

Identifying $I\{*\} = \{*\} \times \mathbb{I} = \mathbb{I}$, the structural maps of the standard interval coincide with the components of the transformations (5.29) on the singleton.

The transformations ∂^α, e, r give rise to the *faces* $\varphi\partial^\alpha$ of a homotopy, the *trivial homotopy* $0_f = fe$ of a map and the *reversed homotopy* $-\varphi = \varphi r$. (More precisely, the homotopy $-\varphi\colon g \to f$ is represented by the map $(-\varphi)^\hat{} = \hat{\varphi} r\colon IX \to Y$.)

A path $a\colon \mathbb{I} \to X$ is the same as a homotopy $a\colon x \to x'$ between its endpoints (always identifying $I\{*\} = \mathbb{I}$).

Two consecutive homotopies $\varphi\colon f \to g$ and $\psi\colon g \to h$ have a *concatenation*, or *vertical composition*

$$\varphi + \psi\colon f \to g, \qquad (5.30)$$

that extends the concatenation of paths (in (5.23)).

Again, this can be formally expressed noting that the *concatenation pushout of the cylinder* – pasting two copies of a cylinder IX, 'one on top of the other' – can be realised as the cylinder itself

$$\begin{array}{ccc}
X & \xrightarrow{\partial^+} & IX \\
{\scriptstyle\partial^-}\Big\downarrow & \swarrow{\scriptstyle c^-} & \Big\downarrow{\scriptstyle c^-} \\
IX & \xrightarrow[c^+]{} & IX
\end{array} \qquad
\begin{aligned}
& c^-(x, t) = (x, t/2), \\
& \\
& \qquad\qquad\qquad (5.31) \\
& \\
& c^+(x, t) = (x, (t + 1)/2).
\end{aligned}$$

Indeed, the subspaces

$$c^-(IX) = X \times [0, 1/2], \qquad c^+(IX) = X \times [1/2, 1],$$

form a finite closed cover of IX, so that a mapping defined on IX is continuous if and only if this holds for its restrictions to these subspaces.

The concatenated homotopy $\varphi + \psi \colon f \to h$ is represented by the map $(\varphi + \psi)\hat{\ } \colon IX \to Y$ which reduces to φ on c^-, and to ψ on c^+.

Here also, there is a 'second degree' structure, with three natural transformations which involve the iterated cylinder $I^2(X) = I(I(X)) = X \times \mathbb{I}^2$ and are used to construct higher homotopies (homotopies of homotopies)

$$
\begin{aligned}
g^\alpha \colon I^2 X \to IX, \quad & g^\alpha(x, t, t') = (x, g^\alpha(t, t')) \quad \text{(connections)}, \\
s \colon I^2 X \to I^2 X, \quad & s(x, t, t') = (x, t', t) \qquad \text{(transposition)}.
\end{aligned}
\tag{5.32}
$$

For instance, given a homotopy $\varphi \colon f^- \to f^+ \colon X \to Y$, we need the transposition to construct a homotopy $I\varphi \colon If^- \to If^+$

$$
\begin{aligned}
(I\varphi)\hat{\ } &= I(\hat{\varphi}).sX \colon I^2 X \to IY, \\
I(\hat{\varphi}).sX.\partial^\alpha(IX) &= I(\hat{\varphi}).I(\partial^\alpha X) = If^\alpha,
\end{aligned}
\tag{5.33}
$$

modifying the mapping $I(\hat{\varphi})$, *which does not have the correct faces.*

The fact that 'pasting two copies of the cylinder gives back the cylinder' is rather *peculiar of spaces.* For instance, it does not hold for chain complexes, where the concatenation of homotopies is based on a more general procedure: an assigned concatenation morphism $c \colon IX \to IX +_X IX$.

5.2.3 Cocylinder and homotopies

We conclude this brief review of the formal bases of classical homotopy recalling that a homotopy $\varphi \colon f \to g$, represented by a map $\hat{\varphi} \colon IX \to Y$, also has a dual representation as a map $\check{\varphi} \colon X \to PY$ with values in the *path-space* $PY = Y^{\mathbb{I}}$.

In fact, as we know from 5.1.1, the cylinder functor $I = - \times \mathbb{I}$ has a right adjoint

$$P \colon \mathsf{Top} \to \mathsf{Top}, \qquad P(Y) = Y^{\mathbb{I}}, \tag{5.34}$$

called the *cocylinder*, or *path functor*: PY is the space of paths $\mathbb{I} \to Y$, with the compact-open topology (also called the topology of uniform convergence on $\mathbb{I}$). The counit is also called *evaluation* (of paths)

$$\mathrm{ev} \colon PY \times \mathbb{I} \to Y, \qquad (a, t) \mapsto a(t). \tag{5.35}$$

The functor P inherits from the interval $\mathbb{I}$, contravariantly, a *dual* structure, which we write with the same symbols.

The structure consists of a basic, first degree part:

$$\partial^\alpha \colon PY \to Y, \qquad \partial^\alpha(a) = a(\alpha), \qquad\qquad (faces),$$

$$e \colon Y \to PY, \qquad e(y)(t) = y \qquad\qquad (degeneracy), \qquad (5.36)$$

$$r \colon PY \to PY, \qquad r(a)(t) = a(1-t) \qquad\qquad (reversion),$$

and a second degree part: two *connections* and a *transposition*

$$g^\alpha \colon PY \to P^2Y, \qquad g^\alpha(a)(t,t') = a(g^\alpha(t,t')),$$

$$s \colon P^2Y \to P^2Y, \qquad s(a)(t,t') = a(t',t). \qquad\qquad (5.37)$$

In this description, the faces of a homotopy $\check{\varphi} \colon X \to PY$ are defined as $\partial^\alpha \check{\varphi} \colon X \to Y$.

Concatenation of homotopies can now be performed with the *concatenation pullback* QY (which can be realised as *the object of pairs of consecutive paths*) and the concatenation map c

$$
\begin{array}{ccc}
QY & \xrightarrow{\;c^+\;} & PY \\
{\scriptstyle c^-}\downarrow & \nearrow & \downarrow{\scriptstyle \partial^-} \\
PY & \xrightarrow[\;\partial^+\;]{} & Y
\end{array}
\qquad (5.38)
$$

$$Q(Y) = \{(a,b) \in PY \times PY \mid \partial^+(a) = \partial^-(b)\},$$

$$c \colon QY \to PY, \qquad c(a,b) = a+b \qquad (concatenation\ map).$$

Again, as a property of topological spaces, the natural transformation c is invertible (by splitting a path into its two halves), and we can also realise QY as PY. (For chain complexes one has a non-invertible concatenation map.)

5.2.4 The homotopy category

Two parallel maps $f, g \colon X \to Y$ in Top are said to be *homotopic*, written as $f \simeq g$, if there exists a homotopy $\varphi \colon f \to g$. This is an equivalence relation in $\mathsf{Top}(X, Y)$, because of trivial homotopies, reversed homotopies and their concatenation.

Furthermore, homotopies have a *whisker composition* with maps

$$
X' \xrightarrow{\;h\;} X \underset{g}{\overset{f}{\underset{\Downarrow\varphi}{\rightrightarrows}}} Y \xrightarrow{\;k\;} Y' \qquad (5.39)
$$

$$k\varphi k \colon kfh \to kgh \colon X' \to Y',$$

where $k\varphi h$ is represented on the cylinder by the map $k.\check{\varphi}.I(h) \colon IX' \to Y'$

(or on the cocylinder by $P(k).\breve{\varphi}.h\colon X' \to PY'$). As a consequence, the homotopy relation is a congruence of categories in Top.

The quotient category $\mathsf{hoTop} = \mathsf{Top}/\simeq$ (already considered in 1.4.5) is called the *homotopy category of topological spaces*, and is important in Algebraic Topology. Its objects are the topological spaces, and a morphism $[f]\colon X \nrightarrow Y$ is a homotopy class of continuous mappings $X \to Y$.

An isomorphism in hoTop is the same as a *homotopy equivalence*: namely a pair of maps $f\colon X \rightleftarrows Y \colon g$ such that $gf \simeq \mathrm{id}X$ and $fg \simeq \mathrm{id}Y$. In particular, a subspace $A \subset X$ is a *deformation retract* of X if the inclusion $i\colon A \to X$ has a (continuous) retraction $p\colon X \to A$ such that $ip \simeq \mathrm{id}X$ (while, of course, $pi = \mathrm{id}A$).

As a basic notion of Algebraic Topology, a functor $F\colon \mathsf{Top} \to \mathsf{C}$ with values in an arbitrary category is said to be *homotopy invariant* if $F(f) = F(g)$ whenever $f \simeq g$ in Top, so that F induces a functor $\mathsf{hoTop} \to \mathsf{C}$. Typically, this is the case of the homology and homotopy functors with values in Ab, Gp, etc.

Exercises and complements. (a) All euclidean spaces $\mathbb{R}^n$ are contractible, i.e. homotopy equivalent to a singleton space.

(b) The sphere $\mathbb{S}^n$ is a deformation retract of the 'pierced space' $X = \mathbb{R}^{n+1} \setminus \{0\}$.

5.2.5 Cones and suspension for spaces

Let X be a topological space. Its *upper cone* C^+X is defined by the following pushout, where $\top = \{*\}$ is the terminal object

$$
\begin{array}{ccc}
X & \xrightarrow{\ p\ } & \{*\} \\
{\scriptstyle\partial^+}\downarrow & & \downarrow{\scriptstyle v^+} \\
IX & \xrightarrow[\ \gamma\ ]{} & C^+X
\end{array}
\qquad C^+X = (IX + \{*\})/(\partial^+X + \{*\}), \qquad (5.40)
$$

and the quotient is defined in 5.1.8 (viewing numerator and denominator as pointed at $*$). When X is not empty, the cone C^+X is the quotient IX/∂^+X, collapsing the upper basis of the cylinder to a point; otherwise $C^+(\emptyset) = \{*\}$.

The cone comes equipped with an *upper vertex* $v^+\colon \{*\} \to X$ and a homotopy

$$
\gamma\colon u^- \to v^+p\colon X \to C^+X, \qquad u^- = \gamma\partial^-\colon X \to C^+X, \qquad (5.41)
$$

between the embedding u^- of X as the (lower) basis of the cone and the constant mapping at the (upper) vertex. The points of C^+X will be written as equivalence classes $[x, s]$, with $[x, 1] = v^+ = v^+(*)$, for $x \in X$ and $s \in \mathbb{I}$ (adding the vertex $v^+ = v^+(*)$ when $X = \emptyset$).

The cone C^+X is always a *contractible space*, homotopy equivalent to the point, by the following homotopy between the identity and the constant endomap ending at its vertex

$$h \colon C^+X \times \mathbb{I} \to C^+X, \qquad h([x, s], t) = [x, s \vee t]. \tag{5.42}$$

The higher structure of the cylinder I gives rise to a monad (C, u, g) on the upper cone functor $C = C^+$. The unit is the lower basis $u \colon 1 \to C$ and the multiplication $g \colon C^2 \to C$ is induced by $g^- \colon \mathbb{I}^2 \to \mathbb{I}$. (A proof can be found in [G8], Theorem 4.8.3.)

Symmetrically, one has the *lower cone*

$$C^-X = (IX + \{*\})/(\partial^-X + \{*\}),$$

with a lower vertex v^- and an upper basis u^+.

The *suspension* ΣX is defined by a colimit (of the solid diagram)

$$\tag{5.43}$$

which collapses, independently, the bases of IX to a lower and an upper vertex, v^- and v^+; or adds these points when X is empty: $\Sigma(\emptyset) \cong \mathbb{S}^0$.

The suspension ΣX can also be obtained by the following pushout: the pasting of both cones $C^\alpha X$ of X, along their bases

$$\tag{5.44}$$

(Again, this coincidence is *not* a 'general' fact, in homotopy theory, but rests on the 'pasting property' of the topological cylinder.)

It is not difficult to prove that $\Sigma \mathbb{S}^n$ is homeomorphic to $\mathbb{S}^{n+1}$ (see Exercise 5.2.7(a)), so that all spheres are constructed over the 0-dimensional one: $\mathbb{S}^n \cong \Sigma^n \mathbb{S}^0$.

As we have seen, the cone and suspension functors of Top do not preserve the initial object $\bot = \emptyset$ (nor sums, since the result is always a connected space); therefore they do not have right adjoints.

But in the category Top. of pointed spaces, the cylinder, cone and suspension functors have each a right adjoint, described below.

5.2.6 Pointed cylinder, cone and suspension

In Top. one works with *pointed homotopies*, which preserve the base point. They are represented by the *pointed cylinder* endofunctor

$$I \colon \mathsf{Top}_\bullet \to \mathsf{Top}_\bullet, \qquad I(X, x_0) = (IX/I\{x_0\}, [x_0, t]), \tag{5.45}$$

defined as the quotient of the unpointed cylinder IX which collapses the fibre of the base point x_0 (providing thus the new base point).

The upper cone and suspension are defined by the same colimits as in (5.40) and (5.43), in the category Top.. They can also be obtained from the corresponding unpointed constructions, by collapsing the fibre of the base point

$$C^+(X, x_0) = I(X, x_0)/\partial^+(X, x_0) = (C^+(X)/I\{x_0\}, [x_0, t]),$$
$$\Sigma(X, x_0) = ((\Sigma X)/I\{x_0\}, [x_0, t]). \tag{5.46}$$

5.2.7 Exercises and complements (Suspension of a sphere)

(a) In Top, the suspension of a sphere gives

$$\Sigma \mathbb{S}^n \cong \mathbb{S}^{n+1} \qquad\qquad (n \geqslant 0), \tag{5.47}$$

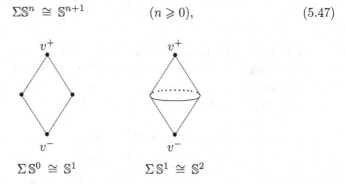

$$\Sigma \mathbb{S}^0 \cong \mathbb{S}^1 \qquad\qquad \Sigma \mathbb{S}^1 \cong \mathbb{S}^2$$

Hints. Project the cylinder $I(\mathbb{S}^n)$ onto $\mathbb{S}^{n+1}$, to obtain an induced homeomorphism as above.

*(b) This is also true for the pointed suspension, in Top.. Choosing (for instance) the base-point $x_n = (1, 0, ..., 0) \in \mathbb{S}^n$, we have

$$\Sigma(\mathbb{S}^n, x_n) \cong (\mathbb{S}^{n+1}, x_{n+1}) \qquad\qquad (n \geqslant 0). \tag{5.48}$$

Giving an analytic expression of this homeomorphism is more complicated than in the unpointed case. The reader might be satisfied with an intuitive idea, based on the previous exercise. (The solution in Chapter 8 is kept at this level.)

5.2.8 Pointed cocylinder, cocone and loop functor

The pointed cylinder, cone and suspension of 5.2.6 have a right adjoint

$$P\colon \mathsf{Top.} \to \mathsf{Top.}, \qquad I \dashv P \qquad (path\ functor),$$

$$E^+\colon \mathsf{Top.} \to \mathsf{Top.}, \qquad C^+ \dashv E^+ \qquad (upper\ cocone), \qquad (5.49)$$

$$\Omega\colon \mathsf{Top.} \to \mathsf{Top.}, \qquad \Sigma \dashv \Omega \qquad (loop\ functor).$$

The pointed path object $P(Y, y_0)$ is simply the unpointed path space $P(Y)$, pointed at the constant path $0_{y_0}(t) = y_0$

$$P(Y, y_0) = (P(Y), 0_{y_0}). \qquad (5.50)$$

The adjunction $I \dashv P$ is induced by the corresponding adjunction of topological spaces

$$\varphi\colon \mathsf{Top.}(I(X, x_0), (Y, y_0)) \to \mathsf{Top.}((X, x_0), P(Y, y_0)),$$
$$\varphi(f)\colon (X, x_0) \to P(Y, y_0), \qquad \varphi(f)(x)(t) = f[x, t], \qquad (5.51)$$

since a map $f\colon I(X, x_0) \to (Y, y_0)$ is the same as a map $f'\colon IX \to Y$ that sends the fibre $I\{x_0\}$ to y_0.

Finally, the upper cocone and the loop object in $\mathsf{Top.}$ are constructed as limits (of the solid diagrams below)

$$
\begin{array}{ccc}
& \Omega(Y, y_0) \dashrightarrow 0 & \\
E^+(Y, y_0) \dashrightarrow 0 & \Big\downarrow \ \searrow^{\omega} & \Big\downarrow \\
\Big\downarrow \quad \Big\downarrow & \quad P(Y, y_0) \longrightarrow (Y, y_0) & (5.52) \\
P(Y, y_0) \xrightarrow[\partial^+]{} (Y, y_0) & \Big\downarrow & \\
& 0 \longrightarrow (Y, y_0) &
\end{array}
$$

$$E^+(Y, y_0) = (\{(y, b) \in Y \times PY \mid b(0) = y, b(1) = y_0\}, (y_0, 0_{y_0})),$$

$$\Omega(Y, y_0) = (\{b \in PY \mid b(0) = y_0 = b(1)\}, 0_{y_0}),$$

with the topology of a subspace of $Y \times PY$ or PY, respectively.

$\Omega(Y, y_0)$ is indeed the space of loops of Y at the base point, with compact-open topology.

5.2.9 *The fundamental groupoid

The reader has likely seen the construction of the fundamental group $\pi_1(X, x)$ of a pointed space, where an element is the class $[a]$ of a loop $a \in \Omega(X, x)$, *up to homotopy with fixed endpoints.*

The operation is induced by path-concatenation:

$$[a] * [b] = [a + b],$$

which becomes a group law in the quotient: the unit is the class $[0_x]$ of the constant loop at x, and the inverse is obtained by reversing a representative path. (We do not review here the easy but tiresome construction of the homotopies required to prove all this.)

This gives a functor

$$\pi_1 \colon \mathsf{Top}_{\bullet} \to \mathsf{Gp}, \tag{5.53}$$

where a pointed map $f \colon (X, x) \to (Y, y)$ is sent to the homomorphism

$$f_* \colon \pi_1(X, x) \to \pi_1(Y, y), \qquad f_*([a]) = [fa].$$

Working on an unpointed space X, one gets a small groupoid $\Pi_1(X)$ whose objects (or vertices) are the points of X, while an arrow $[a] \colon x \to y$ is a class of paths $a \colon \mathbb{I} \to X$ from x to y, *up to homotopy with fixed endpoints*. Composition with a consecutive arrow $[b] \colon y \to z$ is again by concatenation: $[a] * [b] = [a + b]$.

This becomes indeed a groupoid law in the quotient; the proof (for associativity, identities and inverses) is a straightforward extension of the corresponding proof for the fundamental group.

We have now a functor with values in the category Gpd of *small groupoids and functors*

$$\Pi_1 \colon \mathsf{Top} \to \mathsf{Gpd}. \tag{5.54}$$

A map $f \colon X \to Y$ is sent to the functor

$$f_* \colon \Pi_1(X) \to \Pi_1(Y), \qquad f_*(x) = f(x), \quad f_*([a]) = [fa].$$

For every $x \in X$, the fundamental group $\pi_1(X, x)$ is the group of automorphisms $\Pi_1(X)(x, x)$ of $\Pi_1(X)$, at the vertex x.

Exercises and complements. (a) (*Homotopy equivariance*) The functor Π_1 has a canonical enrichment: a homotopy $\varphi \colon f \to g \colon X \to Y$ gives a homotopy of groupoids, i.e. an (invertible) natural transformation

$$\varphi_* \colon f_* \to g_* \colon \Pi_1(X) \to \Pi_1(Y). \tag{5.55}$$

(b) (*Corollary*) A homotopy equivalence of spaces is turned by Π_1 into a categorical equivalence of groupoids.

5.3 Simplicial sets

After describing simplicial objects by faces and degeneracies, we show that every topological space S has an associated *singular simplicial set* $\mathrm{Smp}(S)$ (in 5.3.4), which will be used to construct the singular homology groups of S (in Section 5.5).

On the other hand, a simplicial set can be viewed as a combinatorial description of a topological space of a rather simple type. This space is produced by the geometric realisation functor $\mathcal{R}\colon \mathsf{SmpSet} \to \mathsf{Top}$ (in 5.3.5), by pasting tetrahedra of any dimension, according to the rules 'written' in X. The functor $\mathcal{R}$ is left adjoint to the singular simplicial functor $\mathrm{Smp}\colon \mathsf{Top} \to \mathsf{SmpSet}$.

5.3.1 Simplicial objects

As already said in 1.6.3, a *simplicial object* in the category C is a functor $X\colon \underline{\Delta}^{\mathrm{op}} \to \mathsf{C}$ defined on the category $\underline{\Delta}$ of finite positive ordinals (and increasing maps). A *morphism of simplicial objects* $f\colon X \to Y$ is a natural transformation of such functors; they form the category $\mathrm{Smp}(\mathsf{C}) = \mathsf{C}^{\underline{\Delta}^{\mathrm{op}}}$ of simplicial objects in C.

In particular we have the categories SmpSet of *simplicial sets*, SmpGp of *simplicial groups*, etc. The category $\underline{\Delta}$, as the 'domain' of simplicial objects, is also called the *simplicial site*.

We also consider the *augmented simplicial site* $\underline{\Delta}^{\sim}$ of all finite ordinals; an *augmented simplicial object* is a functor $X\colon (\underline{\Delta}^{\sim})^{\mathrm{op}} \to \mathsf{C}$.

In [M4], Section VII.5, the main subject is the category $\underline{\Delta}^{\sim}$, written as $\underline{\Delta}$, while the subcategory $\underline{\Delta}$ is written as $\underline{\Delta}^{+}$. Here we follow the notation used in Topology. Simplicial objects in [M4] are the same as here, the classical ones.

5.3.2 Faces and degeneracies

The simplicial site $\underline{\Delta}$ can be described by simple generators.

$\underline{\Delta}$ has objects $[n] = \{0, ..., n\}$, starting with the singleton $[0] = \{0\}$. We are interested in the following morphisms. The *face* δ_i (or δ_{ni})

$$\delta_i\colon [n-1] \to [n],$$
$$0, ..., i-1, i, ..., n-1 \;\mapsto\; 0, ..., i-1, i+1, ..., n \quad (0 \leqslant i \leqslant n),$$

(5.56)

is the injective map which misses i; the *degeneracy* ε_i (or ε_{ni}) is the surjective map which repeats i

$$\varepsilon_i\colon [n] \to [n-1],$$
$$0, ..., i, i+1, ..., n \;\mapsto\; 0, ..., i, i, ..., n-1 \quad (0 \leqslant i \leqslant n-1).$$

(5.57)

These mappings (also called *cofaces* and *codegeneracies*) satisfy the *cosimplicial relations*

$$\delta_j \delta_i = \delta_{i+1} \delta_j, \qquad \text{for } j \leqslant i,$$

$$\varepsilon_i \varepsilon_j = \varepsilon_j \varepsilon_{i+1}, \qquad \text{for } j \leqslant i,$$

$$\varepsilon_j \delta_i = \begin{cases} \delta_{i-1} \varepsilon_j, & \text{for } j < i - 1, \\ \text{id}, & \text{for } j = i - 1, i, \\ \delta_i \varepsilon_{j-1}, & \text{for } j > i. \end{cases} \qquad (5.58)$$

Plainly, every injective increasing mapping $\delta \colon [m] \to [n]$ can be uniquely factorised as a composite of faces, indexed by $\{i_1, ..., i_k\} = [n] \setminus \delta([m])$, the missing elements of its image:

$$\delta = \delta_{i_k} ... \delta_{i_1} \colon [m] \to [n] \qquad (0 \leqslant i_1 < i_2 < ... < i_k \leqslant n). \qquad (5.59)$$

Every surjective increasing mapping $\varepsilon \colon [m] \to [n]$ can be uniquely factorised as a composite of degeneracies, indexed by the 'stationary points' j, where $\varepsilon(j) = \varepsilon(j+1)$:

$$\varepsilon = \varepsilon_{j_1} ... \varepsilon_{j_h} \colon [m] \to [n] \qquad (0 \leqslant j_1 < j_2 < ... < j_h \leqslant m - 1). \qquad (5.60)$$

Finally an increasing mapping $\varphi \colon [m] \to [n]$ has a unique epi-mono factorisation $\varphi = \delta \varepsilon$, as a composite

$$\varphi = \delta_{i_k} ... \delta_{i_1} \varepsilon_{j_1} ... \varepsilon_{j_h}$$
$$(0 \leqslant j_1 < j_2 < ... < j_h < m, \quad 0 \leqslant i_1 < i_2 < ... < i_h \leqslant n), \qquad (5.61)$$

with $m - h = n - k$ (the cardinal of the image of φ).

All this shows that the category $\underline{\Delta}$ is generated by faces and degeneracies, under the cosimplicial relations (applying the Special Form Lemma 5.3.7).

Similar facts hold for the augmented simplicial site $\underline{\Delta}^{\sim}$, adding a new face $\delta_0 \colon \emptyset \to [0]$ to the generators (and in the cosimplicial relations).

5.3.3 Simplicial objects by faces and degeneracies

As a consequence, a simplicial object $X \colon \underline{\Delta}^{\mathrm{op}} \to \mathsf{C}$ can be equivalently assigned as a family $(X_n)_{n \geqslant 0}$ of objects of C, linked by *faces* and *degeneracies*

$$\partial_i \colon X_n \to X_{n-1} \qquad (0 \leqslant i \leqslant n),$$
$$e_i \colon X_{n-1} \to X_n \qquad (0 \leqslant i \leqslant n - 1), \qquad (5.62)$$

$$X_0 \underset{e_0}{\overset{\partial_i}{\rightleftarrows}} X_1 \underset{e_i}{\overset{\partial_i}{\rightleftarrows}} X_2 \overset{\partial_i}{\underset{e_i}{\rightleftarrows}} X_3 \, ...$$

that satisfy the *simplicial relations* (dual to the cosimplicial ones of (5.58))

$$\partial_i \partial_j = \partial_j \partial_{i+1}, \qquad \text{for } j \leqslant i,$$

$$e_j e_i = e_{i+1} e_j, \qquad \text{for } j \leqslant i,$$

$$\partial_i e_j = \begin{cases} e_j \partial_{i-1}, & \text{for } j < i-1, \\ \mathrm{id}, & \text{for } j = i-1, i, \\ e_{j-1} \partial_i, & \text{for } j > i. \end{cases} \qquad (5.63)$$

When useful we write the faces as $\partial_{ni} \colon X_n \to X_{n-1}$.

5.3.4 Singular simplices

The category $\underline{\Delta}$ can be embedded in Top, forming the cosimplicial object of the *standard simplices* Δ^n (with euclidean topology), for $n \geqslant 0$:

$$J \colon \underline{\Delta} \to \mathsf{Top},$$

$$[n] \;\mapsto\; \Delta^n = \{(t_0, ..., t_n) \in \mathbb{R}^{n+1} \mid t_i \geqslant 0, \; \Sigma t_i = 1\},$$

$$\delta_i \colon \Delta^{n-1} \to \Delta^n \qquad (0 \leqslant i \leqslant n),$$

$$\delta_i(t_0, ..., t_{n-1}) = (t_0, ..., t_{i-1}, 0, t_i, ..., t_{n-1}), \qquad (5.64)$$

$$\varepsilon_i \colon \Delta^n \to \Delta^{n-1} \qquad (0 \leqslant i \leqslant n-1),$$

$$\varepsilon_i(t_0, ..., t_i, ..., t_n) = (t_0, ..., t_i + t_{i+1}, ..., t_n),$$

To a topological space S we associate a *singular simplicial set* $\mathrm{Smp}(S)$, obtained by composing the functor $J \colon \underline{\Delta} \to \mathsf{Top}$ with the contravariant functor $\mathsf{Top}(-, S) \colon \mathsf{Top}^{\mathrm{op}} \to \mathsf{Set}$

$$\mathrm{Smp}(S) \colon \underline{\Delta}^{\mathrm{op}} \to \mathsf{Set}, \qquad (\mathrm{Smp}(S))_n = \mathsf{Top}(\Delta^n, S),$$

$$\partial_i(a) = a\delta_i \colon \Delta^{n-1} \to S \qquad (a \in \mathsf{Top}(\Delta^n, S)), \qquad (5.65)$$

$$e_i(a) = a\varepsilon_i \colon \Delta^n \to S \qquad (a \in \mathsf{Top}(\Delta^{n-1}, S)).$$

Globally we have the *singular simplicial* functor

$$\mathrm{Smp} \colon \mathsf{Top} \to \mathsf{SmpSet}, \qquad (5.66)$$

which is used to define the singular homology of a space (see Section 5.5).

5.3.5 The geometric realisation

The *geometric realisation* of a simplicial set $X \colon \underline{\Delta}^{\mathrm{op}} \to \mathsf{Set}$ is a topological space $\mathcal{R}(X)$ constructed by pasting a standard simplex $\Delta^n(x)$ for every $x \in X_n$, under identifications induced by the faces and degeneracies of X.

More precisely, we form the topological sum

$$S = \Sigma_n \, DX_n \times \Delta^n$$

(where $D \colon \mathsf{Set} \to \mathsf{Top}$ equips a set with the discrete topology), and we define the space $\mathcal{R}(X) = S/R$, as the quotient of S modulo the equivalence relation R spanned by the following identifications

$$
\begin{aligned}
&[x, \delta_i t] = [\partial_i x, t] && (x \in X_n,\ t \in \Delta^{n-1}), \\
&[x, \varepsilon_i t] = [e_i x, t] && (x \in X_{n-1},\ t \in \Delta^n).
\end{aligned}
\tag{5.67}
$$

The canonical map $u_n \colon DX_n \times \Delta^n \to S/R$ is the restriction of the canonical projection $S \to S/R$.

We have now, for any simplicial set X, a canonical morphism of simplicial sets

$$\eta X \colon X \to \mathrm{Smp}(\mathcal{R}(X)), \qquad (\eta X)_n(x) = u_n(x, -) \colon \Delta^n \to \mathcal{R}(X). \tag{5.68}$$

Finally, $(\mathcal{R}(X), \eta X)$ is a universal arrow from X to the functor $\mathrm{Smp} \colon \mathsf{Top} \to \mathsf{SmpSet}$, and the geometric realisation is its left adjoint

$$\mathcal{R} \colon \mathsf{SmpSet} \to \mathsf{Top}, \qquad \mathcal{R} \dashv \mathrm{Smp}. \tag{5.69}$$

Remarks. The space $\mathcal{R}(X) = S/R$ is the coend (see 2.8.6)

$$\mathcal{R}(X) = \int^n DX_n \times \Delta^n, \tag{5.70}$$

of the composed functor

$$\underline{\Delta}^{\mathrm{op}} \times \underline{\Delta} \xrightarrow{\ DX \times J\ } \mathsf{Top} \times \mathsf{Top} \xrightarrow{\ -\times-\ } \mathsf{Top}$$

produced by $DX \colon \underline{\Delta}^{\mathrm{op}} \to \mathsf{Top}$ and the embedding $J \colon \underline{\Delta} \to \mathsf{Top}$ of (5.64), which realises $[n]$ as the standard simplex $\Delta^n \subset \mathbb{R}^{n+1}$.

5.3.6 Exercises and complements

(a) (*Faces and degeneracies as adjoints*) The unbounded chain of adjunctions of the ordered integral line, in 1.7.7(e), can be restricted to finite ordinals, giving a finite chain (involving $n + 1$ faces, n degeneracies and $2n$ adjunctions)

$$\delta_i \colon [n-1] \rightleftarrows [n] : \varepsilon_j \qquad (0 \leqslant i \leqslant n, \ 0 \leqslant j \leqslant n-1),$$

$$\delta_n \dashv \varepsilon_{n-1} \dashv \ \ldots \ \delta_{i+1} \dashv \varepsilon_i \dashv \delta_i \dashv \ \ldots \ \varepsilon_0 \dashv \delta_0, \qquad (5.71)$$

$$\varepsilon_i \delta_i = \mathrm{id} = \varepsilon_{i-1} \delta_i, \qquad \delta_{i+1} \varepsilon_i \leqslant \mathrm{id}, \qquad \delta_i \varepsilon_i \geqslant \mathrm{id}.$$

(b) The category $\underline{\Delta}^{\sim}$ of all finite ordinals $n = \{0, ..., n-1\}$ has a strict monoidal structure (see Section 3.4) given by the ordinal sum $m + n$, with identity 0.

*(c) The interested reader can see in [M4], Section VII.5, that the terminal object 1 has a unique structure of internal monoid in $\underline{\Delta}^{\sim}$ (in the sense of 3.4.8). The latter can be characterised as *the universal strict monoidal category equipped with an internal monoid.*

*(d) The augmented *symmetric simplicial site* $!\underline{\Delta}^{\sim}$ is the category *of finite cardinals* (equivalent to the category of finite sets). It also has a description by generators and relations, and can be characterised as *the universal strict monoidal category equipped with an internal symmetric monoid*: see [G5].

5.3.7 *Lemma* (Special Form Lemma)

Let S be a category generated by a subgraph G, whose maps satisfy in S a system of relations Φ.

Then S is freely generated by G under these relations if and only if every S-map can be expressed in a unique special form $f = g_m ... g_1$, as a composite of G-maps, and every G-factorisation $f = g'_m ... g'_1$ in S can be made special by applying the relations Φ, finitely many times.

Proof The necessity of this condition is easily proved by choosing, arbitrarily, one special form in each equivalence class of the quotient of the free category generated by G, under the congruence generated by Φ.

Conversely, we take a graph-morphism $F_0 \colon G \to \mathsf{C}$, with values in an arbitrary category, that satisfies the system of relations Φ; this morphism extends to at most one functor $F \colon \mathsf{S} \to \mathsf{C}$, letting it operate on special forms $F(g_m ... g_1) = F_0(g_m) ... F_0(g_1)$. This construction defines indeed a functor, since any composite $f' f$ in S can be rewritten in special form using relations which are preserved in C. $\square$

5.3.8 Further comments on horror vacui

Finally, let us remark that the categories $\Delta^{\sim}$ and $!\Delta^{\sim}$, of *all* finite ordinals or cardinals, have substantial categorical properties and simple characterisations (recalled in 5.3.6).

Taking out the initial object (the empty ordinal or cardinal) destroys these properties, but gives two categories – namely Δ and $!\Delta$ – which have an important auxiliary role in Combinatorial and Algebraic Topology, as the sites of $\mathsf{Smp}(\mathsf{C})$ and its symmetric version $!\mathsf{Smp}(\mathsf{C})$.

We can therefore refine our comments on *horror vacui*, in 1.1.6: artificial exclusions *can* be of interest, for a category S which is used in an 'auxiliary role' – typically as a domain of functors $X \colon \mathsf{S} \to \mathsf{C}$ with values in a (generally 'well behaved') category C.

In the same perspective we can also note that, when discussing limits in C, *we do not expect the domain of a functor* $\mathsf{S} \to \mathsf{C}$ *to have good categorical properties*; this can even be a hindrance: for instance, if S has an initial object the limit of each functor defined on S is trivial (as we have observed in 2.8.1).

5.4 Cubical sets

After simplicial sets, cubical sets have also been studied in Algebraic Topology, in a parallel way. They are 'modelled' on the standard euclidean cubes $\mathbb{I}^n$, much in the same way as simplicial sets are modelled on the euclidean tetrahedra Δ^n.

The main advantage with respect to the simplicial framework comes from the fact that the cubes $\mathbb{I}^n$ are closed under cartesian products (in Top), while the products of tetrahedra have to be covered with tetrahedra. (The advantage will be evident in the proof of the homotopy invariance of singular homology, in Theorem 5.5.7.)

Here we describe cubical objects as functors $X \colon \underline{\mathbb{I}}^{\mathrm{op}} \to \mathsf{C}$ defined on the basic cubical site, that was introduced and characterised in [GM]. Various extensions, studied in this paper, are mentioned at the end of this section.

Also here the two-valued index α (or β) varies in the cardinal $2 = \{0, 1\} = \{-, +\}$. The singleton $1 = 2^0$ is also written as $\{*\}$.

5.4.1 The basic cubical site

Let $\underline{\mathbb{I}}$ be the subcategory of Set consisting of the *elementary cubes* $2^n = \{0, 1\}^n$ (for $n \geqslant 0$) together with the mappings $f \colon 2^m \to 2^n$ which delete some coordinates and insert some 0's and 1's (without modifying the order of the remaining coordinates).

$\underline{\mathrm{I}}$ is a strict symmetric monoidal category; its tensor product $2^m \square 2^n = 2^{m+n}$ is induced by the cartesian product of Set, but is no longer a cartesian product in the subcategory (where the diagonal $2 \to 2 \times 2$ of Set is not present); the exponent of 2^n denotes now a tensor power.

The object $2 = \{0, 1\}$ is an *internal interval* (both in (Set, $\times$) and $(\underline{\mathrm{I}}, \square)$), with (elementary) *faces* δ^α and *degeneracy* ε (as in 5.2.1, with slightly different notation)

$$\delta^\alpha \colon 1 \to 2, \qquad \varepsilon \colon 2 \to 1, \qquad \varepsilon . \delta^\alpha = \mathrm{id},$$
$$\delta^\alpha(*) = \alpha \qquad (\alpha = 0, 1). \tag{5.72}$$

Higher faces and degeneracies are derived from the previous ones, via the monoidal structure, for $1 \leqslant i \leqslant n$ and $\alpha = 0, 1$

$$\delta_{ni}^\alpha = 2^{i-1} \square \delta^\alpha \square 2^{n-i} \colon 2^{n-1} \to 2^n,$$
$$\delta_{ni}^\alpha(t_1, ..., t_{n-1}) = (t_1, ..., t_{i-1}, \alpha, t_i, ..., t_{n-1}),$$
$$\varepsilon_{ni} = 2^{i-1} \square \varepsilon \square 2^{n-i} \colon 2^n \to 2^{n-1},$$
$$\varepsilon_{ni}(t_1, ..., t_n) = (t_1, ..., t_{i-1}, t_{i+1}, ..., t_n). \tag{5.73}$$

The *cocubical relations* follow easily from the formulas (5.72) (omitting the index n):

$$\delta_j^\beta \, \delta_i^\alpha = \delta_{i+1}^\alpha \, \delta_j^\beta, \qquad \text{for } j \leqslant i,$$
$$\varepsilon_i \, \varepsilon_j = \varepsilon_j \, \varepsilon_{i+1}, \qquad \text{for } j \leqslant i,$$
$$\varepsilon_j \, \delta_i^\alpha = \begin{cases} \delta_{i-1}^\alpha \, \varepsilon_j, & \text{for } j < i, \\ \mathrm{id}, & \text{for } j = i, \\ \delta_i^\alpha \, \varepsilon_{j-1}, & \text{for } j > i. \end{cases} \tag{5.74}$$

5.4.2 Lemma (The canonical form of the cubical site)

Using the cocubical relations (5.74) *as* rewriting rules, *each composite in* Set *of the cubical faces and degeneracies can be rewritten in a unique canonical factorisation (empty for an identity)*

$$\delta_{j_s}^{\alpha_s} \, ... \, \delta_{j_1}^{\alpha_1} \varepsilon_{i_1} \, ... \, \varepsilon_{i_r} \colon 2^m \to 2^{m-r} \to 2^n,$$
$$1 \leqslant i_1 < ... < i_r \leqslant m, \qquad 1 \leqslant j_1 < ... < j_s \leqslant n, \tag{5.75}$$
$$m - r = n - s \geqslant 0,$$

consisting of a surjective composed degeneracy $\varepsilon_{i_1} ... \varepsilon_{i_r}$ *(that deletes the co-ordinates specified by the indices) and an injective composed face* $\delta_{j_s}^{\alpha_s} ... \delta_{j_1}^{\alpha_1}$ *(that inserts 0's and 1's in the specified positions).*

Proof Obvious.   □

5.4.3 Theorem (The basic cubical site)

The category $\underline{I}$ *can be characterised in the following equivalent ways:*

(a) the subcategory of Set *with objects* 2^n, *generated by all* faces *and degeneracies (5.73),*

(b) the subcategory of Set *with objects* 2^n, *closed under the binary-product functor (realised as* $2^m \times 2^n = 2^{m+n}$), *and generated by the* basic faces *(*$\delta^\alpha \colon 1 \to 2$) *and degeneracy (*$\varepsilon \colon 2 \to 1$),*

(c) the category generated by the graph (5.73), subject to the cocubical relations (5.74),

(d) the free strict monoidal category with an assigned internal interval, namely $(2, \delta^\alpha, \varepsilon)$.

Proof The characterisation (a) is obvious: every map of $\underline{I}$ can clearly be factorised as in (5.75), in a unique way. Therefore, (b) follows from the construction of higher faces and degeneracies as tensor products, in (5.73), while (c) follows from the Special Form Lemma, in 5.3.7.

For (d), let $A = (A, \otimes, E)$ be a strict monoidal category with an assigned internal interval $(A, \delta^\alpha, \varepsilon)$. We write a tensor power as $(-)^n$.

Defining the higher faces and degeneracies of A as above, in (5.73)

$$\delta^\alpha_{ni} = A^{i-1} \otimes \delta^\alpha \otimes A^{n-i} \colon A^{n-1} \to A^n,$$
$$\varepsilon_{ni} = A^{i-1} \otimes \varepsilon \otimes A^{n-i} \colon A^n \to A^{n-1}, \tag{5.76}$$

the cocubical relations are satisfied, and there is a unique functor $F \colon \underline{I} \to A$ sending 2^n to A^n and preserving higher faces and degeneracies.

It is now sufficient to prove that this F is strictly monoidal: then, it will be the unique such functor sending 2 to A and preserving $\delta^\alpha, \varepsilon$. As we already know that F is a functor, our result follows from the following formulas

$$F(2^p \,\square\, 2^q) = F(2^{p+q}) = A^{p+q} = A^p \otimes A^q,$$
$$F(2^{i-1} \,\square\, \delta^\alpha \,\square\, 2^{n-i}) = F(\delta^\alpha_{ni}) = \delta^\alpha_{ni} = A^{i-1} \otimes \delta^\alpha \otimes A^{n-i}, \tag{5.77}$$
$$F(2^{i-1} \,\square\, \varepsilon \,\square\, 2^{n-i}) = F(\varepsilon_{ni}) = \varepsilon_{ni} = A^{i-1} \otimes \varepsilon \otimes A^{n-i},$$

since the tensor product of two arbitrary $\underline{I}$-maps $f = f_r...f_1$ and $g = g_s...g_1$, both in canonical form, can be decomposed as

$$f \,\square\, g = (f_r \,\square\, 1)...(f_1 \,\square\, 1).(1 \,\square\, g_s)...(1 \,\square\, g_1).   □$$

5.4.4 Cubical objects

A *cubical object* in the category C is a presheaf $X\colon \underline{\mathbb{I}}^{\mathrm{op}} \to \mathsf{C}$ on the cubical site. A *morphism of cubical objects* $f\colon X \to Y$ is a natural transformation of these functors; they form the category

$$\mathrm{Cub}(\mathsf{C}) = \mathsf{C}^{\underline{\mathbb{I}}^{\mathrm{op}}}$$

of cubical objects in C.

As a consequence, a cubical object $X\colon \underline{\mathbb{I}}^{\mathrm{op}} \to \mathsf{C}$ can be equivalently assigned as a family $(X_n)_{n \geqslant 0}$ of objects of C, linked by *faces* and *degeneracies* (for $1 \leqslant i \leqslant n$ and $\alpha = \pm$)

$$\partial_{ni}^\alpha\colon X_n \to X_{n-1}, \qquad e_{ni}\colon X_{n-1} \to X_n, \tag{5.78}$$

that satisfy the *cubical relations* (dual to the cocubical ones of (5.74))

$$\partial_i^\alpha.\partial_j^\beta = \partial_j^\beta.\partial_{i+1}^\alpha, \qquad \text{for } j \leqslant i,$$

$$e_j.e_i = e_{i+1}.e_j, \qquad \text{for } j \leqslant i,$$

$$\partial_i^\alpha e_j = \begin{cases} e_j.\partial_{i-1}^\alpha, & \text{for } j < i, \\ \mathrm{id}, & \text{for } j = i, \\ e_{j-1}.\partial_i^\alpha, & \text{for } j > i. \end{cases} \tag{5.79}$$

5.4.5 The singular cubical set of a space

In Top (with the cartesian monoidal structure) the standard interval $\mathbb{I} = [0,1]$ of the euclidean line has an obvious structure of internal interval (already considered in 5.2.1)

$$\delta^\alpha\colon \{0\} \to \mathbb{I}, \quad \varepsilon\colon \mathbb{I} \to \{0\}, \qquad \varepsilon.\delta^\alpha = \mathrm{id},$$

$$\delta^\alpha(0) = \alpha \qquad (\alpha = 0,1). \tag{5.80}$$

Higher faces and degeneracies are constructed as above, for $1 \leqslant i \leqslant n$ and $\alpha = 0,1$

$$\delta_{ni}^\alpha = \mathbb{I}^{i-1} \times \delta^\alpha \times \mathbb{I}^{n-i}\colon \mathbb{I}^{n-1} \to \mathbb{I}^n,$$

$$\varepsilon_{ni} = \mathbb{I}^{i-1} \times \varepsilon \times \mathbb{I}^{n-i}\colon \mathbb{I}^n \to \mathbb{I}^{n-1}, \tag{5.81}$$

and satisfy the cocubical relations.

We have thus an embedding

$$J\colon \underline{\mathbb{I}} \to \mathsf{Top}, \qquad 2^n \mapsto \mathbb{I}^n, \tag{5.82}$$

or – in other words – a cocubical space.

For every space S the contravariant functor $\mathsf{Top}(-, S)\colon \mathsf{Top}^{\mathrm{op}} \to \mathsf{Set}$ produces the *singular cubical set*

$$\mathrm{Cub}(S) = \mathsf{Top}(J, S)\colon \underline{I}^{\mathrm{op}} \to \mathsf{Set},$$

$$\mathrm{Cub}(S) = ((\mathrm{Cub}_n(S)), (\partial_i^\alpha), (e_i)), \qquad \mathrm{Cub}_n(S) = \mathsf{Top}(\mathbb{I}^n, S),$$

$$\partial_i^\alpha(a\colon \mathbb{I}^n \to S) = (a.\delta_i^\alpha\colon \mathbb{I}^{n-1} \to S),$$

$$e_i(a\colon \mathbb{I}^{n-1} \to S) = (a.\varepsilon_i\colon \mathbb{I}^n \to S).$$

(5.83)

Acting similarly on each map $f\colon S \to T$ of topological spaces

$$\mathrm{Cub}(f)\colon \mathrm{Cub}(S) \to \mathrm{Cub}(T),$$

$$(\mathrm{Cub}(f))(a\colon \mathbb{I}^n \to S) = (fa\colon \mathbb{I}^n \to T),$$

(5.84)

we have a functor:

$$\mathrm{Cub}\colon \mathsf{Top} \to \mathsf{CubSet}. \qquad (5.85)$$

5.4.6 *The geometric realisation*

Proceeding in the same way as for simplicial sets, in 5.3.5, the *geometric realisation* of a cubical set $X\colon \underline{I}^{\mathrm{op}} \to \mathsf{Set}$ is a topological space $\mathcal{R}(X)$ constructed by pasting a standard cube $\mathbb{I}^n(x)$ for every $x \in X_n$, under the identifications induced by the faces and degeneracies of X.

This is the coend (see 2.8.6)

$$\mathcal{R}(X) = \int^n DX_n \times \mathbb{I}^n, \qquad u_n\colon DX_n \times \mathbb{I}^n \to \mathcal{R}(X), \qquad (5.86)$$

of the functor $\underline{I}^{\mathrm{op}} \times \underline{I} \to \mathsf{Top}$ produced by the composite $DX\colon \underline{I}^{\mathrm{op}} \to \mathsf{Top}$ and the embedding $J\colon \underline{I} \to \mathsf{Top}$ defined in (5.82) (that realises the vertex 2^n of $\underline{I}$ as the standard euclidean cube $\mathbb{I}^n$).

We have now, for any cubical set X, a canonical morphism of cubical sets

$$\eta X\colon X \to \mathrm{Cub}(\mathcal{R}(X)), \qquad (\eta X)_n(x) = u_n(x, -)\colon \mathbb{I}^n \to \mathcal{R}(X), \qquad (5.87)$$

and the geometric realisation is left adjoint to the singular-cubical-set functor $\mathrm{Cub}\colon \mathsf{Top} \to \mathsf{CubSet}$ (with unit η)

$$\mathcal{R}\colon \mathsf{CubSet} \to \mathsf{Top}, \qquad \mathcal{R} \dashv \mathrm{Cub}. \qquad (5.88)$$

5.4.7 *Extended cubical sites*

As already remarked in Section 5.2, the formal study of homotopies requires to add further structure to cubical faces and degeneracies: namely the connections (introduced in Brown–Higgins [BroH1]), together with reversions and transpositions (cf. [G8]).

These structures are embodied in the *intermediate cubical site* $\underline{J}$, the *extended cubical site* $\underline{K}$ and its *symmetric* version $!\underline{K}$, that have been introduced in [GM].

5.5 Chain complexes and singular homology

This new section is a brief introduction to singular homology, the simplest homology theory of topological spaces. The definition is presented in simplicial form in 5.5.4, and in cubical form in 5.5.5 (extending two subsections of the first edition).

The few results we are able to mention here, in 5.5.1 and 5.5.8, should already give an idea of the power of this theory. Relative singular homology will be introduced in Section 6.7.

For the interested reader, there are two elementary, very clear textbooks, by Vick [Vi] and Massey [Mas], following respectively the simplicial and the cubical construction. More advanced books are cited in the Introduction to this chapter.

The trivial group is always written as 0.

5.5.1 Introducing homology

The sphere $\mathbb{S}^2$ and the 'torus' $\mathbb{T}$ (the surface of a life buoy) have a different shape

Sphere and Torus   (5.89)

This looks quite obvious to our intuition; however, proving that these subspaces of $\mathbb{R}^3$ are not homeomorphic is not an elementary thing.

Likely, the simplest way to approach this fact is to think of a 'loop' (a closed path) on these surfaces. On the sphere, every loop can be continuously deformed to a point (a constant path), but on the torus we can easily guess that a tight rope around the life buoy cannot be further contracted (staying in the space we are considering).

All this can be made precise with elementary results of Algebraic Topology: the homology group $H_1(\mathbb{S}^2)$ is trivial, while $H_1(\mathbb{T}) \cong \mathbb{Z}^2$ (as one can see in any book on Algebraic Topology). Since $H_1 \colon \mathsf{Top} \to \mathsf{Ab}$ is a functor, $\mathbb{S}^2$ and $\mathbb{T}$ cannot be isomorphic objects in Top.

The homology groups of a space have a geometrical meaning, which is often fairly intuitive: the two generators of $H_1(\mathbb{T})$ can be realised as two non-trivial loops, around a 'meridian' or an 'equator' of the torus (with respect to the axis of rotation). Very roughly speaking, a non-trivial group $H_n(X)$ detects 'holes' of dimension n in the space X. We shall recall below that $H_n(\mathbb{S}^n) \cong \mathbb{Z}$, for every $n > 0$.

There are diverse homology theories of topological spaces, but all of them coincide on 'reasonably good' spaces. *More precisely, all homology theories satisfying Eilenberg–Steenrod axioms, with integral coefficients, give the same result on every CW-space, up to isomorphism [EiS].*

5.5.2 Chain complexes and chain homology

Before dealing with topological spaces, we have to fix a preliminary algebraic tool, chain complexes of abelian groups. (We are here in the domain of Homological Algebra, at its beginning.)

A (positive) chain complex of abelian groups is a diagram in Ab

$$A = ((A_n)_{n \geqslant 0}, (\partial_n)_{n \geqslant 1}),$$

consisting of a sequence of consecutive homomorphisms

$$\ldots A_{n+1} \xrightarrow{\partial_{n+1}} A_n \xrightarrow{\partial_n} A_{n-1} \ldots \quad A_1 \xrightarrow{\partial_1} A_0 \dashrightarrow 0 \ldots \qquad (5.90)$$

with $\partial_n \partial_{n+1} = 0$ for $n \geqslant 1$.

The homomorphisms ∂_n are called *boundaries* or *differentials*. When useful, all terms are prolonged at the right with trivial groups A_n (for $n < 0$) and zero homomorphisms ∂_n (for $n \leqslant 0$).

A *morphism of chain complexes* $f \colon A \to B$, or *chain morphism*, is a family $(f_n \colon A_n \to B_n)_{n \geqslant 0}$ of homomorphisms in Ab, its components, that commute with the boundaries:

$$
\begin{array}{ccc}
A_n & \xrightarrow{\partial_n} & A_{n-1} \\
{\scriptstyle f_n}\downarrow & & \downarrow{\scriptstyle f_{n-1}} \\
B_n & \xrightarrow[\partial_n]{} & B_{n-1}
\end{array}
\qquad \partial_n f_n = f_{n-1}\partial_n \qquad (n > 0). \qquad (5.91)
$$

Again, these components are extended with zero homomorphisms in negative degree, when useful; note also that we are using the same notation for the differentials of A and B.

With the componentwise composition we get a category $\mathsf{Ch}_+(\mathsf{Ab})$ of *chain complexes of abelian groups*. (It is a full subcategory of $\mathsf{Cat}(\mathsf{N}^{\mathrm{op}}, \mathsf{Ab})$.)

For a chain complex A, an element $a \in A_n$ is called a *chain of degree n* (or dimension n, when appropriate). We are interested in two subgroups of A_n, the subgroup $Z_n(A)$ *of n-cycles* and the subgroup $B_n(A)$ *of n-boundaries*

$$Z_n(A) = \mathrm{Ker}\,(\partial_n \colon A_n \to A_{n-1}) \subset A_n,$$
$$B_n(A) = \mathrm{Im}\,(\partial_{n+1} \colon A_{n+1} \to A_n) \subset A_n. \tag{5.92}$$

This includes $Z_0(A) = \mathrm{Ker}\,(\partial_0 \colon A_0 \to 0) = A_0$. The condition $\partial_n \partial_{n+1} = 0$ is equivalent to saying that $B_n(A) \subset Z_n(A)$ (for $n \geqslant 0$): *every boundary is a cycle.*

We have thus a quotient, the *n-th homology group* of the complex A

$$H_n(A) = Z_n(A)/B_n(A) = \mathrm{Ker}\,(\partial_n)/\mathrm{Im}\,(\partial_{n+1}) \quad (n \geqslant 0). \tag{5.93}$$

Every cycle $z \in Z_n(A)$ has a *homology class* $[z] \in H_n(A)$, which annihilates if and only if z is a boundary. Saying that $H_n(A) = 0$ amounts to the condition $\mathrm{Im}\,\partial_{n+1} = \mathrm{Ker}\,\partial_n$, also expressed saying that the sequence (5.90) is *exact* at A_n, or – more precisely – in degree n.

It is now easy to see that we have functors

$$Z_n \colon \mathsf{Ch}_+(\mathsf{Ab}) \to \mathsf{Ab},$$
$$B_n \colon \mathsf{Ch}_+(\mathsf{Ab}) \to \mathsf{Ab} \qquad (n \geqslant 0), \tag{5.94}$$

where $Z_n(f) \colon Z_n(A) \to Z_n(B)$ and $B_n(f) \colon B_n(A) \to B_n(B)$ are restrictions of the homomorphism $f_n \colon A_n \to B_n$ (cf. 2.4.2). The chain morphism $f = (f_n) \colon A \to B$ gives thus a commutative diagram (for each $n \geqslant 0$)

$$
\begin{array}{ccccc}
B_n(A) & \rightarrowtail & Z_n(A) & \twoheadrightarrow & H_n(A) \\
{\scriptstyle B_n(f)}\downarrow & & {\scriptstyle Z_n(f)}\downarrow & & {\scriptstyle H_n(f)}\downarrow \\
B_n(B) & \rightarrowtail & Z_n(B) & \twoheadrightarrow & H_n(B)
\end{array}
\qquad H_n(f)[z] = [f_n(z)], \tag{5.95}
$$

where the rows are formed of inclusions and projections, while the left and right columns are induced homomorphisms, in the sense of (2.28). *(The rows are 'short exact sequences', see 6.3.4.)*

We have constructed the *chain homology functor* of degree n

$$H_n \colon \mathsf{Ch}_+(\mathsf{Ab}) \to \mathsf{Ab}, \qquad H_n(A) = (\mathrm{Ker}\,\partial_n)/(\mathrm{Im}\,\partial_{n+1}). \tag{5.96}$$

The homomorphism $H_n(f) \colon H_n(A) \to H_n(B)$ is usually written as f_{*n}.

The whole construction can be extended to every abelian category A, giving an abelian category $\mathsf{Ch}_+(\mathsf{A})$, with homology functors $H_n \colon \mathsf{Ch}_+(\mathsf{A}) \to \mathsf{A}$.

5.5.3 Chain homotopy

The category $\mathrm{Ch}_+(\mathsf{Ab})$ has also a notion of homotopy, well related to topological homotopy (as we shall see).

For two morphisms $f, g\colon A \to B$ in $\mathrm{Ch}_+(\mathsf{Ab})$, a *homotopy* $\varphi\colon f \to g$ is a sequence of homomorphisms $\varphi_n\colon A_n \to B_{n+1}$ $(n \geqslant 0)$ such that

$$\partial_{n+1}\varphi_n + \varphi_{n-1}\partial_n = g_n - f_n, \qquad (5.97)$$

where, again, we are letting $\varphi_{-1} = 0$.

Then the morphisms $f, g\colon A \to B$ are said to be *homotopic* in $\mathrm{Ch}_+(\mathsf{Ab})$, written as $f \simeq g$.

More precisely, we should define the homotopy as the triple (f, φ, g); this becomes evident in Exercise (c), below.

Exercises and complements. (a) The homotopy relation $f \simeq g$ is a congruence of categories, in $\mathrm{Ch}_+(\mathsf{Ab})$.

(b) (*Homotopy invariance of chain homology*) If two arrows $f, g\colon A \to B$ in $\mathrm{Ch}_+(\mathsf{Ab})$ are homotopic, then $H_n(f) = H_n(g)$, for all $n \geqslant 0$.

*(c) (*The cylinder functor*) Homotopies of chain complexes can be defined by a cylinder endofunctor $I\colon \mathrm{Ch}_+(\mathsf{Ab}) \to \mathrm{Ch}_+(\mathsf{Ab})$

$$(IA)_n = A_n \oplus A_{n-1} \oplus A_n,$$

$$\partial_n(a, x, b) = (\partial_n a - x, -\partial_{n-1}x, \partial_n b + x), \qquad (5.98)$$

$$(If)_n = f_n \oplus f_{n-1} \oplus f_n,$$

with *faces*

$$\partial^\alpha\colon A \to IA, \qquad \partial^-(a) = (a, 0, 0), \quad \partial^+(a) = (0, 0, a).$$

Prove that a morphism $\Phi = [f, \varphi, g]\colon IA \to B$ in $\mathrm{Ch}_+(\mathsf{Ab})$ is the same as a homotopy from $f = \Phi\partial^-$ to $g = \Phi\partial^+$.

The functor I has a structure analogous to that of the cylinder functor of Top, in Section 5.2. This is studied in detail in [G8], Section 4.4, for chain complexes over an additive category A.

5.5.4 Singular homology, I

We define the *chain functor* of simplicial sets

$$\mathrm{Ch}_+\colon \mathsf{SmpSet} \to \mathrm{Ch}_+(\mathsf{Ab}), \qquad \mathrm{Ch}_+(K) = ((F(K_n)), (\partial_n)), \qquad (5.99)$$

where, for a simplicial set K, the component $\mathrm{Ch}_n(K)$ is the free abelian group $F(K_n) = \mathbb{Z}K_n$. The differential is defined on the basis K_n as a linear combination of the faces $\partial_{ni}\colon K_n \to K_{n-1}$

$$\partial_n(a) \;=\; \Sigma_i \,(-1)^i \,\partial_{ni}(a) \;\in\; F(K_{n-1}), \qquad\qquad (5.100)$$

for $i = 0, ..., n$. The only non-trivial point is proving that $\partial_n \partial_{n+1} = 0$, as a consequence of the simplicial relations of the faces; it will be the subject of a nice exercise, in 5.5.6(a).

On a morphism $f\colon K \to L$ of simplicial sets, the components of the morphism $\mathrm{Ch}_+(f)$ of chain complexes

$$\mathrm{Ch}_+(f)\colon \mathrm{Ch}_+(K) \to \mathrm{Ch}_+(L),$$
$$\mathrm{Ch}_n(f) = F(f_n)\colon \Sigma \lambda_i\, a_i \mapsto \Sigma \lambda_i f_n(a_i) \qquad (n \geqslant 0), \qquad (5.101)$$

are the linear extensions of the components f_n.

The n-th *singular homology functor*

$$H_n\colon \mathsf{Top} \to \mathsf{SmpSet} \to \mathrm{Ch}_+(\mathsf{Ab}) \to \mathsf{Ab},$$
$$H_n(-) = H_n(\mathrm{Ch}_+(\mathrm{Smp}(-))), \qquad\qquad (5.102)$$

is now defined by composing three functors:

- the singular simplicial functor $\mathrm{Smp}\colon \mathsf{Top} \to \mathsf{SmpSet}$ of (5.66),
- the chain functor $\mathrm{Ch}_+\colon \mathsf{SmpSet} \to \mathrm{Ch}_+(\mathsf{Ab})$ defined above,
- the chain homology functor $H_n\colon \mathrm{Ch}_+(\mathsf{Ab}) \to \mathsf{Ab}$ defined in (5.96).

For a topological space X, a (formal) linear combination of singular simplices $a_i\colon \Delta^n \to X$, with coefficients in $\mathbb{Z}$

$$c \;=\; \Sigma_i \lambda_i\, a_i \;\in\; \mathrm{Ch}_n(\mathrm{Smp}(X)) = F(\mathsf{Top}(\Delta^n, X)) \qquad (5.103)$$

is called a *singular (simplicial) chain* of the space X, of dimension n. Again, it is an n-*cycle* if $\partial_n(c) = 0$, and an n-*boundary* if $c \in \mathrm{Im}\,\partial_{n+1}$. Every cycle c has a homology class $[c] \in H_n(S)$, which annihilates if and only if c is a boundary.

The differential (5.100) can be viewed as a morphism $K_n \nrightarrow K_{n-1}$ of the Kleisli category Set_T of sets and $\mathbb{Z}$-weighted mappings, considered in 3.7.1(b).

5.5.5 Singular homology, II

Singular homology can be equivalently constructed using the cubical models $\mathbb{I}^n$ instead of the simplicial models Δ^n. (A proof of the equivalence of these constructions is given in [HiW], Section 8.4, applying the technique of 'acyclic models', due to Eilenberg and Mac Lane.)

The singular homology of the space X is defined here as the homology of the (normalised) chain complex associated to the cubical set $\mathrm{Cub}(X)$

$$H_n\colon \mathsf{Top} \to \mathsf{Ab}, \qquad H_n(X) = H_n(\mathrm{Ch}_+(\mathrm{Cub}(X))). \qquad (5.104)$$

To make sense of this, we must construct the chain complex $\mathrm{Ch}_+(K)$ associated to a cubical set K; this is more complicated than the simplicial analogue: here we have to quotient the free abelian group $F(K_n) = \mathbb{Z}K_n$ modulo the subgroup spanned by the 'degenerate elements' (otherwise we would not get the correct homology groups).

First, an n-cube $a \in K_n$ is degenerate if $a = e_j(b)$, for some cube $b \in K_{n-1}$ (and $n > 0$)

$$\mathrm{Deg}_n K = \bigcup_j \mathrm{Im}\,(e_j\colon K_{n-1} \to K_n), \qquad \mathrm{Deg}_0 K = \emptyset. \qquad (5.105)$$

Because of the cubical relations, we have (for $i = 1, ..., n$)

$$a \in \mathrm{Deg}_n K \ \Rightarrow\ (\partial_i^\alpha a \in \mathrm{Deg}_{n-1} K \ \text{or}\ \partial_i^- a = \partial_i^+ a). \qquad (5.106)$$

Now the chain complex $\mathrm{Ch}_+(K)$ is defined as follows:

$$\mathrm{Ch}_n(K) = (\mathbb{Z}K_n)/(\mathbb{Z}\mathrm{Deg}_n K),$$
$$\partial_n\colon \mathrm{Ch}_n(K) \to \mathrm{Ch}_{n-1}(K), \qquad (5.107)$$
$$\partial_n(\hat{a}) = \Sigma_{i,\alpha}\,(-1)^{i+\alpha}\,(\partial_i^\alpha a)\hat{\ } \qquad (a \in K_n).$$

Here $\hat{a}$ is the class of the n-cube a, up to degenerate cubes; the differential is legitimate because of (5.106). Again, the only non-trivial point is proving that $\partial_n \partial_{n+1} = 0$, by applying the cubical relations of the faces (Exercise 5.5.6(b)). In computations, we shall usually write the class $\hat{a}$ as a; of course one as to check that a definition involving this class is invariant, up to degenerate cubes.

On a morphism $f\colon K \to L$ of cubical sets, the n-component of the chain morphism $\mathrm{Ch}_+(f)$ is induced by the linear extension of f_n

$$\mathrm{Ch}_n(f)\colon \mathrm{Ch}_n(K) \to \mathrm{Ch}_n(L), \qquad \Sigma \lambda_i a_i \mapsto \Sigma \lambda_i f_n(a_i), \qquad (5.108)$$

which is legitimate, because a degenerate n-cube $a = e_j b$ is sent to the degenerate cube $f_n(a) = e_j f_n(b)$.

The heavier definition of the chain complex $\mathrm{Ch}_+(K)$ is compensated by an effective simplification of all the issues related to cartesian products of spaces, from the Homotopy Invariance Theorem of singular homology (in 5.5.7) to more complex topics, like the homology of a product $X \times Y$ of spaces (see [Mas]).

5.5.6 Exercises and complements (Basic results)

(a) Prove that $\partial_n \partial_{n+1} = 0$, for the differential of the chain complex of a simplicial set K, defined in (5.100). *Hints:* use the simplicial relations of the faces.

(b) Same exercise for the differential of the chain complex of a cubical set K, defined in (5.107).

(c) A few very elementary computations of homology groups can be done directly, out of the definition. Begin by proving the following formulas for the homology of the empty space and the singleton, using the simplicial definition of 5.5.4

$$H_k(\emptyset) = 0 \qquad (k \geqslant 0), \tag{5.109}$$

$$H_0(\{*\}) \cong \mathbb{Z}, \qquad H_k(\{*\}) = 0 \quad (k > 0). \tag{5.110}$$

(d) Same exercise, using the cubical form of 5.5.5.

(e) (*Path-connected spaces*) For a non-empty path-connected space X we have

$$H_0(X) \cong \mathbb{Z}. \tag{5.111}$$

More precisely, $H_0(X)$ is the free abelian group generated by the homology class $[x]$ of any point $x \in X$. (All these classes coincide.)

(f) In general, $H_0(X)$ is isomorphic to the free abelian group $\mathbb{Z}I$ generated by the set I of path-connected components of X, namely the direct sum $\bigoplus_{i \in I} \mathbb{Z}$. More precisely, $H_0(X)$ has a canonical basis $\xi_i = [x_i]$ ($i \in I$), where x_i is any point of the path-connected component i.

Loosely speaking, H_0 'counts' the path-components of the space.

(g) Compute the singular homology of a space X whose path-connected components (if any) are singletons.

This includes all discrete spaces and the rational line $\mathbb{Q}$. Note that singular homology does not distinguish the euclidean spaces $\mathbb{Q}$ and $\mathbb{Z}$.

5.5.7 Homotopy Invariance Theorem

(a) *The functors H_n: Top $\to$ Ab are homotopy invariant: if $f \simeq g \colon X \to Y$ in Top, then $f_{*n} = g_{*n} \colon H_n(X) \to H_n(Y)$, for all $n \geqslant 0$.*

(b) *If the spaces X, Y are homotopy equivalent, then $H_n(X) \cong H_n(Y)$, for all $n \geqslant 0$.*

Note. Contractible spaces have thus the same homology as the singleton (in 5.5.6(c)). In particular, this applies to all euclidean spaces $\mathbb{R}^m$.

Proof It is sufficient to prove (a). Using the cubical form, we want to prove that a topological homotopy $\varphi \colon f \to g \colon X \to Y$ (as defined in 5.2.2) between two maps has an associated chain homotopy Φ between the corresponding morphisms of chain complexes, written as $f_\sharp$ and $g_\sharp$

$$\Phi \colon f_\sharp \to g_\sharp \colon \mathrm{Ch}_+(\mathrm{Cub}(X)) \to \mathrm{Ch}_+(\mathrm{Cub}(Y)),$$

$$(\partial_{n+1}\Phi_n + \Phi_{n-1}\partial_n)(a) = ga - fa \qquad \text{(for } a \colon \mathbb{I}^n \to X). \tag{5.112}$$

This will imply that $H_n(f) = H_n(g)$ (for $n \geqslant 0$), by the homotopy invariance of chain homology (in Exercise 5.5.3(b)).

We write the homotopy $\varphi \colon f \to g \colon X \to Y$ as a continuous mapping $\varphi \colon \mathbb{I} \times X \to Y$. For every $n \geqslant 0$ we define the homomorphism

$$\Phi_n \colon \mathrm{Ch}_n(\mathrm{Cub}(X)) \to \mathrm{Ch}_{n+1}(\mathrm{Cub}(Y)), \tag{5.113}$$

sending any n-cube $a \colon \mathbb{I}^n \to X$ to the composite $\varphi\,(\mathbb{I} \times a) \colon \mathbb{I}^{n+1} \to Y$, an $(n+1)$-cube of Y. This is legitimate, because a degenerate cube $a = b\varepsilon_i \colon \mathbb{I}^n \to \mathbb{I}^{n-1} \to X$ is sent to the cube

$$\varphi\,(\mathbb{I} \times b\varepsilon_i) = \varphi\,(\mathbb{I} \times b)\varepsilon_{i+1} \colon \mathbb{I}^{n+1} \to Y,$$

which is also degenerate. Finally (with $\alpha = 0, 1$ in each indexed sum)

$$\begin{aligned}
\partial_{n+1}\Phi_n(a) = \partial_{n+1}(\varphi\,(\mathbb{I} \times a)) &= \textstyle\sum_{1 \leqslant i \leqslant n+1}(-1)^{i+\alpha}\,\varphi\,(\mathbb{I} \times a)\,\delta_i^\alpha \\
&= \textstyle\sum_{i=1}(-1)^{i+\alpha}\,\varphi\,(\delta_i^\alpha \times a) + \sum_{2 \leqslant i \leqslant n+1}(-1)^{i+\alpha}\,\varphi\,(\mathbb{I} \times a\delta_{i-1}^\alpha) \\
&= ga - fa - \textstyle\sum_{1 \leqslant i \leqslant n}(-1)^{i+\alpha}\,\varphi\,(\mathbb{I} \times a\delta_i^\alpha) \\
&= ga - fa - \Phi_{n-1}\partial_n(a).
\end{aligned}$$

Remarks. (i) If we use the cylinder $X \times \mathbb{I}$ (instead of $\mathbb{I} \times X$) the computation is slightly more complex: we should pick out the last faces (instead of the first ones), and apply a sign-change depending on the degree n.

(ii) The reader will note that the construction of Φ is easy because the cylinder functor $\mathbb{I} \times - \colon \mathsf{Top} \to \mathsf{Top}$ takes cubes to cubes (from dimension n to dimension $n+1$).

Using the simplicial form of singular homology, the construction would be more complex, as one can see in [Vi]: the product $\mathbb{I} \times \Delta^1$ is a square, and has to be 'covered' with two triangles which meet at an edge; in general, $\mathbb{I} \times \Delta^n$ is covered with $n+1$ tetrahedra homeomorphic to Δ^{n+1}, setting up a convenient geometrical machinery. $\qquad\square$

5.5.8 *An overview of some applications

Singular homology allows us to prove important topological facts.

(a) (*Homology of the spheres*) Computing the homology groups of the spheres $\mathbb{S}^n$ (after the trivial case of dimension 0) requires the Mayer–Vietoris exact sequence, and we can only write here the result

$$H_0(\mathbb{S}^0) \cong \mathbb{Z}^2, \qquad\qquad H_k(\mathbb{S}^0) = 0 \quad (k > 0),$$
$$H_0(\mathbb{S}^n) \cong H_n(\mathbb{S}^n) \cong \mathbb{Z}, \qquad H_k(\mathbb{S}^n) = 0 \quad (k \neq 0, n), \tag{5.114}$$

(for $n > 0$) as one can find in [Vi, Mas], or any book on Algebraic Topology.

Roughly speaking, these groups say that $\mathbb{S}^0$ has two connected components, and each higher sphere $\mathbb{S}^n$ is path-connected, with one 'hole' of dimension n.

(b) This has many important consequences. For instance, these results prove that the sphere $\mathbb{S}^n$ is not homeomorphic to any other sphere $\mathbb{S}^m$ (for $m \neq n$): we simply apply the fact that any functor preserves isomorphisms.

(c) (*Theorem of Topological Dimension*) As a consequence, if the euclidean spaces $\mathbb{R}^m$ and $\mathbb{R}^n$ are homeomorphic, the same is true of their one-point compactifications $\mathbb{S}^m$ and $\mathbb{S}^n$, *and we conclude that $m = n$.*

Note that we cannot get this result applying the homology functors to $\mathbb{R}^m$ and $\mathbb{R}^n$, directly: these spaces are contractible (Exercise 5.2.4(a)), and a homotopy-invariant functor cannot distinguish them.

(d) Applying (5.114) we also conclude that the sphere $\mathbb{S}^n$ (for $n > 0$) is not a retract of $\mathbb{R}^{n+1}$, or of any contractible space, because $H_n(\mathbb{S}^n) \cong \mathbb{Z}$ is not a retract of the trivial group.

For $n = 0$ the argument can be adapted: $H_0(\mathbb{S}^0) \cong \mathbb{Z}^2$ is not a retract of $H_0(\mathbb{R}) \cong \mathbb{Z}$. But the Intermediate Value Theorem already proves that $\mathbb{S}^0$ is not a retract of $\mathbb{R}$.

5.5.9 *Comonads and homology

A comonad on a category C can be used to construct simplicial objects on C, and then homology or cohomology theories on this category.

Mac Lane gives a brief, clear presentation of group (co)homology in this way, in [M4], Section VII.6. This is an excellent introduction to a field also called 'triple cohomology' or 'cohomology on standard constructions' [Bc, BarB, In].

The starting point is a comonad (S, ε, δ) on the category C (as defined in 3.6.9)

$$S: \mathsf{C} \to \mathsf{C}, \qquad \varepsilon: S \to 1, \qquad \delta: S \to S^2. \tag{5.115}$$

Exercises and complements. (a) Prove that an object X of C defines an augmented simplicial object in C (see 5.3.1)

$$S_*(X) \colon (\underline{\Delta}^\sim)^{\mathrm{op}} \to \mathsf{C},$$

$$S_*(X) = ((S^n(X))_{n \geqslant 0}, (\partial_i), (e_i)),$$

$$\partial_i = S^i \varepsilon S^{n-i} \colon S^{n+1}(X) \to S^n(X) \qquad (0 \leqslant i \leqslant n),$$

$$e_i = S^i \delta S^{n-i-1} \colon S^n(X) \to S^{n+1}(X) \qquad (0 \leqslant i \leqslant n-1).$$

(5.116)

(b) Taking out the component $S^0(X)$, we have a simplicial object $S_+(X)$ in C, whose components can be renamed as $S_n(X) = S^{n+1}(X)$, for $n \geqslant 0$. Any functor $F \colon \mathsf{C} \to \mathsf{Ab}$ produces a simplicial abelian group $F_+S_+(X)$, with components $A_n = F(S^{n+1}(X))$. (The category Ab can be replaced with any abelian one.)

(c) Define a chain complex with the same components $A_n = F(S^{n+1}(X))$, and differential as in (5.100).

Its algebraic homology gives a homology theory on C; or a cohomology theory, if we use a functor $F \colon \mathsf{C} \to \mathsf{Ab}^{\mathrm{op}}$.

5.6 *Smash product and colax monoidal categories

While the smash product of pointed sets gives a symmetric monoidal closed structure (in 3.4.2(c) and 3.4.7(d)), the smash product of Top. is symmetric but not associative.

We show here that it has a natural extension to an n-ary operation, defined by the universal multi-pointed mapping

$$X_1 \times \ldots \times X_n \dashrightarrow X_1 \wedge \ldots \wedge X_n.$$

This interesting structure makes Top. into a colax monoidal category.

For particular objects the 'comparisons' of the colax structure are invertible, and produce an isomorphism $X \wedge (Y \wedge Z) \to (X \wedge Y) \wedge Z$. The main case is when the underlying spaces of X and Z are exponentiable in Top, for the cartesian product (see 5.1.1). (Then X and Z are also exponentiable in Top., for the smash product: see 5.6.7.)

Examining this structure may have a general significance: *lax* and *colax* monoidal structures might be viewed as the primitive notion, and the associative case a particular situation – if an important one – and a consequence of exponentiability. A reader interested in the general theory of lax monoidal categories is referred to Leinster's book [Le].

5.6.1 Kernels and cokernels of pointed mappings

Kernels and cokernels in a general pointed category will be examined in Section 6.1. Here we introduce them for the categories Set. and Top., of pointed sets and pointed spaces.

We recall from 2.1.9 that these categories have a zero object, written here as $\{*\}$. An object of these categories will be simply written as $X, Y, ...$; the base-point of X is written as 0_X, or also as 0. The zero morphism $0_{XY} \colon X \to \{*\} \to Y$ is the constant mapping at 0_Y.

The *kernel* of a morphism $f \colon X \to Y$ is defined as the equaliser of f and $0_{XY} \colon X \to Y$. This is easily computed, as the inclusion of the pointed subset, or subspace, where f annihilates

$$\ker f \colon \operatorname{Ker} f \to X,$$
$$\operatorname{Ker} f = \{x \in X \mid f(x) = 0_Y\} = f^{-1}\{0_Y\}. \tag{5.117}$$

Dually, the *cokernel* of $f \colon X \to Y$ is the coequaliser of f and 0_{XY}. It can be realised as the projection of the set, or space, Y on the quotient that collapses the image $f(X)$ on the base point (see 5.1.8)

$$\operatorname{cok} f \colon Y \to \operatorname{Cok} f, \qquad \operatorname{Cok} f = Y/f(X). \tag{5.118}$$

Every pointed subset, or subspace, A of X is a *normal subobject*, i.e. a kernel of a morphism (in particular, of the projection $X \to X/A$). On the other hand, a projection $Y \to Y/R$ on a quotient pointed set, or space, is a *normal quotient*, i.e. a cokernel (of its kernel) if and only if every equivalence class $[y]_R \neq [0_Y]_R$ (if any) is a singleton.

We also recall that the categorical sum $X \vee Y$, in Set. and Top., can be realised as a subset or subspace of the cartesian product, as we have seen in 2.1.6(d), (e)

$$X \vee Y = (X \times \{0_Y\}) \cup (\{0_X\} \times Y) \subset X \times Y. \tag{5.119}$$

5.6.2 Reviewing the smash product of pointed sets

The category Set. of pointed sets has a symmetric monoidal closed structure, described in 3.4.2(c), which we review now as produced by a universal property similar to that of the tensor product of R-modules.

A *bipointed mapping* $\varphi \colon X \times Y \dashrightarrow Z$ is a mapping which is pointed in each variable:

$$\varphi(x, 0) = 0 = \varphi(0, y) \qquad (\text{for } x \in X,\ y \in Y). \tag{5.120}$$

These mappings will be denoted by dot-marked arrows.

They are closed under an appropriate form of whisker composition

$$k\varphi(h \times h'): X' \times Y' \to X \times Y \twoheadrightarrow Z \to Z', \qquad (5.121)$$

where $h: X' \to X$, $h': Y' \to Y$ and $k: Z \to Z'$ are pointed mappings.

Now, the smash product $X \wedge Y$ can be defined by means of the universal bipointed mapping

$$\eta: X \times Y \twoheadrightarrow X \wedge Y, \qquad (5.122)$$

characterised by the fact that every bipointed mapping $\varphi: X \times Y \twoheadrightarrow Z$ factorises through η, by a unique pointed mapping h

$$\begin{array}{ccc}
X \times Y & \xrightarrow{\;\;\eta\;\;} & X \wedge Y \\
 & \searrow\raisebox{0.5ex}{φ} & \downarrow{\scriptstyle h} \\
 & & Z
\end{array} \qquad (5.123)$$

The solution is the cokernel of the inclusion $u: X \vee Y \to X \times Y$, i.e. the projection on the quotient pointed set $(X \times Y)/(X \vee Y)$, which collapses all points of $X \vee Y$ to the base point.

We shall write $x \wedge y = \eta(x, y) = [x, y]$, so that $x \wedge 0 = 0 = 0 \wedge y$.

We fix the identity as the two-point set $\mathbb{S}^0 = \{-1, 1\}$, pointed at 1. The associator

$$\kappa: X \wedge (Y \wedge Z) \to (X \wedge Y) \wedge Z, \qquad x \wedge (y \wedge z) \mapsto (x \wedge y) \wedge z, \qquad (5.124)$$

will be obtained in 5.6.4(b), as a by-product of the topological smash product.

The internal hom is computed as the set $\mathsf{Set}_\bullet(X, Y)$, pointed at the zero-mapping $X \to Y$. The exponential law is the natural bijection

$$\varphi^A_{XY}: \mathsf{Set}_\bullet(X \wedge A, Y) \to \mathsf{Set}_\bullet(X, \mathsf{Set}_\bullet(A, Y)),$$
$$(f: X \wedge A \to Y) \mapsto (g: X \to \mathsf{Set}_\bullet(A, Y)), \qquad (5.125)$$
$$g(x) = f(x \wedge -): A \to Y.$$

5.6.3 Smash product of pointed spaces

The category $\mathsf{Top}_\bullet$ of pointed spaces has a similar smash product, which is symmetric but no longer associative.

Again, $X \wedge Y$ is given by the universal bi-pointed map (5.122); it can be obtained as the cokernel of the inclusion $u: X \vee Y \to X \times Y$, i.e. the projection on the quotient pointed space $(X \times Y)/(X \vee Y)$, which collapses the subset $X \vee Y$ on the base point.

The discrete two-point space $\mathbb{S}^0 = \{-1, 1\}$, pointed at 1, acts as an identity (up to canonical isomorphism).

A counterexample to associativity, found by D. Puppe, is based on the triple of spaces $\mathbb{Q}$, $\mathbb{Q}$, $\mathbb{N}$. A proof of the associativity failure in this case can be found in [MayS].

5.6.4 A colax monoidal structure

(a) This non-associative smash product can be extended, to make Top. into a symmetric *colax monoidal* category (as defined in [Le], Section 3.1).

The n-ary smash product $X_1 \wedge ... \wedge X_n$ is obtained as the universal n-pointed mapping (i.e. pointed in each variable)

$$\eta \colon X_1 \times ... \times X_n \twoheadrightarrow X_1 \wedge ... \wedge X_n, \quad \eta(x_1, ..., x_n) = x_1 \wedge ... \wedge x_n. \quad (5.126)$$

The solution is the cokernel of the embedding of the subspace H of coordinate hyperplanes

$$H \to X_1 \times ... \times X_n, \qquad H = \{(x_i) \in \Pi X_i \mid \exists i \colon x_i = 0\} \qquad (5.127)$$

(equal to the subspace $\vee X_i$ of coordinate axes when $n = 2$). The term $x_1 \wedge ... \wedge x_n$ annihilates if and only if at least one coordinate x_i is zero. Let us note that η is the projection on a topological quotient.

As described in Leinster [Le] (in the dual lax case), we have a coherent system of associativity-comparisons which need not be invertible. In particular there are two of them, for a triple of spaces

$$\begin{aligned} \gamma' \colon X \wedge Y \wedge Z \to X \wedge (Y \wedge Z), \quad & \gamma'(x \wedge y \wedge z) = x \wedge (y \wedge z), \\ \gamma'' \colon X \wedge Y \wedge Z \to (X \wedge Y) \wedge Z, \quad & \gamma''(x \wedge y \wedge z) = (x \wedge y) \wedge z, \end{aligned} \quad (5.128)$$

each of them being produced by the universal property of the tri-pointed mapping $\eta \colon X \times Y \times Z \twoheadrightarrow X \wedge Y \wedge Z$.

(b) Forgetting topologies, the category Set. has a similar colax monoidal structure, which actually is an *unbiased monoidal structure* [Le]: all its associativity-comparisons are invertible, i.e. bijective pointed mappings.

In particular the associator κ_{XYZ} of Set. is determined by a commutative diagram of isomorphisms

$$\begin{array}{ccc} X \wedge Y \wedge Z & \xrightarrow{\ \gamma'\ } & X \wedge (Y \wedge Z) \\ & {\gamma''}\searrow & \ \downarrow{\kappa} \\ & & (X \wedge Y) \wedge Z \end{array} \qquad (5.129)$$

(c) The forgetful functor $\mathsf{Top}_\bullet \to \mathsf{Set}_\bullet$ preserves the colax monoidal structure. All the associativity-comparisons of $\mathsf{Top}_\bullet$ (including the previous γ' and γ'') are thus bijective pointed maps; generally, they are not invertible.

However there are particular cases where these comparisons are invertible, *and there is an associator* $\kappa\colon X \wedge (Y \wedge Z) \to (X \wedge Y) \wedge Z$. The main case is exposed in the following proposition.

Saying that the pointed space X *is exponentiable in* Top we always mean that the corresponding unpointed space is exponentiable for the cartesian structure of Top; we have seen that this property is enjoyed by all locally compact spaces, and characterised as core compactness (in 5.1.1).

5.6.5 Proposition

If the pointed space X is exponentiable in Top, *the bijective pointed map* $\gamma'\colon X \wedge Y \wedge Z \to X \wedge (Y \wedge Z)$ *(in (5.128)) is a homeomorphism.*

By symmetry, if the pointed space Z is exponentiable in Top, *the same is true of* $\gamma''\colon X \wedge Y \wedge Z \to (X \wedge Y) \wedge Z$.

Proof We form a commutative square in $\mathsf{Top}_\bullet$

$$
\begin{array}{ccc}
X \times Y \times Z & \xrightarrow{\ \eta\ } & X \wedge Y \wedge Z \\
{\scriptstyle X \times \eta'}\big\downarrow & & \big\downarrow{\scriptstyle \gamma'} \\
X \times (Y \wedge Z) & \xrightarrow[\ \eta''\ ]{} & X \wedge (Y \wedge Z)
\end{array}
\qquad
\begin{array}{ccc}
(x,y,z) & \longmapsto & x \wedge y \wedge z \\
\big\downarrow & & \big\downarrow \\
(x, y \wedge z) & \longmapsto & x \wedge (y \wedge z)
\end{array}
\qquad (5.130)
$$

The mappings η, η' and η'' are projections on a topological quotient, by definition. The mapping $X \times \eta'$ is also a topological projection (i.e. a surjective continuous mapping with final topology on the codomain), because the functor $X \times -\colon \mathsf{Top} \to \mathsf{Top}$ preserves coequalisers, as a left adjoint. Therefore the bijective continuous map γ' is a homeomorphism.

Let us note that, in Top, the functor $X \times -$ preserves coequalisers *only if* X is exponentiable [GiS]. $\square$

5.6.6 Corollary

If X, Y, Z are pointed spaces and X, Z are core compact, the following commutative triangle in $\mathsf{Top}_\bullet$

$$
\begin{array}{ccc}
X \wedge Y \wedge Z & \xrightarrow{\ \gamma'\ } & X \wedge (Y \wedge Z) \\
 & {\scriptstyle \gamma''}\searrow & \big\downarrow{\scriptstyle \kappa} \\
 & & (X \wedge Y) \wedge Z
\end{array}
\qquad (5.131)
$$

gives an invertible associator κ_{XYZ}, whose underlying mapping is the associator of Set. (in (5.129)).

5.6.7 Lemma

Let A be a pointed space whose corresponding unpointed space is exponentiable in Top, *for the cartesian structure. Then A is exponentiable in* Top. *for the smash product.*

For a pointed space Y, the internal hom $[A, Y]$ is the set Top.(A, Y) *of pointed maps, with the subspace topology of the exponential Y^A in* Top.

Note. In fact a pointed space A is exponentiable for the smash product *if and only if* its underlying space is exponentiable in Top for the cartesian structure, as proved in [Ca] (independently of the base point of A). Here we are only interested in one implication.

Proof The natural bijection of the exponential law in Top

$$\mathsf{Top}(X \times A, Y) \to \mathsf{Top}(X, Y^A)$$

restricts to a natural bijection

$$\varphi_{XY}^A \colon \mathsf{Top}_{\bullet}(X \wedge A, Y) \to \mathsf{Top}_{\bullet}(X, [A, Y]),$$

$$(f \colon X \wedge A \to Y) \mapsto (g \colon X \to [A, Y]), \qquad (5.132)$$

$$g(x) = f(x \wedge -) \colon A \to Y.$$

In particular, if A is locally compact, the space $[A, Y]$ has the compact-open topology. $\qquad\square$

5.7 *Hints at Directed Algebraic Topology

Directed Algebraic Topology is a recent subject which arose in the 1990's, on the one hand in abstract settings for homotopy theory, like [G3], and on the other hand in investigations in the theory of concurrent processes, like [FGR1, FGR2]. Its general aim can be stated as 'modelling non-reversible phenomena'. The subject has a deep relationship with category theory.

The domain of Directed Algebraic Topology is distinguished from the domain of classical Algebraic Topology by the principle that *directed spaces have privileged directions*, and their (directed) paths need not be reversible.

While Topology and Algebraic Topology deal with reversible worlds, where a path can always be travelled backwards, the study of non-reversible phenomena requires more structured worlds, where a directed space *can* have non-reversible paths. (It is important to develop the directed theory so that it *extends* the reversible one.)

The classical tools, consisting of:

- ordinary homotopies, fundamental groups and fundamental n-groupoids, are now replaced by (possibly) non-reversible versions:

- *directed homotopies, fundamental monoids* and *fundamental n-categories.*

Similarly, the homological theories of the directed setting take values in 'directed' algebraic structures, like *preordered* abelian groups (with indiscrete preorder in the reversible case).

Here we give a few hints at the starting points of this subject, taken from the book [G8]. An initial study of directed manifolds can now be found in [G12], Section 4.4.

The prefix $\uparrow$ denotes a directed enrichment of a reversible structure.

5.7.1 *Preordered topological spaces*

The simplest topological setting where one can study directed paths and directed homotopies is likely the category pTop of *preordered topological spaces* and *preorder-preserving continuous mappings.*

(We are not requiring any particular relationship between topology and preorder.)

Here, the *standard directed interval* $\uparrow\mathbb{I} = \uparrow[0, 1]$ has the euclidean topology and the natural order. A (directed) *path* in a preordered space X is – by definition – a map $a\colon \uparrow\mathbb{I} \to X$ (continuous and monotone).

The category pTop has all limits and colimits, constructed as for topological spaces and equipped with the initial or final preorder for their structural maps; for instance, in a product $X = \Pi X_j$, we have the product preorder: $(x_j) \prec_X (x'_j)$ if and only if, for each index j, $x_j \prec x'_j$ in X_j.

The forgetful functor $U\colon \text{pTop} \to \text{Top}$ has both a left and a right adjoint, $D \dashv U \dashv C$ where DS (resp. CS) is the space S equipped with the *discrete* order (resp. the *chaotic,* or *indiscrete* preorder).

The standard embedding of Top in pTop will be the one given by the *indiscrete preorder*, so that all (ordinary) paths in S are directed in CS.

Our category is not cartesian closed, of course; but it is easy to transfer here the classical result for topological spaces, recalled above (in 5.1.1).

Thus, every preordered space A that has a *locally compact topology* and an *arbitrary preorder* is exponentiable in pTop, with Y^A consisting of the set pTop$(A, Y) \subset$ Top(UA, UY) of *monotone continuous mappings,* equipped with the (induced) compact-open topology and the *pointwise preorder*

$$f \prec g \ \text{ if } \ (\forall x \in A, \ f(x) \prec_Y g(x)). \tag{5.133}$$

The main setting studied in [G8] (and many papers, by several authors) is in fact a richer one, the category dTop of d-spaces, or *spaces with distinguished paths*, where an object can have *vortices* (i.e. non-reversible loops); this framework will not be recalled here.

5.7.2 Comments

As to our choice of the category pTop, we note two points – in the line of the remarks in 1.1.5(b).

First, *if we replace preordered spaces with the ordered ones we miss the embedding* C: Top → pTop by the chaotic preorder, and can no longer view classical Algebraic Topology within the directed one. Moreover, colimits of ordered spaces are 'different' from the topological ones (in the same way as the colimits of ordered sets do not agree with those of sets).

Second, if we require that in a preordered space X the graph of the preorder is closed in $X \times X$ (as is often done), then the discrete (pre)order would only be allowed in Hausdorff spaces (which are characterised by having a closed diagonal), and the previous embedding D: Top → pTop would fail. Now we would complicate limits in pTop.

Finer connections between topology and preorder, determined by directed paths or locality, are investigated in [G12], Sections 4.3 and 4.5.

5.7.3 The basic structure of the directed interval

The standard directed interval $\uparrow\mathbb{I} = \uparrow[0,1]$, in the category pTop of preordered topological spaces, has a structure *partially* similar to the ordinary one, in (5.22).

Faces and *degeneracy* are as in the classical case

$$\partial^\alpha : \{*\} \rightleftarrows \uparrow\mathbb{I} : e,$$
$$\partial^-(*) = 0, \quad \partial^+(*) = 1, \quad e(t) = *, \tag{5.134}$$

where $\uparrow\mathbb{I}^0 = \{*\}$ is now an ordered space, with the unique order-relation on the singleton.

On the other hand, the classical reversion $r(t) = 1 - t$ is not an *endomap* of $\uparrow\mathbb{I}$, but becomes a map which we prefer to call *reflection*

$$r : \uparrow\mathbb{I} \to \uparrow\mathbb{I}^{op}, \qquad r(t) = 1 - t, \tag{5.135}$$

as it takes values in the *opposite* preordered space $\uparrow\mathbb{I}^{op}$, with the opposite preorder.

Again, the *standard concatenation pushout* can be realised as $\uparrow\mathbb{I}$ itself

$$
\begin{array}{ccc}
\{*\} & \xrightarrow{\ \partial^+\ } & \uparrow\mathbb{I} \\
\partial^- \downarrow & \ \ \ \ \downarrow c^- & \\
\uparrow\mathbb{I} & \xrightarrow{\ c^+\ } & \uparrow\mathbb{I}
\end{array}
\qquad
\begin{array}{l}
c^-(t) = t/2, \\[1em]
c^+(t) = (t+1)/2,
\end{array}
\tag{5.136}
$$

since a mapping $a\colon \uparrow\mathbb{I} \to X$ with values in a preordered space is a *map* (continuous and monotone) if and only if its two restrictions ac^α (to the first or second half of the interval) are maps.

5.7.4 Cylinder, cocylinder and homotopies

The directed interval produces the cylinder functor of preordered topological spaces, where the product $I(X) = X \times \uparrow\mathbb{I}$ has the product topology and the product preorder:

$$
I\colon \mathsf{pTop} \to \mathsf{pTop}, \qquad I(X) = X \times \uparrow\mathbb{I},
$$
$$
(x,t) \prec (x',t') \ \text{ if } \ (x \prec x' \text{ in } X \text{ and } t \leqslant t' \text{ in } \uparrow\mathbb{I}).
\tag{5.137}
$$

The cylinder functor has a first-degree structure, formed of four natural transformations. Faces and degeneracy are as in the classical case (see (5.29)), while the reflection r, again, has to be expressed via the *reversor*, i.e. the involutive endofunctor R which reverses the preorder-relation

$$
\partial^\alpha\colon 1 \to I, \qquad\quad e\colon I \to 1, \qquad r\colon IR \to RI,
$$
$$
R\colon \mathsf{pTop} \to \mathsf{pTop}, \qquad R(X) = X^{\mathrm{op}} \qquad (reversor), \tag{5.138}
$$
$$
r\colon I(X^{\mathrm{op}}) \to (IX)^{\mathrm{op}}, \quad r(x,t) = (x, 1-t) \qquad (reflection).
$$

Since $\uparrow\mathbb{I}$ is exponentiable in pTop (by 5.7.1), we also have a path functor, right adjoint to the cylinder functor, where the preordered space $P(Y)$ has the compact-open topology and the pointwise preorder (for $a,b\colon \uparrow\mathbb{I} \to Y$)

$$
P\colon \mathsf{pTop} \to \mathsf{pTop}, \qquad P(Y) = Y^{\uparrow\mathbb{I}},
$$
$$
a \prec b \ \text{ if } \ (\forall t \in [0,1],\ a(t) \prec_Y b(t)). \tag{5.139}
$$

The path functor is equipped with a dual (first-degree) structure, formed of four natural transformations (with the same reversor R as above)

$$
\partial^\alpha\colon P \to 1, \qquad e\colon 1 \to P, \qquad r\colon RP \to PR. \tag{5.140}
$$

A (directed) *homotopy* $\varphi\colon f^- \to f^+\colon X \to Y$ is defined by a map

$$
\hat\varphi\colon X \times \uparrow\mathbb{I} \to Y, \qquad \hat\varphi\,\partial^\alpha = f^\alpha,
$$

or, equivalently, by a map

$$\check{\varphi}\colon X \to Y^{\uparrow I}, \qquad \partial^{\alpha}\check{\varphi} = f^{\alpha}.$$

It yields a *reflected* homotopy between the opposite spaces

$$\varphi^{\mathrm{op}}\colon Rf^{+} \to Rf^{-}\colon X^{\mathrm{op}} \to Y^{\mathrm{op}},$$

$$(\varphi^{\mathrm{op}})\widehat{} = R(\hat{\varphi}).rX\colon IRX \to RIX \to RY. \tag{5.141}$$

Extending the construction of the fundamental groupoid of a space, in 5.2.9, we get here the *fundamental category* $\uparrow\Pi_1(X)$ of a preordered space, and a functor

$$\uparrow\Pi_1\colon \mathsf{pTop} \to \mathsf{Cat} \tag{5.142}$$

which sends the opposite space to the opposite category.

The fundamental monoids

$$\uparrow\pi_1(X, x) = \uparrow\Pi_1(X)(x, x)$$

are often of little importance, even in richer settings where vortices are possible.

5.7.5 Some examples

We end with some hints at notions and applications dealt with in [G8], without any proof and still in the elementary framework of preordered spaces.

Consider the following order relation in the plane

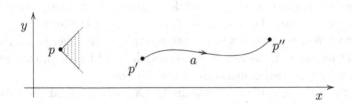

$$(x, y) \leqslant (x', y') \quad \Leftrightarrow \quad |y' - y| \leqslant x' - x. \tag{5.143}$$

The picture shows the 'cone of the future' at a point p (i.e. the set of points which follow it) and a *directed path* from p' to p'', i.e. a continuous mapping $a\colon [0, 1] \to \mathbb{R}^2$ which is (weakly) *increasing*, with respect to the natural order of the standard interval and the previous order of the plane: if $t \leqslant t'$ in $[0, 1]$, then $a(t) \leqslant a(t')$ in the plane.

Take now the following (compact) subspaces X, Y of the plane, with the induced order (the cross-marked open rectangles are taken out). A directed path in X or Y satisfies the same conditions as above

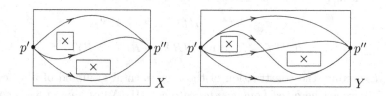

Then – as displayed in the figures above and easy to guess – there are, respectively, 3 or 4 'homotopy classes' of directed paths from the point p' to the point p'', in the fundamental categories $\uparrow\Pi_1(X)$, $\uparrow\Pi_1(Y)$; in both cases there are none from p'' to p', and every loop is constant.

First, we can view each of these 'directed spaces' as a stream with two islands, and the induced order as an upper bound for the relative velocity feasible in the stream.

Secondly, one can interpret the horizontal coordinate as (a measure of) time, the vertical coordinate as position in a 1-dimensional physical medium, and the order as the possibility of going from (x, y) to (x', y') with velocity $\leqslant 1$ (with respect to a 'rest frame' of the medium). The two forbidden rectangles are now linear obstacles in the medium, with a bounded duration in time.

Thirdly, our figures can be viewed as execution paths of concurrent automata subject to some conflict of resources, as in [FGR2], fig. 14.

In all these cases, the fundamental category distinguishes between obstructions (islands, temporary obstacles, conflict of resources) which intervene essentially together (in the earlier diagram on the left) or one after the other, *in a discernible way* (on the right). On the other hand, the underlying topological spaces are homeomorphic, and topology, or algebraic topology, cannot distinguish these two situations.

Again, all the fundamental monoids $\uparrow\pi_1(X, x_0)$ are trivial: as a striking difference to the classical case, the fundamental monoids can carry a very minor part of the information of the fundamental category $\uparrow\Pi_1(X)$.

5.7.6 From directed to weighted algebraic topology

The final chapter of [G8] investigates spaces where paths have a 'weight', or 'cost', expressing length, duration, price, energy, etc. The general aim is now: measuring the cost of (possibly non-reversible) phenomena.

The weight function takes values in the interval $[0, \infty]$ of extended real numbers. 'Weighted algebraic topology' can thus be developed as an enriched version of directed algebraic topology, where illicit paths are penalised with an infinite cost, and the licit ones are measured. Its algebraic counterpart consists of 'weighted algebraic structures', equipped with a sort of directed seminorm.

A prime structure for this purpose is given by Lawvere generalised metric spaces (see 7.3.4). A more general framework, *w-spaces* or *spaces with weighted paths*, is a natural enrichment of d-spaces.

6

Applications in Homological Algebra

In this chapter we introduce a particular approach to Homological Algebra, exposed in the book [G9] and further extended in [G10]. This presentation cannot be self-contained, and various proofs will be referred to [G9].

After the classical text on Homological Algebra by Cartan–Eilenberg [CE], in 1956, a crucial contribution to this discipline came from the theory of abelian categories, introduced by Buchsbaum and Grothendieck [Bu1, Bu2, Gt]. This extension covers new domains, and allows one to 'duplicate' theorems by duality. Yet some aspects of the theory are better viewed in the more general context of Puppe-exact categories [Pu, Mit], or p-exact categories for short.

In fact, as shown in [G9] and briefly suggested here in Section 6.6, the problem of coherence of induced morphisms (particularly relevant in complex systems like those that generate spectral sequences) can only be solved by working in auxiliary 'distributive' p-exact categories. This cannot be done in the setting of abelian categories, which are only distributive when trivial (i.e. equivalent to the category $\mathbf{1}$).

Here, after a general definition of kernels and cokernels, we show in Section 6.2 that particular categories of lattices are of help in studying exactness properties in Ab and $R\,\mathsf{Mod}$. The natural framework of this approach is Puppe-exact categories, since we need from the very beginning a *nonabelian* p-exact category Mlc, of modular lattices and modular connections.

Puppe-exact and abelian categories are presented in Sections 6.3–6.5. In Section 6.6 we introduce the category of relations of a p-exact category, as the basis for studying its subquotients, their induced relations and the coherence problem of the latter.

We end with a new Section 6.7, hinting at an extension of exactness to unpointed categories, and interpreting relative singular homology as a connected sequence of functors, after [G10].

Two-dimensional categories can also be useful in Homological Algebra, as we shall see in Sections 7.5 and 7.6.

Different perspectives on the use of category theory in Homological Algebra can be found, for instance, in [BoB, In].

6.1 Kernels and cokernels

Kernels and cokernels are the basic notions in classical Homological Algebra, the starting point to define exact sequences and exact functors. They can be defined in an arbitrary pointed category C, generalising what we have seen for pointed sets and pointed spaces, in 5.6.1.

Kernels and cokernels can be further extended to a far more general framework, beyond pointed categories: see Section 6.7.

6.1.1 Pointed categories

We assume that we are working in a *pointed* category C. As already said in 2.1.9, this means that there is a *zero object* 0, that is both initial and terminal in C: for every object A there is a unique morphism $0 \to A$ and a unique morphism $A \to 0$. *The* zero object is determined up to isomorphism; as usual, one of them is chosen.

For instance Ab, R Mod, Gp, Set., gRng, Ban and Ban$_1$ are pointed categories, with zero object given by the (adequate) singleton.

Given two objects A, B in C, the composite $A \to 0 \to B$ is called the *zero morphism* from A to B, and written as $0_{AB} \colon A \to B$, or just 0.

Plainly, the morphism $0 \to A$ is mono, also written as $0_A \colon 0 \rightarrowtail A$ and called the *zero subobject* of A; dually the morphism $A \to 0$ is epi, written as $0^A \colon A \twoheadrightarrow 0$ and called the *zero quotient* of A.

6.1.2 Kernels and cokernels

In the pointed category C the *kernel* of a morphism $f \colon A \to B$ is defined by a universal property, as an object Ker f equipped with a morphism ker $f \colon \mathrm{Ker}\, f \to A$, such that $f.(\mathrm{ker}\, f) = 0$ and

(i) every map h such that $fh = 0$ factorises uniquely through ker f

$$\mathrm{Ker}\, f \xrightarrow{\ \mathrm{ker}\, f\ } A \xrightarrow{\ f\ } B$$

(6.1)

which means that there is a unique morphism u such that $h = (\mathrm{ker}\, f)u$.

Equivalently ker f is the equaliser of f and the zero morphism $A \to B$. A *normal monomorphism* is any kernel of a morphism, and is always a regular mono. The existence of kernels does not require the existence of all equalisers, as many examples will show, in 6.3.3.

Dually, the *cokernel* of f is the coequaliser of f and the zero morphism $A \to B$, i.e. an object Cok f equipped with a morphism cok $f \colon B \to \text{Cok}\, f$ such that $(\text{cok}\, f).f = 0$ and

(i*) every map h such that $hf = 0$ factorises uniquely through cok f

$$A \xrightarrow{\ f\ } B \xrightarrow{\ \text{cok}\, f\ } \text{Cok}\, f$$

(6.2)

i.e. there is a unique morphism u such that $h = u(\text{cok}\, f)$.

A *normal epimorphism* is any cokernel of a morphism, and is always a regular epi.

6.1.3 Exercises and complements, I

Let $f \colon A \to B$ be a morphism in a pointed category. This is a list of solved exercises, which only require a straightforward verification.

(a) The kernel ker f is the same as the preimage $f^*(0)$ (see 2.4.8) of the zero subobject $0 \rightarrowtail B$, while the cokernel cok f amounts to the image of the zero quotient $A \twoheadrightarrow 0$.

(b) Normal monos (resp. epis) are closed under preimages (resp. direct images).

(c) In $R\,\mathsf{Mod}$ or $\mathsf{Set_{\bullet}}$, or $\mathsf{Top_{\bullet}}$, or Gp, or gRng, or Ban the natural 'kernel-object' is the usual subobject Ker f (the preimage of the zero subobject) and ker f is its embedding in the domain of f.

Let us note that, if $u \colon K \to \text{Ker}\, f$ is an isomorphism, also the composite $(\text{ker}\, f)u \colon K \to A$ is *a kernel* of f. The preservation of kernels by functors holds in this sense, generally.

(d) In $R\,\mathsf{Mod}$ and $\mathsf{Set_{\bullet}}$ every subobject is normal, while in $\mathsf{Top_{\bullet}}$ (resp. Gp, gRng, Ban) the normal subobjects correspond to the pointed subspaces (resp. invariant subgroups, bilateral ideals, closed linear subspaces).

Normal monomorphisms are *not* closed under composition in Gp, nor in gRng, but they are in Ban.

(e) In $R\,\mathsf{Mod}$ the natural cokernel-object is the quotient $B/f(A)$, and cok f is the canonical projection $B \to B/f(A)$.

In Gp (resp. gRng) the cokernel-object is the quotient of B modulo the invariant subgroup (resp. the bilateral ideal) generated by $f(A)$.

In these categories all regular epimorphisms are normal.

(f) We have already seen that cokernels exist in Set, and Top,, and there are non-normal epimorphisms (in 5.6.1). *The same holds in Ban.*

(g) Let us note that in R Mod the image $f(A)$ *is determined as* $\operatorname{Ker}(\operatorname{cok} f)$, i.e. *the kernel-object of the morphism* $\operatorname{cok} f$, while $\operatorname{Cok}(\ker f) = A/\operatorname{Ker} f$ gives the *coimage* of f, an isomorphic object. The other categories we have been considering above do not behave in this 'simple' way, which holds in all abelian categories, and – loosely speaking – characterises the Puppe-exact ones (see Section 6.3).

6.1.4 Exercises and complements, II

Let C be a pointed category.

(a) Define a functor $C \to C^2$ which has a right adjoint $K\colon C^2 \to C$ if and only if C has kernels; describe the action of K on objects and arrows.

(a*) Define a functor $C \to C^2$ dually related to cokernels.

6.2 Categories of lattices in Homological Algebra

We show, *in an informal way*, how the category Mlc of modular lattices and modular (Galois) connections can be used, dealing with subobjects. For simplicity, we work in an elementary context, the category R Mod of (left) modules on a (unital) ring R.

This section is also a preparation for the next: introducing the non-abelian category Mlc and analysing its exactness properties will prepare the definition of Puppe-exact categories in Section 6.3.

6.2.1 An outline

Direct and inverse images of *subsets* have been dealt with in 1.7.5. Images and preimages of *submodules* will be analysed in this section, according to the following layout.

We already know that, for every object A in R Mod, the ordered set Sub$A$ of its subobjects (i.e. submodules) is a modular lattice (see 1.2.6), with maximum $1_A\colon A \to A$ and minimum $0_A\colon 0 \to A$. For submodules $X, X' \subset A$ the meet and the join are

$$X \wedge X' = X \cap X',$$
$$X \vee X' = X + X' = \{x + x' \mid x \in X, \, x' \in X'\}. \tag{6.3}$$

A homomorphism $f\colon A \to B$ gives two monotone mappings, called *direct* and *inverse image*, or image and preimage

$$f_* \colon \mathrm{Sub}A \rightleftarrows \mathrm{Sub}B \colon f^*,$$
$$f_*(X) = f(X), \qquad f^*(Y) = f^{-1}(Y) \qquad (X \subset A, \, Y \subset B), \tag{6.4}$$

with $\mathrm{Ker}\, f = f^*(0_B)$ and $\mathrm{Im}\, f = f_*(1_A)$.

These mappings form a Galois connection $f_* \dashv f^*$ (as in 1.7.1), but here we have a stronger property

$$f^*f_*(X) = X \vee f^*(0_B) \supset X,$$
$$f_*f^*(Y) = Y \wedge f_*(1_A) \subset Y, \tag{6.5}$$

which we describe saying that the pair $(f_*, f^*)\colon \mathrm{Sub}A \to \mathrm{Sub}B$ is a *modular connection*.

All this will be used to define a *transfer functor* for subobjects of $R\,\mathsf{Mod}$

$$\mathrm{Sub}_R \colon R\,\mathsf{Mod} \to \mathsf{Mlc}, \tag{6.6}$$

with values in the category Mlc of modular lattices and modular connections – a subcategory of the category AdjOrd of ordered sets and Galois connections (see Section 1.7). This functor will be seen to be exact in 6.3.8, i.e. a functor which preserves kernels and cokernels (or equivalently exact sequences, as defined in 6.3.4).

6.2.2 *Modular lattices and modular connections*

We begin now to study the category Mlc of *modular lattices and modular connections*, that we have seen to abstract the properties of direct and inverse images for $R\,\mathsf{Mod}$ (and will similarly work for a wide class of categories).

An object is a modular lattice (with 0 and 1). A morphism, called a *modular connection*

$$f = (f_\bullet, f^\bullet)\colon X \to Y,$$

is a pair where

(i) $f_\bullet \colon X \to Y$ and $f^\bullet \colon Y \to X$ are monotone mappings,

(ii) $f^\bullet f_\bullet(x) = x \vee f^\bullet 0, \quad f_\bullet f^\bullet(y) = y \wedge f_\bullet 1 \qquad$ (for $x \in X$, $y \in Y$).

As a consequence $f^\bullet f_\bullet \geqslant \mathrm{id}X$ and $f_\bullet f^\bullet \leqslant \mathrm{id}Y$, and we have an adjunction $f_\bullet \dashv f^\bullet$ between ordered sets, i.e. a covariant Galois connection (see 1.7.1).

As we have seen, $f_\bullet$ determines $f^\bullet$, and conversely

$$f^\bullet(y) = \max\{x \in X \mid f_\bullet(x) \leqslant y\},$$
$$f_\bullet(x) = \min\{y \in Y \mid f^\bullet(y) \geqslant x\}. \tag{6.7}$$

Moreover $f_\bullet$ preserves all the existing joins (including $0 = \vee\emptyset$), $f^\bullet$ preserves all the existing meets (including $1 = \wedge\emptyset$), and

$$f_\bullet f^\bullet f_\bullet = f_\bullet, \qquad f^\bullet f_\bullet f^\bullet = f^\bullet. \tag{6.8}$$

Condition (ii) can be equivalently rewritten in a seemingly stronger form:

(ii$'$) $f^\bullet(f_\bullet x \vee y) = x \vee f^\bullet y, \qquad f_\bullet(f^\bullet y \wedge x) = y \wedge f_\bullet x \qquad (x \in X, y \in Y).$

Indeed, from (ii) *and* the modularity of lattices, we have

$$x \vee f^\bullet y = (x \vee f^\bullet y) \vee f^\bullet 0 = f^\bullet f_\bullet(x \vee f^\bullet y) = f^\bullet(f_\bullet x \vee f_\bullet f^\bullet y)$$
$$= f^\bullet(f_\bullet x \vee (y \wedge f_\bullet 1)) = f^\bullet((f_\bullet x \vee y) \wedge f_\bullet 1)$$
$$= f^\bullet(f_\bullet f^\bullet(f_\bullet x \vee y)) = f^\bullet(f_\bullet x \vee y).$$

This makes evident that modular connections are closed under composition in the category AdjOrd of ordered sets and Galois connections (introduced in 1.7.4),

$$(g_\bullet, g^\bullet) \cdot (f_\bullet, f^\bullet) = (g_\bullet f_\bullet, f^\bullet g^\bullet), \tag{6.9}$$

and form a subcategory Mlc of the latter.

Mlc inherits a (contravariant) involution from AdjOrd, and an order consistent with composition and involution (see 1.7.4)

$$X \mapsto X^{\mathrm{op}}, \quad ((f_\bullet, f^\bullet): X \to Y) \mapsto ((f^\bullet, f_\bullet): Y^{\mathrm{op}} \to X^{\mathrm{op}}),$$
$$f \leqslant g \iff f_\bullet \leqslant g_\bullet \iff f^\bullet \geqslant g^\bullet. \tag{6.10}$$

Mlc is thus selfdual. We are also interested in the category Dlc of *distributive lattices and modular connections*, a full subcategory of Mlc.

6.2.3 Exercises and complements

(a) The morphism $f: X \to Y$ of Mlc is invertible if and only if the following equivalent conditions hold:

(i) $f^\bullet f_\bullet = 1_X, \qquad f_\bullet f^\bullet = 1_Y,$

(ii) $f^\bullet 0 = 0, \qquad f_\bullet 1 = 1,$

(iii) $f_\bullet$ is a bijective mapping,

(iii*) $f^\bullet$ is a bijective mapping,

(iv) $f_\bullet$ is an isomorphism of ordered sets (hence of lattices, in Mlh),

(iv*) $f^\bullet$ is an isomorphism of ordered sets (hence of Mlh).

(b) In this way, the isomorphisms of the category Mlc correspond to the isomorphisms of Mlh, namely the usual bijective homomorphisms of modular lattices; the isomorphism relation in these categories is the same.

(c) Both of the following faithful functors reflect the isomorphisms (where $|X|$ denotes the underlying set of X)

$$U: \mathsf{Mlc} \to \mathsf{Set}, \qquad X \mapsto |X|, \qquad (f_\bullet, f^\bullet) \mapsto f_\bullet, \qquad (6.11)$$

$$V: \mathsf{Mlc} \to \mathsf{Set}^{\mathrm{op}}, \qquad X \mapsto |X|, \qquad (f_\bullet, f^\bullet) \mapsto f^\bullet. \qquad (6.12)$$

U will be called the *forgetful functor* of Mlc.

(d) Let $f = (f_\bullet, f^\bullet): X \to Y$ be a modular connection between distributive lattices. Prove that the mappings $f_\bullet$ and $f^\bullet$ are homomorphisms of *quasi lattices* (in the sense of 1.2.4): in other words, $f_\bullet$ also preserves binary meets (but need not preserve the maximum, of course), while (dually) $f^\bullet$ also preserves binary joins (but need not preserve the minimum).

Hints: one can begin by writing $f_\bullet(x \wedge x') = f_\bullet f^\bullet f_\bullet(x \wedge x')$.

6.2.4 Exactness properties

We prove now that the category Mlc of modular lattices and modular connections satisfies exactness conditions that will make it Puppe-exact, or p-exact for short (according to a definition made precise below, in 6.3.2).

The reader is warned that Mlc is not abelian, and cannot even be exactly embedded in an abelian category (this will be proved in Theorem 6.5.5).

First there is a *zero object* (both terminal and initial, see 6.1.1), namely the one-point lattice $0 = \{*\}$, since every object X has unique morphisms t and s:

$$X \overset{t}{\longrightarrow} \{*\} \overset{s}{\longrightarrow} X \qquad (t^\bullet(*) = 1, \ s_\bullet(*) = 0). \qquad (6.13)$$

The zero morphism $0_{XY}: X \to \{*\} \to Y$ is defined by $x \mapsto 0$ and $y \mapsto 1$.

Every morphism $f = (f_\bullet, f^\bullet): X \to Y$ has a *kernel* $\ker f$ and a *cokernel* $\operatorname{cok} f$ (see 6.1.2)

$$
\begin{aligned}
m &= \ker f: \downarrow f^\bullet 0 \to X, & m_\bullet(x) &= x, & m^\bullet(x) &= x \wedge f^\bullet 0, \\
p &= \operatorname{cok} f: Y \to \uparrow f_\bullet 1, & p_\bullet(y) &= y \vee f_\bullet 1, & p^\bullet(y) &= y.
\end{aligned}
\qquad (6.14)
$$

Verifying the universal property of m is straightforward: if $h: Z \to X$ is annihilated by f, then for every $z \in Z$ we have:

$$f^\bullet 0 = f^\bullet(f_\bullet h_\bullet(z)) = h_\bullet(z) \vee f^\bullet 0 \geqslant h_\bullet(z),$$

so that $h_\bullet$ takes values in $\downarrow f^\bullet 0$ (and we can restrict it). Dually for p.

As a consequence of these properties, the morphism f has a canonical ternary factorisation $f = ngq$, through $q = \text{cok}\,(\ker f)$ (the *coimage* of f) and $n = \ker(\text{cok}\,f)$ (the *image* of f), by a unique central morphism g

$$\downarrow f^\bullet 0 \xrightarrow{\;m\;} X \xrightarrow{\;f\;} Y \xrightarrow{\;p\;} \uparrow f_\bullet 1$$

$$q \downarrow \qquad\qquad \uparrow n \qquad\qquad\qquad (6.15)$$

$$\uparrow f^\bullet 0 \xrightarrow{\;g\;} \downarrow f_\bullet 1$$

$$q_\bullet(x) = x \vee f^\bullet 0, \quad q^\bullet(x) = x, \qquad n_\bullet(y) = y, \quad n^\bullet(y) = y \wedge f_\bullet 1,$$

$$g_\bullet(x) = f_\bullet(x), \qquad g^\bullet(y) = f^\bullet(y).$$

Moreover g is an *isomorphism* of Mlc (as characterised above), because

$$g^\bullet g_\bullet(x) = f^\bullet f_\bullet(x) = x, \qquad g_\bullet g^\bullet(y) = f_\bullet f^\bullet(y) = y,$$

for $x \geqslant f^\bullet 0$ and $y \leqslant f_\bullet 1$.

All this will mean that the category Mlc is Puppe-exact.

As an easy consequence of the canonical factorisation, each monomorphism is normal (i.e. a kernel of some arrow) and each epimorphism is normal (i.e. a cokernel).

6.2.5 Subobjects and quotients

We also note that every element $a \in X$ determines a *subobject* and a *quotient* of X

$$m\colon \downarrow a \rightarrowtail X, \qquad m_\bullet(x) = x, \quad m^\bullet(x) = x \wedge a, \qquad (6.16)$$

$$p\colon X \twoheadrightarrow \uparrow a, \qquad p_\bullet(x) = x \vee a, \quad p^\bullet(x) = x, \qquad (6.17)$$

where $m = \ker p$ and $p = \text{cok}\,m$.

The correspondence $a \mapsto m$ establishes an isomorphism between the *lattice* X and the ordered set Sub$X$ of *subobjects* of X in Mlc, while the correspondence $a \mapsto p$ gives an anti-isomorphism of X with the ordered set of *quotients* of X.

The subobject m and the quotient p determined by the element $a \in X$ form a *short exact sequence* (cf. 6.3.4), i.e. a pair (m, p) meeting the condition

$$\downarrow a \rightarrowtail X \twoheadrightarrow \uparrow a, \qquad m \sim \ker p, \quad p \sim \text{cok}\,m. \qquad (6.18)$$

Conversely, every short exact sequence (m, p) in Mlc with central object X is isomorphic to a unique sequence of this type, with $a = m_\bullet 1 = p^\bullet 0$.

6.2.6 Exercises and complements (The transfer functor)

(a) Coming back to 6.2.1, the reader can prove that we have a functor

$$\mathrm{Sub}_R \colon R\,\mathsf{Mod} \to \mathsf{Mlc}, \qquad A \mapsto \mathrm{Sub}(A), \quad f \mapsto (f_*, f^*). \qquad (6.19)$$

(b) This functor is *exact*, i.e. preserves kernels and cokernels.

6.3 Puppe-exact categories

We review the main definitions about p-exact categories, exact sequences and exact functors. Then we deal with the category Mlc of modular lattices and modular connections, introduced in the previous section.

We show now that this category, which is p-exact and not abelian, abstracts the behaviour of direct and inverse images of subobjects *for all p-exact categories* (including the abelian ones); p-exact categories form thus the natural setting of this analysis.

Subquotients, a crucial tool of Homological Algebra, are introduced in 6.3.9. Some proofs are only referred to the book [G9].

6.3.1 Three forms of exactness

The term 'exact category' has assumed different meanings in category theory, *and three of them are still in use* (listed as (b), (c), (d) below).

(a) (*Abelian categories*) The term 'exact category' was first used in 1955–56 by Buchsbaum [Bu1, Bu2], essentially meaning what is now called an abelian category (even if the existence of finite biproducts was seemingly deferred to an additional axiom).

The name subsisted – in this sense – in various papers of the 1950's and 1960's (by Atiyah, Hilton, Heller, etc.), together with the term 'abelian category' that had been introduced by Grothendieck in his Tôhoku paper of 1957 [Gt]. Gradually the last term was universally accepted in the literature, including the first books on category theory, by Freyd [Fr1] (1964) and Mitchell [Mit] (1965) – both focused on the embedding of abelian categories in categories of modules.

(b) (*Puppe-exact categories*) Meanwhile, in 1962, Puppe [Pu] had introduced a more general notion, called a 'quasi exact category', still selfdual but not additive, and based on the axioms (AB1), (AB2) of Grothendieck.

This framework was investigated by Tsalenko [T1, T2] (also transliterated as 'Calenko') in 1964 and 1967, for the construction of the category of relations, and by many researchers for diagram lemmas. It became an 'exact category' in Mitchell's book [Mit] (1965), where abelian categories are defined as additive exact categories. The name is kept, in this sense, in subsequent works by Brinkmann and Puppe [Bri, BriP], in 1969, and by Herrlich and Strecker [HeS], in 1973. The books [AHS, FrS] still use in 1990 exact categories in the sense of Puppe and Mitchell.

(c) (*Barr-exact categories*) Later, in 1971, Barr [Bar] used the term 'exact category' for a different generalisation of abelian categories, based on regular categories and not selfdual (see Chapter 4).

This setting became popular in category theory and has been extended in various forms (see [Bou1, BoB, JaMT, Bo4, Bou2]). As we have seen, it contains all varieties of algebras, and can give a general categorical presentation of Universal Algebra.

(d) (*Quillen-exact categories*) In 1973 a paper by Quillen [Qu] on higher K-theory introduced another notion of 'exact category', as a full additive subcategory E of an abelian category A, closed under extensions.

This means that, for every short exact sequence $A \rightarrowtail C \twoheadrightarrow B$ of A, if A and B are in E, so is C. This notion, already used elsewhere, can be defined intrinsically, without reference to an abelian environment, as an additive category equipped with a class of 'short exact sequences' satisfying some axioms.

The cases (b), (c), (d) are thus distinguished as: *Puppe-exact, Barr-exact* and *Quillen-exact* categories. It is often remarked that the first two notions satisfy the 'equation': exact + additive = abelian.

A Puppe-exact category is abelian if and only if it has finite products, or equivalently finite sums (by Theorem 6.4.6). Therefore the intersection of each pair of these three frameworks only contains the abelian categories.

6.3.2 Exact and abelian categories

A *Puppe-exact*, or *p-exact*, category E has to satisfy two selfdual axioms:

(pex.1) E is pointed and every morphism has a kernel and a cokernel,

(pex.2) in the *canonical factorisation* of a morphism $f \colon A \to B$ through its *coimage* and its *image*

$$\mathrm{Ker}\, f \overset{k}{\rightarrowtail} A \xrightarrow{\ f\ } B \overset{c}{\twoheadrightarrow} \mathrm{Cok}\, f$$
$$q \downarrow \qquad\qquad \uparrow n \qquad\qquad\qquad (6.20)$$
$$\mathrm{Coim}\, f \underset{g}{\longrightarrow} \mathrm{Im}\, f$$

$$\mathrm{Coim}\, f = \mathrm{Cok}\,(\mathrm{ker}\, f), \qquad \mathrm{Im}\, f = \mathrm{Ker}\,(\mathrm{cok}\, f),$$
$$q = \mathrm{coim}\, f = \mathrm{cok}\,(\mathrm{ker}\, f), \qquad n = \mathrm{im}\, f = \mathrm{ker}\,(\mathrm{cok}\, f),$$

the unique morphism g such that $f = ngq$ is an *isomorphism*.

The existence of the factoristion is proved in Exercise (a), below. These axioms correspond to the axioms (AB1), (AB2) of Grothendieck [Gt]; their formulation is redundant, but clearer and often simpler to check than other, more concise ones: see Exercise (d).

As an easy consequence of the axioms, each monomorphism is normal, i.e. a kernel of some arrow, and each epimorphism is normal, i.e. a cokernel (see Exercise (b)). Every morphism has an essentially unique epi-mono factorisation, the canonical one, as in diagram (6.20), – which explains our names of coimage and image, above. E is a *balanced* category, i.e. epi and mono implies iso.

Following the notation of 2.4.1, we write as $\mathrm{Sub}_E A$ and $\mathrm{Quo}_E A$ the (possibly large) ordered sets of subobjects and quotients of an object A.

Cokernels and kernels give two *antitone* mappings

$$\mathrm{cok}\colon \mathrm{Sub}_E A \rightleftarrows \mathrm{Quo}_E A \colon \mathrm{ker} \qquad (\textit{kernel duality}), \qquad (6.21)$$

that are easily proved to be inverse to each other (see Exercise (c)), using the fact that all monos and epis are normal

$$\mathrm{ker}\,(\mathrm{cok}\,m) = m, \qquad \mathrm{cok}\,(\mathrm{ker}\,p) = p. \qquad (6.22)$$

(On the other hand, *categorical* duality links $\mathrm{Sub}_E A$ and $\mathrm{Quo}_{E^{\mathrm{op}}} A$, preserving the ordering. Working with both dualities simplifies many proofs.)

From now on we assume that a p-exact category is well powered (see 2.4.1), i.e. that all sets $\mathrm{Sub}A$ (and $\mathrm{Quo}A$, as a consequence) are small. We shall see in 6.3.8 that all $\mathrm{Sub}A$ and $\mathrm{Quo}A$ are modular lattices, anti-isomorphic by (6.21).

A p-exact category is said to be *trivial* if all its objects are zero objects; or, in other words, if it is equivalent to the singleton category **1**.

An *abelian category* can be defined as a p-exact category having finite products, or equivalently finite sums, or finite limits and colimits; this equivalence will be proved in Theorem 6.4.6, but we shall use from now the notion of abelian category, much better known than that of a p-exact category.

Exercises and complements. (a) Assuming the axiom (pex.1), prove the existence of the factorisation (6.20), with a unique morphism g (which need not be invertible).

(b) In a p-exact category every monomorphism is normal. Dually, every epimorphism is normal.

(c) Prove the kernel duality (6.21).

*(d) A pointed category is p-exact if and only if every morphism factorises as a normal epi followed by a normal mono.

6.3.3 Examples and complements

The pointed categories Gp, gRng and Ban are not p-exact, since there are non-normal monomorphisms; Set. and Top. are not p-exact, since there are non-normal epimorphisms.

The categories Set., Top., Ban are homological, in the sense of [G10]; this is also true of the category of 'pairs' of spaces (or sets, or groups), with respect to an assigned ideal of null morphisms: see Section 6.7.

The following examples of p-exact categories are taken from [G9], Subsection 1.5.6.

(a) Every abelian category.

(b) Every non-empty full subcategory of a p-exact category that is closed under subobjects and quotients; for instance:

- cyclic groups, *or* finite cyclic groups,

- abelian groups of cardinal (or rank) lower than a fixed integer,

- vector spaces on a fixed field, of dimension lower than a fixed integer.

(Replacing a 'fixed integer' with a 'fixed infinite cardinal' just gives an abelian subcategory.)

(c) The category Mlc of modular lattices and modular connections, as proved in 6.2.2–6.2.4. Mlc abstracts the behaviour of direct and inverse images of subobjects in (abelian or) p-exact categories (as we shall see in 6.3.8). It is *not* abelian and cannot even be exactly embedded in an abelian category (as will be proved in Theorem 6.5.5).

Its full subcategory Dlc of *distributive lattices* (also defined in 6.2.2) will play the same role for *distributive p-exact categories*, i.e. those p-exact categories whose lattices of subobjects are distributive. For instance, the category of cyclic groups is distributive p-exact, by Exercise 1.2.7(c).

(d) The category $\mathcal{I}$ of sets and partial bijections, introduced in 1.8.7. A morphism $f\colon X \rightarrowtail Y$ in $\mathcal{I}$ is a bijection between a subset (Def f) of X and a subset (Val f) of Y, or equivalently a single-valued, injective relation; they compose as relations in Rel (Set). $\mathcal{I}$ is selfdual; the zero object is $\emptyset$, and

$$\text{Ker}\, f = X \setminus \text{Def}\, f, \qquad \text{Cok}\, f = Y \setminus \text{Val}\, f.$$

This category is distributive p-exact (with Sub $X = \mathcal{P}X$), and 'universal' in this domain, in the sense that every small category of this kind has an exact embedding in $\mathcal{I}$ (cf. [G9], 4.6.7(b)).

(e) The *projective category* Pr E associated to any p-exact category E, that will be studied in Section 6.5. In particular, if E $= K$ Vct is the abelian category of vector spaces over the field K, $\mathsf{Pr}\,\mathsf{E}$ 'is' the (p-exact) category of projective spaces and projective linear maps over K.

*(f) The *distributive expansion* $\mathsf{Dst}\,\mathsf{E}$ of any p-exact category E, defined in [G9], Section 2.8. It is a distributive p-exact category, and is never abelian – unless E (and therefore $\mathsf{Dst}\,\mathsf{E}$) is trivial.

Its objects are the pairs (A, X) consisting of an object A of E and a distributive sublattice $X \subset \mathrm{Sub}A$. A morphism $f \colon (A, X) \to (B, Y)$ comes from an E-morphism $f \colon A \to B$ which carries the subobjects of X into Y and those of Y into X (by direct and inverse images, see 6.3.7).

(g) Every category of functors E^{S}, where E is a p-exact category and S is a small category. This includes all cartesian powers of E (when S is a discrete category), the category of morphisms of E (when $\mathsf{S} = \mathbf{2}$, the arrow category), the category of commutative squares of E (when $\mathsf{S} = \mathbf{2} \times \mathbf{2}$), etc. The proof that E^{S} is also p-exact is straightforward.

Similarly, if E is abelian, so is E^{S}.

6.3.4 Exact functors and exact sequences

A functor $F \colon \mathsf{E} \to \mathsf{E}'$ between p-exact categories is said to be *exact* if it preserves kernels and cokernels, in the usual sense of preserving limits, that is – in the present case – up to equivalence of monos and epis, respectively.

As a consequence it also preserves the zero object (that is the kernel and cokernel of any identity), canonical factorisations and exact sequences.

The latter are defined in the usual way: in the p-exact category E the sequence

$$A \xrightarrow{\ f\ } B \xrightarrow{\ g\ } C \tag{6.23}$$

is said to be *exact* (in B) if $\mathrm{im}\,f = \ker g$, or equivalently $\mathrm{cok}\,f = \mathrm{coim}\,g$. A sequence of consecutive morphisms is said to be *exact* if it is in all locations where the condition makes sense.

Short exact sequences are of particular interest (see 6.3.5). By definition, this is a sequence of the following form, that is exact (in A, B, C)

$$0 \longrightarrow A \xrightarrow{\ m\ } B \xrightarrow{\ p\ } C \longrightarrow 0 \tag{6.24}$$

Plainly, this means that the sequence satisfies the following equivalent conditions:

(i) m is a monomorphism, p is an epimorphism and $\mathrm{im}\,m = \ker p$,

(ii) $m \sim \ker p$ (as a mono in B) and $p \sim \mathrm{cok}\,m$ (as an epi from B).

This sequence will also be written in the form $A \rightarrowtail B \twoheadrightarrow C$.

A functor F between p-exact categories is *left exact* (resp. *right* exact) if it preserves kernels (resp. cokernels), or equivalently exact sequences of the form $0 \to A \to B \to C$ (resp. $A \to B \to C \to 0$). More generally, F is said to be *half exact* if it takes any sort exact sequence $0 \to A \to B \to C \to 0$ to an exact sequence $FA \to FB \to FC$; note that this property is not closed under composition.

A functor is exact if and only if it is left and right exact. A left adjoint functor between p-exact categories is necessarily right exact. We shall see in 6.4.7 that any exact (or even half exact) functor between abelian categories preserves finite products and sums, and the additive structure.

6.3.5 *Exercises and complements*

(a) In a p-exact category, a sequence (f, g) of consecutive morphisms is exact if and only if it can be inserted in the commutative diagram below, with a slanting short exact sequence

$$A \xrightarrow{\quad f \quad} B \xrightarrow{\quad g \quad} C \qquad\qquad (6.25)$$

(b) An object X is a zero object if and only if the pair $(1_X, 1_X)$ is an exact sequence, if and only if it is a short exact sequence.

(c) A functor between p-exact categories is exact if and only if it preserves short exact sequences.

6.3.6 *Lemma* (Pullbacks and pushouts in p-exact categories)

A p-exact category E *has pullbacks of monos and pushouts of epis along arbitrary maps. The following points are a more detailed formulation.*

(a) Along a map f, the pullback (or preimage) $m = f^*(n)$ of a monomorphism n and the pushout (or direct image) $q = f_\circ(p)$ of an epimorphism p always exist, and are computed as follows

$$f^*(n) = \ker((\operatorname{cok} n).f), \qquad\qquad f_\circ(p) = \operatorname{cok}(f.(\ker p)), \qquad (6.26)$$

(b) In the same situation, if in the left diagram f is epi, then so is g and the square is also a pushout (a bicartesian square*). Dually, if in the right diagram f is mono, then so is g and the square is also a pullback.*

(c) Given the following commutative diagram with (short) exact rows

$$
\begin{array}{ccccc}
A & \xrightarrow{\ m\ } & B & \xrightarrow{\ p\ } & C \\
\scriptstyle u\downarrow & & \scriptstyle v\downarrow & & \scriptstyle w\downarrow \\
A' & \xrightarrow[\ m'\ ]{} & B' & \xrightarrow[\ p'\ ]{} & C'
\end{array}
\tag{6.27}
$$

the left square is a pullback if and only if w is mono. The right square is a pushout if and only if u is epi.

Proof The reader can prove the statement for a category of modules, or try the general case. The latter can be found in [G9], Lemma 2.2.4. □

6.3.7 Direct and inverse images

In a p-exact category E every morphism $f\colon A \to B$ gives rise to two monotone mappings between ordered sets, called – respectively – *direct and inverse image* of subobjects along f

$$
\begin{aligned}
f_*\colon \mathrm{Sub}A \to \mathrm{Sub}B, && f_*(m) = \mathrm{im}\,(fm) = \ker\,(\mathrm{cok}\,(fm)), \\
f^*\colon \mathrm{Sub}B \to \mathrm{Sub}A, && f^*(n) = \ker\,((\mathrm{cok}\,n).f),
\end{aligned}
\tag{6.28}
$$

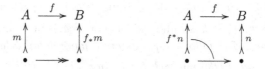

The left square is commutative and the right one is a pullback (by the previous lemma).

The theorem below proves that the pair (f_*, f^*) is a *modular connection between modular lattices*, i.e. a morphism of the category Mlc studied in Section 6.2.

In particular:

$$
f_*0 = 0, \quad f_*1 = \mathrm{im}\,f, \qquad f^*0 = \ker f, \quad f^*1 = 1.
\tag{6.29}
$$

6.3.8 Theorem and Definition (The transfer functor)

(a) Every (well-powered) p-exact category E *has an exact functor with values in the p-exact category* Mlc *of modular lattices and modular connections*

$$\mathrm{Sub_E}\colon \mathsf{E} \to \mathsf{Mlc},$$

$$A \mapsto \mathrm{Sub}A, \qquad f \mapsto \mathrm{Sub}(f) = (f_*, f^*).$$

(6.30)

This functor also reflects exactness: a sequence of E *is exact if and only if its image in* Mlc *is exact.*

$\mathrm{Sub_E}$ *will be called the* transfer functor *(for subobjects), or also the* projective functor *of* E. *(The name is explained by the projective category* Pr E, *that will be studied in Section 6.5.)*

(b) The lattice operations and the ordering of SubA *can be described as follows, for* $m, n \in \mathrm{Sub}A$ *and* $p = \mathrm{cok}\, m$, $q = \mathrm{cok}\, n$ *(the corresponding quotients)*

$$m \wedge n = m_* m^*(n) = n_* n^*(m),$$

$$m \vee n = p^* p_*(n) = q^* q_*(m),$$

(6.31)

$$n \leqslant m \;\Leftrightarrow\; n^*(m) = 1.$$

(c) An exact functor $F\colon \mathsf{E} \to \mathsf{E}'$ *preserves direct and inverse images of subobjects, as well as their finite joins and meets. Therefore it induces a homomorphism of lattices, for every* A *in* E

$$(\mathrm{Sub}_F)A\colon \mathrm{Sub_E}A \to \mathrm{Sub_{E'}}(FA), \qquad \mathrm{Sub}_F(m) = \mathrm{im}\,(Fm),$$

(6.32)

(where $\mathrm{im}\,(Fm)$ *is the* subobject *associated to the monomorphism* Fm*).*

Proof The particular case $\mathsf{E} = R\,\mathrm{Mod}$ has been studied in 6.2.1, 6.2.6. The general case is proved in [G9], Theorem 2.2.6. □

6.3.9 Subquotients

As a prime example, the homology group

$$H_n(A) = \mathrm{Ker}\,\partial_n / \mathrm{Im}\,\partial_{n+1}$$

of a chain complex A of abelian groups (see 5.5.2) is a subquotient of the component A_n. Complex systems of subquotients arise, for instance, in the study of spectral sequences (see [M2, HiW, G9]).

In general, a *subquotient* $S = M/N$ of an object A in a p-exact category E is a quotient of a subobject (M) of A, or equivalently a subobject of a quotient (A/N) of A.

It is determined by a decreasing pair of subobjects of A

$$m: M \rightarrowtail A, \quad n: N \rightarrowtail A \quad (n \prec m),$$

via the commutative square produced by the epi-mono factorisation of $pm: M \to A \to A/N$, where $p = \operatorname{cok} n$

$$
\begin{array}{ccc}
M & \xrightarrow{\;\;m\;\;} & A \\
{\scriptstyle p'}\big\downarrow & & \big\downarrow{\scriptstyle p} \\
S & \xrightarrow[\;\;m'\;\;]{} & A/N
\end{array}
\qquad
\begin{array}{l}
p = \operatorname{cok} n, \\[2em]
n = \ker p \prec m.
\end{array}
\qquad (6.33)
$$

This square is bicartesian, i.e. pullback and pushout at the same time.

(In fact, it is a pullback by (6.26), because $p^*(m') = p^*p_*(m) = m \vee \ker p = m$; it is a pushout by 6.3.6(b), or by duality.)

Conversely, every bicartesian square in E formed of two 'parallel' monos and two 'parallel' epis is of this type, up to isomorphism.

6.4 Additive and abelian categories

We now review additive and abelian categories, with a more complete analysis of biproducts – after 2.1.9 and 3.2.8.

The reader can find further results in [M4], Chapter VIII, and in [Gt]. The embedding of abelian categories in categories of modules was studied in the books [Mit, Fr1], right at the beginning of category theory.

6.4.1 Preadditive categories and biproducts

A *preadditive*, or $\mathbb{Z}$-*linear*, category is a category C where every hom-set $\mathsf{C}(A, B)$ is equipped with a structure of abelian group, so that composition is bilinear over $\mathbb{Z}$.

More generally, we are also interested in $\mathbb{N}$-*linear categories*, where every hom-set $\mathsf{C}(A, B)$ is equipped with a structure of abelian monoid, and composition is bilinear over the semiring $\mathbb{N}$. The zero element of $\mathsf{C}(A, B)$ is generally written as $0_{AB}: A \to B$, or simply 0. (Both these structures are related to 'enrichment', see 7.3.3.)

By definition, the *biproduct* of two objects A, B in an arbitrary *pointed* category is an object C equipped with four maps u, v, p, q, so that

$$
\begin{array}{ccccc}
A & & & & B \\
{\scriptstyle 1}\big\downarrow & \searrow{\scriptstyle u} & & \swarrow{\scriptstyle v} & \big\downarrow{\scriptstyle 1} \\
& & C & & \\
\big\downarrow & \swarrow{\scriptstyle p} & & \searrow{\scriptstyle q} & \big\downarrow \\
A & & & & B
\end{array}
\qquad (6.34)
$$

- $pu = \mathrm{id}A, \quad qv = \mathrm{id}B, \qquad qu = 0, \quad pv = 0,$

- $(C; p, q)$ is the product of A, B and $(C; u, v)$ is their sum.

The object C is often written as $A \oplus B$; the maps p, q are called *projections* while u, v are called *injections*.

Similarly one defines the biproduct $\bigoplus_i A_i$ of any family of objects; the biproduct of the empty family is the zero object. A morphism $C \to \bigoplus_i A_i$ will often be written as (f_i), by its *components* $f_i \colon C \to A_i$, while a morphism $\bigoplus_i A_i \to D$ can be written as $[f_i]$, by its *co-components* $f_i \colon A_i \to D$ (as in our notation for products and coproducts, in 2.1.1 and 2.1.5).

As we know, all categories of modules, Abm and Ban have finite biproducts; other examples will be examined in 6.4.5, including some cases of arbitrary biproducts.

6.4.2 Lemma

Let C *be an* N-*linear category.*

The following conditions on an object Z are equivalent:

(a) Z *is terminal,*

(a*) Z *is initial,*

(b) Z *is the zero object,*

(c)   C(Z, Z) *is the trivial group,*

(d) $\mathrm{id}Z = 0_{ZZ}.$

Secondly, the following conditions on the diagram (6.34) *are equivalent (using the notation by components and co-components, recalled above):*

(i) $(C; p, q)$ *is the product of A, B and* $u = (\mathrm{id}A, 0)$, $v = (0, \mathrm{id}B)$,

(i*) $(C; u, v)$ *is the sum of A, B and* $p = [\mathrm{id}A, 0]$, $q = [0, \mathrm{id}B]$,

(ii) $pu = \mathrm{id}A$, $qv = \mathrm{id}B$, $up + vq = \mathrm{id}C$,

(iii) p, q *are jointly mono,* $pu = \mathrm{id}A$, $qv = \mathrm{id}B$, $pv = 0$, $qu = 0$,

(iii*) u, v *are jointly epi,* $pu = \mathrm{id}A$, $qv = \mathrm{id}B$, $pv = 0$, $qu = 0$.

Therefore in an N-*linear category (a fortiori in a preadditive category), the existence of binary (or finite) products is equivalent to the existence of binary (or finite) sums, which are then biproducts.*

Proof Well-known and straightforward. We only write down the proof of the second part, taking advantage of categorical duality.

The implication (i) $\Rightarrow$ (iii) is obvious. (iii) $\Rightarrow$ (ii) We have $p(up+vq) = p$ and $q(up+vq) = q$, whence $up+vq = \mathrm{id}\,C$. (ii) $\Rightarrow$ (i) The maps $f\colon X \to A$ and $g\colon X \to B$ are given. If $h\colon X \to C$ satisfies $ph = f$ and $qh = g$, then $h = (up + vq)h = uf + vg$; conversely, the morphism $h = uf + vg\colon X \to C$ does have f, g as components. $\square$

6.4.3 Theorem and Definition (Semiadditive categories)

Let C *be a category.*

(a) C *is said to be* semiadditive *if it satisfies the following equivalent conditions:*

(*i*) C *is an* $\mathbb{N}$-*linear category with finite products,*

(*i**) C *is an* $\mathbb{N}$-*linear category with finite sums,*

(*ii*) C *is a pointed category with finite biproducts.*

When these conditions hold, the sum $f + g$ of two maps $f, g\colon A \to B$ is determined by the categorical structure

$$A \xrightarrow{d} A \oplus A \xrightarrow{f \oplus g} B \oplus B \xrightarrow{\partial} B \qquad f + g = \partial(f \oplus g)d, \qquad (6.35)$$

where $d = (1,1)\colon A \to A \oplus A$ is the diagonal of the product and $\partial = [1,1]\colon B \oplus B \to B$ is the codiagonal of the coproduct.

(b) A functor $F\colon$ C $\to$ C$'$ between semiadditive categories is said to be additive *if it satisfies the following equivalent conditions:*

(*i*) *F preserves finite products,*

(*i**) *F preserves finite sums,*

(*ii*) *F preserves finite biproducts,*

(*iii*) *F preserves (finite) sums of parallel maps.*

Proof (a) We already know, from the previous lemma, that (i) and (i*) are equivalent. Furthermore, if they hold, C has finite biproducts and the binary ones satisfy the conditions of 6.4.2.

Therefore, for $f, g\colon A \to B$, the composite (6.35) is computed as follows (letting p', q' be the projections of $B \oplus B$):

$$\partial(f \oplus g)d = (p' + q')(u'fp + v'gq)(u + v) = f + g.$$

Now we suppose that C is pointed and has finite biproducts (as defined in 6.4.1), and define the sum of parallel maps as in (6.35). One verifies that this is indeed an enrichment over Abm (see [M4], Section VIII.2, Exercise 4a).

Moreover, in diagram (6.34), the map $up + vq \colon Z \to Z$ must be the identity, because:

$$p(up + vq) = pup + pvq = p, \qquad q(up + vq) = qup + qvq = q.$$

(b) It is a straightforward consequence. $\square$

6.4.4 Definition (Additive categories)

As a consequence of the previous theorem, we can define an *additive category* C as a preadditive category (i.e. a $\mathbb{Z}$-linear category) with finite products, or equivalently with finite sums.

Equivalently, we can say that C is a pointed category with finite biproducts, such that the $\mathbb{N}$-linear structure defined in (6.35) (and determined by the categorical structure) is actually $\mathbb{Z}$-linear, i.e. all abelian monoids $C(A, B)$ are groups.

It is easy to see that a $\mathbb{Z}$-linear category is finitely complete if and only if it is additive and has kernels. In fact, in any category the existence of finite limits is equivalent to the existence of finite products and equalisers (by Theorem 2.2.4). Furthermore, if C is $\mathbb{Z}$-linear, the equaliser of two maps $f, g \colon A \to B$ amounts to the kernel of the morphism $f - g$, while the kernel of f amounts to the equaliser of f and 0_{AB}.

6.4.5 Exercises and complements

The following exercises only require easy verifications, except (c), whose solution can be found in Chapter 8.

(a) We know that all categories $R\,\mathsf{Mod}$ and Ban are additive.

As a less obvious example, the reader can prove that the category Set_T of sets and $\mathbb{Z}$-weighted maps (see 3.7.1(b)) is additive, with the obvious sum of maps; a finite biproduct is a disjoint union $X = \sum X_i$ (a sum in Set), with the usual injections $u_i \colon X_i \to X$ and suitable projections $p_i \colon X \twoheadrightarrow X_i$ (that are not mappings).

But also this example becomes obvious, if we replace Set_T with the equivalent category of free abelian groups – a full subcategory of Ab closed under finite biproducts (see 3.7.1(c)).

(b) Abm is obviously a semiadditive, not additive, category. The examples below consist of selfdual categories with finite (or arbitrary) biproducts.

(c) The category Ltc of *lattices and Galois connections* is a full subcategory of AdjOrd. It has finite biproducts, realised as cartesian products ΠX_i of lattices. On the other hand, the p-exact category Mlc lacks finite products and finite sums (or it would be abelian, by Theorem 6.4.6).

(d) Rel Set is also semiadditive, and has arbitrary biproducts realised as disjoint unions (see 2.2.6(c)). The sum of a family of parallel relations $a_i \colon X \nrightarrow Y$ is their join $\cup a_i$, a union of subsets of $X \times Y$.

6.4.6 Theorem and definition (Abelian categories)

Let E *be a Puppe-exact category.*

We say that E *is* abelian *if it satisfies the following equivalent properties:*

(*i*)   E *has finite products,*

(i^*)   E *has finite sums,*

(*ii*)   E *has finite biproducts (in the sense of 6.4.1),*

(*iii*)   E *has pullbacks,*

(iii^*) E *has pushouts,*

(*iv*)   E *has finite limits,*

(iv^*) E *has finite colimits,*

(*v*)   E *is additive.*

The sum of parallel maps is then determined by the categorical structure, as in (6.35).

Note. A preadditive Puppe-exact category need not be abelian: see the examples of 6.3.3(b).

Proof See [G9], Theorem 2.1.5. The essential part of this result was stated in [HeS] and proved in Freyd–Scedrov [FrS]. A weaker result based on the existence of biproducts was already proved in [Mit].   □

6.4.7 Biproducts in abelian categories

We already know that, in any $\mathbb{N}$-linear category C, the biproduct $A \oplus B$ can be characterised as a diagram (6.34) satisfying the condition 6.4.2(iii):

(iii) p, q are jointly mono, $pu = \text{id}A$, $qv = \text{id}B$, $pv = 0$, $qu = 0$.

It is now easy to see that, if C is *preadditive* (i.e. $\mathbb{Z}$-linear) and *p-exact*, the biproduct $A \oplus B$ can also be characterised as a diagram (6.34) satisfying:

(iv) the diagram is commutative, with (short) exact 'diagonals'.

Indeed, if (iv) holds and a morphism $f \colon X \to C$ gives $pf = 0$, $qf = 0$, it must factorise through $v = \ker p$, as $f = vg$. It follows that $g = qvg = qf = 0$ and $f = 0$. For two parallel morphisms $f_i \colon A \to Z$ on which p, q coincide, take $f = f_1 - f_2$. Conversely, if (iii) is satisfied, let us prove that $v = \ker p$. Take

$f: X \to C$ such that $pf = 0$; since we know that $\mathrm{id}C = up + vq$, it follows that $f = (up + vq)f = vqf$ factorises through v (that is mono).

Let $F: \mathsf{E} \to \mathsf{E}'$ be a functor *between abelian categories*. As an immediate consequence of the previous point, if F is exact (in the previous sense, i.e. it preserves kernels and cokernels), then it preserves biproducts, the sum of parallel maps, all finite limits and finite colimits.

In fact, the preservation of all biproducts, including the sum of maps, *also holds if F is just half exact* (as defined in 6.3.4) because, in the diagonals of (6.34), all monos and epis are split, and preserved by any functor.

6.4.8 Exercises and complements (Squares in abelian categories)

We start from a square in the abelian category E, which is not supposed to commute

$$
\begin{array}{ccc}
A & \xrightarrow{\ f\ } & X \\
{\scriptstyle g}\downarrow & & \downarrow{\scriptstyle h} \\
Y & \xrightarrow{\ k\ } & B
\end{array}
\tag{6.36}
$$

and form the associated sequence

$$
0 \to A \xrightarrow{\ (f,g)\ } X \oplus Y \xrightarrow{\ [h,-k]\ } B \to 0
\tag{6.37}
$$

(a) The square (6.36) is commutative if and only if the sequence (6.37) is *of order two* (i.e. its composed morphisms are null).

(b) The square (6.36) is a pullback if and only if the sequence (6.37) is exact in A and $X \oplus Y$.

(b*) The square (6.36) is a pushout if and only if the sequence (6.37) is exact in B and $X \oplus Y$.

(c) More generally the square (6.36) is said to be *exact* in the sense of Hilton [Hi] if the sequence is exact in the central object $X \oplus Y$, which means that

$$
\mathrm{im}\,(f,g) = \ker[h,-k].
\tag{6.38}
$$

(d) If $\mathsf{E} = R\,\mathsf{Mod}$ this is the same as condition 2.3.8(ii) and is equivalent to saying that *the square* (6.36) *becomes bicommutative in* $\mathrm{Rel}\,\mathsf{E}$. *In fact this equivalence holds for all abelian categories E (cf. [Hi]).* More general characterisations of this fact in p-exact categories will be given in 6.6.4.

(e) The exactness of square (6.36) is also equivalent to each of the following (dual) conditions:

  (i) the pair (h,k) and the pushout of (f,g) have the same pullback,

  (i*) the pair (f,g) and the pullback of (h,k) have the same pushout.

6.5 Projective p-exact categories and projective spaces

Categories of projective spaces are rarely considered in category theory; the reason might be that one needs the (generally overlooked) theory of p-exact categories. The present analysis first appeared in the article [G1], and later in the book [G9].

We show here that every p-exact category E has an associated *projective* p-exact category $\Pr E$, with a faithful projective functor. If E is the abelian category $K \mathsf{Vct}$ of vector spaces on the field K, $\Pr E$ can be identified with the category $K \mathsf{Prj}$ of projective spaces and projective linear maps on K (see 6.5.4(a)). This p-exact category is *not* abelian, generally, and cannot even be exactly embedded in an abelian category (see Theorem 6.5.5). But it inherits from $K \mathsf{Vct}$ a symmetric monoidal closed structure (see 6.5.6).

A characterisation of the projective categories associated to abelian categories was given in [CaG].

K is always a (commutative) field and K^* the multiplicative group of its non-zero elements.

6.5.1 *The associated projective category*

We say that a p-exact category is *projective* if its transfer functor $\mathsf{Sub_E}$: $E \to \mathsf{Mlc}$ is faithful. Let us recall that $\mathsf{Sub_E}$, defined in 6.3.8, is also called the projective functor of E.

We construct the projective p-exact category *associated* to an arbitrary p-exact category E as the quotient of E modulo the *projective congruence* $\sim_S$, namely the congruence of categories associated to the functor $\mathsf{Sub_E}$.

The objects of E are unchanged, but we identify two parallel maps $f, g\colon A \to B$ of E when they have the same modular connection $(f_*, f^*) = (g_*, g^*)\colon \mathrm{Sub}A \to \mathrm{Sub}B$

$$\Pr E = E/\!\!\sim_S,$$
$$(f \sim_S g) \ \Leftrightarrow \ (f_* = g_*) \ \Leftrightarrow \ (f^* = g^*). \tag{6.39}$$

One easily verifies that the category $\Pr E$ is p-exact, with the same zero object, the same kernel and cokernel objects and the induced kernel and cokernel morphisms

$$\ker [f] = [\ker_E f]\colon \mathrm{Ker}\, f \rightarrowtail A,$$
$$\mathrm{cok}\,[f] = [\mathrm{cok}_E f]\colon B \twoheadrightarrow \mathrm{Cok}\, f, \tag{6.40}$$

which means that the canonical projection

$$P\colon E \to \Pr E, \qquad A \mapsto A, \quad f \mapsto [f], \tag{6.41}$$

preserves and reflects exactness (including mono-, epi- and isomorphisms).

The projective functor $\mathrm{Sub_E}$ obviously factorises through P, by a faithful functor $S\colon \mathrm{Pr\,E} \to \mathrm{Mlc}$

$$\mathrm{Sub_E} = SP\colon \mathrm{E} \to \mathrm{Pr\,E} \to \mathrm{Mlc},$$
$$S(A) = \mathrm{Sub}(A), \qquad S[f] = \mathrm{Sub}(f) = (f_*, f^*). \tag{6.42}$$

The functor S is isomorphic to the projective functor $\mathrm{Sub_{E'}}$ of $\mathrm{E'} = \mathrm{Pr\,E}$, by a natural isomorphism $\sigma\colon S \to \mathrm{Sub_{E'}}$

$$\sigma A\colon S(A) \to \mathrm{Sub_{E'}}(A), \qquad \sigma A(m) = [m]. \tag{6.43}$$

This shows that $\mathrm{Pr\,E}$ is projective and $\mathrm{Pr(Pr\,E)}$ can be identified with $\mathrm{Pr\,E}$; it will be called the *projective* p-exact category associated to E.

$\mathrm{Pr\,E}$ is determined up to isomorphism of categories by an obvious *universal property*: every exact functor $F\colon \mathrm{E} \to \mathrm{E'}$ with values in a projective p-exact category factorises in a unique way as $F = GP$, by a functor $G\colon \mathrm{Pr\,E} \to \mathrm{E'}$, necessarily exact.

(Loosely speaking, in the category of p-exact categories and exact functors, the full subcategory of projective p-exact categories is reflective, with reflector Pr and unit P.)

6.5.2 Proposition (The projective congruence of vector spaces)

In the abelian category $\mathrm{E} = K\mathrm{Vct}$ *of vector spaces over the (commutative) field* K, *the projective congruence* (6.39) *is characterised as:*

$$(f \sim_S g) \quad \Leftrightarrow \quad (\exists \lambda \in K^*\colon f = \lambda g). \tag{6.44}$$

Therefore the category $K\mathrm{Vct}$ *is projective if and only if* K *is a two-element field (isomorphic to* $\mathbb{Z}/2$*).*

Note. Property (6.44) is essentially well known, in different forms.

Proof The right-hand condition in (6.44) obviously implies that $f_* = g_*$. Conversely, let us suppose that $f_* = g_*$ (and $f^* = g^*$). Let $m = f_*(1) = g_*(1)$ and $p = \mathrm{cok}\,(f^*(0)) = \mathrm{cok}\,(g^*(0))$. There are unique isomorphisms u, v such that $f = mup$, $g = mvq$ and

$$u_* = (m^*m_*)u_*(p_*p^*) = m^*f_*p^* = m^*g_*p^* = v_*.$$

Therefore we only have to prove our property for two isomorphisms u, v: $A \to B$ with $u_* = v_*$. For every $x \in A$, $x \neq 0$, the mappings u_* and v_* coincide on the subspace $\langle x \rangle \subset A$ spanned by x, hence there is a unique scalar $\lambda_x \in K$ such that $u(x) = \lambda_x.v(x)$, and $\lambda_x \neq 0$.

Let $y \in A$, $y \neq 0$. If $y = \lambda x$ for some non-zero scalar λ, we have

$$\lambda_y.v(y) = u(y) = u(\lambda x) = \lambda \lambda_x.v(x) = \lambda_x.v(y)$$

and $\lambda_y = \lambda_x$. Otherwise, x and y are linearly independent and

$$\lambda_x.v(x) + \lambda_y.v(y) = u(x) + u(y) = u(x+y) = \lambda_{x+y}.v(x) + \lambda_{x+y}.v(y).$$

But $v(x)$ and $v(y)$ are also linearly independent, so that λ_x and λ_y coincide with λ_{x+y}. $\qquad\square$

6.5.3 Projective spaces and projective maps

The associated projective category of K Vct

$$K\,\mathsf{Prj} = \mathrm{Pr}(K\,\mathsf{Vct}) = K\,\mathsf{Vct}/\!\!\sim_S, \tag{6.45}$$

will be called the category *of projective spaces and projective linear maps*.

It is p-exact, with a projection functor $P\colon K\,\mathsf{Vct} \to K\,\mathsf{Prj}$ that preserves and reflects exactness (hence also the isomorphisms). We show below (in Theorem 6.5.5) that $K\,\mathsf{Prj}$ is not abelian, except in 'one' case, i.e. if K is the two-element field, when obviously $K\,\mathsf{Prj} = K\,\mathsf{Vct}$ 'is' the abelian category of abelian groups satisfying the axiom $2x = 0$.

Our quotient $K\,\mathsf{Prj}$ admits various concrete representations, in terms of faithful functors, that allow us to recognise the 'usual projective spaces' and their morphisms. The third, below, is the classical one.

(a) *Projective spaces as lattices of linear subspaces of vector spaces.* We use the composite

$$\begin{aligned} F_1 = US\colon K\,\mathsf{Prj} \to \mathsf{Mlc} \to \mathsf{Set}, \\ F_1(X) = |\mathsf{Sub}_\mathsf{E}(X)|, \qquad F_1[f] = f_*, \end{aligned} \tag{6.46}$$

of the transfer functor S, in (6.42), and the forgetful functor U of Mlc, in (6.11).

(b) *Projective spaces as* pointed *sets, quotients of vector spaces.* We use the functor

$$\begin{aligned} F_2\colon K\,\mathsf{Prj} \to \mathsf{Set}_\bullet, \\ F_2(X) = (|X|/\!\sim, [0]), \qquad (F_2[f])[x] = [f(x)], \end{aligned} \tag{6.47}$$

where $F_2(X)$ is the quotient of the underlying set $|X|$ modulo the usual equivalence relation

$$x \sim y \quad \Leftrightarrow \quad (\exists \lambda \in K^*\colon x = \lambda y), \tag{6.48}$$

pointed at the singleton class $[0] = \{0\}$.

F_2 is faithful because every linear subspace is the join of its 1-dimensional subspaces, and direct images preserve joins.

(c) *Projective spaces as* unpointed *sets.* We use the composed functor

$$F_3 = RF_2 \colon K\,\mathsf{Prj} \to \mathsf{Set}_{\bullet} \to \mathcal{S}, \qquad F_3(X) = (|X| \setminus \{0\})/{\sim}, \qquad (6.49)$$

where

$$R \colon \mathsf{Set}_{\bullet} \to \mathcal{S}, \qquad R(S, x_0) = S \setminus \{x_0\},$$

is the equivalence of categories between $\mathsf{Set}_{\bullet}$ and the category $\mathcal{S}$ of sets and partial mappings (see (1.75)).

$K\,\mathsf{Prj}$ is thus isomorphic to the category $K\,\mathsf{Prj}'$ whose objects are the K-vector spaces, while a morphism $\varphi \colon X \nrightarrow Y$ is a *partial mapping* of sets

$$F_3(f) \colon (|X| \setminus \{0\})/{\sim} \;\nrightarrow\; (|Y| \setminus \{0\})/{\sim}, \qquad (6.50)$$

induced by some linear mapping $f \colon X \to Y$, and defined on the set-theoretic complement of $\operatorname{Ker} f$ (see the exercises below).

6.5.4 Exercises and complements

(a) The reader can check that the realisation $K\,\mathsf{Prj}'$ of $K\,\mathsf{Prj}$ considered above amounts to the construction of 'projective linear mappings' as partial mappings, in Bourbaki [Bor1], Section II.9.10.

(b) Other authors use the term 'projective linear transformation' to mean an *isomorphism* of $K\,\mathsf{Prj}'$; this is a bijection

$$(|X| \setminus \{0\})/{\sim} \;\to\; (|Y| \setminus \{0\})/{\sim},$$

induced by some isomorphism $X \to Y$ of vector spaces. (Such isomorphisms form a disconnected groupoid, with poor categorical properties.)

(c) For an arbitrary category E, the group $\pi = \operatorname{Aut}(\mathrm{id}E)$ of natural automorphisms of the identity functor defines a congruence: two morphisms $f, g \colon A \to B$ are π-equivalent if

$$(f \sim_\pi g) \quad \Leftrightarrow \quad (\exists \lambda \in \pi \colon\; f = g.\lambda_A = \lambda_B.g). \qquad (6.51)$$

If E is p-exact, this congruence plainly implies the projective one. One can prove that the quotient $\mathsf{E}/{\sim_\pi}$ is p-exact, the projection $\mathsf{E} \to \mathsf{E}/{\sim_\pi}$ is an exact functor, and the projection $\mathsf{E} \to \mathsf{Pr}\,\mathsf{E}$ factorises through the former.

If $\mathsf{E} = K\,\mathsf{Vct}$, π can be identified with the multiplicative group K^* and the congruence (6.51) coincides with the projective one, by 6.5.2; thus, $\mathsf{E}/{\sim_\pi} = \mathsf{Pr}\,\mathsf{E}$.

(d) On the other hand, if $E = Ab = \mathbb{Z} \, Mod$, π 'is' the multiplicative group of invertible integers, $\{-1, 1\}$.

Verify that, in the hom-set $Hom(\mathbb{Z}, \mathbb{Z}/5)$, the relations $\sim_\pi$ and $\sim_S$ are different, proving that $Ab/\sim_\pi$ is not projective.

*(e) Find similar results replacing $\mathbb{Z}/5$ with a cyclic group $\mathbb{Z}/p$, for a prime number p.

6.5.5 *Theorem* (The lack of products)

Let the field K have more than two elements.

(a) The p-exact category K Prj lacks binary products and sums, and cannot have an exact embedding, or even a faithful half exact functor (see 6.3.4), with values in an abelian category.

(b) The biproduct $A \oplus B$ of K Vct does not induce a two-variable functor of projective spaces.

(c) Similarly, the p-exact category Mlc cannot have an exact embedding, or even a faithful half exact functor, with values in an abelian category.

Proof (a) The argument is based on the fact that the canonical injections

$$u_i \colon K \to K^2 \qquad (i = 1, 2),$$

give two projective linear maps $[u_i]$ that are *not* jointly epi in K Prj.

Indeed, let us choose some $\lambda \in K$ different from 0 and 1. The linear automorphisms

$$f = \mathrm{id} K^2, \qquad g \colon K^2 \to K^2, \quad g(x, y) = (x, \lambda y), \qquad (6.52)$$

are not related by the projective congruence $\sim_S$ (because of Proposition 6.5.2); therefore $[f] \neq [g]$, but $[fu_i] = [gu_i]$, for $i = 1, 2$.

Suppose now that we have a half exact functor $F \colon K$ Prj $\to$ A with values in an abelian category. Then the composite $FP \colon K$ Vct $\to K$ Prj $\to$ A is a half exact functor between abelian categories and preserves biproducts (by 6.4.7), whence the pair $F[u_i]$ is jointly epi in A. Since a faithful functor reflects such pairs, F cannot be faithful.

As a consequence, the p-exact category K Prj is not abelian and cannot have binary products, nor sums.

(b) In K Vct the linear mappings (6.52) can be obtained as

$$f = f_1 \oplus f_1 \colon K^2 \to K^2, \qquad g = f_1 \oplus f_\lambda \colon K^2 \to K^2,$$

where $f_1 = \mathrm{id} K$ and $f_\lambda \colon K \to K$ is the multiplication by $\lambda \neq 0, 1$. But $f_1 \sim_S f_\lambda$ while f and g are not equivalent modulo $\sim_S$.

Therefore, even fixing one variable, the biproduct $A \oplus B$ of K Vct does not induce a functor $A \oplus - : K\,\mathsf{Prj} \to K\,\mathsf{Prj}$.

Let us note that the butterfly diagram (6.34) with exact diagonals subsists in $K\,\mathsf{Prj}$, but – without a $\mathbb{Z}$-linear structure – does not give a biproduct.

(c) It is sufficient to recall that the projective functor $\mathrm{Sub}\colon K\,\mathsf{Prj} \to \mathsf{Mlc}$ is exact and faithful. $\qquad\qquad\qquad\qquad\qquad\qquad\qquad\qquad\qquad\qquad\square$

6.5.6 The exponential law of projective spaces

The tensor product and the Hom functor of vector spaces

$$\otimes_K\colon K\,\mathsf{Vct} \times K\,\mathsf{Vct} \ \to\ K\,\mathsf{Vct},$$
$$\mathrm{Hom}_K\colon K\,\mathsf{Vct}^{\mathrm{op}} \times K\,\mathsf{Vct} \ \to\ K\,\mathsf{Vct}, \tag{6.53}$$

have been considered in 3.4.6, more generally for modules.

Both functors are easily seen to be consistent with the projective congruence, characterised in 6.5.2: if $f = \lambda f'$ and $g = \mu g'$, then

$$f \otimes g = \lambda\mu.f' \otimes g', \qquad \mathrm{Hom}(f,g) = \lambda\mu.\mathrm{Hom}(f',g').$$

Therefore, there are induced functors for $\mathsf{E} = K\,\mathsf{Prj}$ (that act on the objects as the previous ones):

$$\otimes_K\colon \mathsf{E} \times \mathsf{E} \to \mathsf{E}, \qquad\qquad [f] \otimes [g] = [f \otimes g],$$
$$\mathrm{Hom}_K\colon \mathsf{E}^{\mathrm{op}} \times \mathsf{E} \to \mathsf{E}, \qquad \mathrm{Hom}_K([f],[g]) = [g].-.[f]. \tag{6.54}$$

The original adjunction of vector spaces

$$\varphi_{XY}\colon K\,\mathsf{Vct}(X \otimes_K A, Y) \ \to\ K\,\mathsf{Vct}(X, \mathrm{Hom}_K(A,Y)),$$
$$\varphi_{XY}(u)(x) = u(x \otimes -)\colon A \to Y, \tag{6.55}$$

(for $u\colon X \otimes_K A \to Y$ and $x \in X$) is also consistent with the projective congruence: if $u = \lambda v$ then $\varphi(u) = \lambda.\varphi(v)$.

We thus have an induced adjunction $F \dashv G$ between endofunctors of $K\,\mathsf{Prj}$

$$F(X) = X \otimes A, \qquad G(Y) = \mathrm{Hom}_K(A,Y) = Y^A, \tag{6.56}$$

where $\mathrm{Hom}_K(A,Y)$ *is still the original vector space.*

This gives a symmetric monoidal closed structure on $K\,\mathsf{Prj}$. Its identity is K, i.e. the one-point projective space in the classical representation 6.5.3(c). The canonical forgetful functor $K\,\mathsf{Prj}(K,-)$ 'is' the composite of $F_2\colon K\,\mathsf{Prj} \to \mathsf{Set}_\bullet$ (see (6.47)) with the forgetful functor $\mathsf{Set}_\bullet \to \mathsf{Set}$.

6.6 Relations and induction for p-exact categories

A consistent part of Homological Algebra, from the homology of chain complexes to the theory of spectral sequences, is about *subquotients* and *induced morphisms* between them; the latter are defined through the categories of relations, since a subquotient in $R\,\mathsf{Mod}$ is simply a subobject in $\mathrm{Rel}\,(R\,\mathsf{Mod})$ (as shown in 6.6.2(f)).

The problem arises of the coherence of induction with composition, as already noted in Mac Lane's book 'Homology' ([M2], Section II.6). The problem can be apparently complicated, when we are working in complex systems. Yet, it has a clear solution, in the domain of p-exact categories and their relations.

The extension of relations from abelian to p-exact categories was built by Tsalenko [T1, T2], then studied by Brinkmann and Puppe [Bri, BriP]. Here we briefly recall this construction. (The interested reader will find the proofs and further information in these references, or in [G9].)

We end by considering subquotients and induced relations, with some hints at our Coherence Theorem for induction.

6.6.1 Relations of modules revisited

The category $\mathrm{Rel}\,(R\,\mathsf{Mod})$ of relations between modules has been reviewed in 1.8.5; we give now a more detailed analysis, to prepare the construction of relations for all p-exact categories.

A relation $u\colon A \nrightarrow B$ is a submodule of the direct sum $A \oplus B$. It can be viewed as a 'partially defined, multi-valued homomorphism', that sends an element $x \in A$ to the subset $\{y \in B \mid (x,y) \in u\}$ of B.

The composite of u with $v\colon B \nrightarrow C$ is

$$vu = \{(x,z) \in A \oplus C \mid \exists y \in B\colon (x,y) \in u \text{ and } (y,z) \in v\}.$$

$\mathrm{Rel}\,(R\,\mathsf{Mod})$ is an involutive ordered category. For two parallel relations $u, u'\colon \ A \nrightarrow B$, $u \leqslant u'$ means that $u \subset u'$ as submodules of $A \oplus B$. The *opposite relation* of $u\colon A \nrightarrow B$ is obtained by reversing pairs, and written as $u^{\sharp}\colon B \nrightarrow A$. The involution $u \mapsto u^{\sharp}$ is *regular* in the sense of von Neumann, i.e. $uu^{\sharp}u = u$ for all relations u.

As a consequence, a *monorelation*, i.e. a monomorphism in the category $\mathrm{Rel}\,(R\,\mathsf{Mod})$, is characterised by the condition $u^{\sharp}u = 1$ (and is a split mono); an *epirelation* by the dual condition $uu^{\sharp} = 1$.

$\mathrm{Rel}\,(R\,\mathsf{Mod})$ is balanced.

The category $R\,\mathsf{Mod}$ is embedded in the category of relations $\mathrm{Rel}(R\,\mathsf{Mod})$, identifying a homomorphism $f\colon A \to B$ with its graph

$$\{(x,y) \in A \oplus B \mid y = f(x)\}.$$

A relation $u\colon A \nrightarrow B$ determines two submodules of A, called *definition* and *annihilator*

$$\operatorname{Def} u = \{x \in A \mid \exists y \in B\colon (x,y) \in u\},$$
$$\operatorname{Ann} u = \{x \in A \mid (x,0) \in u\}, \tag{6.57}$$

and two submodules of B, called *values* and *indeterminacy*

$$\operatorname{Val} u = \{y \in B \mid \exists x \in A\colon (x,y) \in u\} = \operatorname{Def} u^{\sharp},$$
$$\operatorname{Ind} u = \{y \in B \mid (0,y) \in u\} = \operatorname{Ann} u^{\sharp}. \tag{6.58}$$

We note that, if $(a,b) \in u$:

$$(a',b) \in u \iff a - a' \in \operatorname{Ann} u, \quad (a,b') \in u \iff b - b' \in \operatorname{Ind} u. \tag{6.59}$$

The ordered set $\operatorname{Rel}(R\operatorname{Mod})(A,B)$ has a least element ω_{AB} and a greatest element Ω_{AB}

$$\omega_{AB} = \{0,0\}, \qquad \Omega_{AB} = A \times B. \tag{6.60}$$

The submodule $\operatorname{Val} u$ of B is called the 'image' of u in [M2]. Yet, the epi-mono factorisation of u, in (6.61), will give a different image.

6.6.2 Exercises and complements

(a) A relation $u\colon A \nrightarrow B$ is a homomorphism if and only if it is *everywhere defined* (i.e. $\operatorname{Def} u = A$) and *single-valued* (i.e. $\operatorname{Ind} u = 0$); in this case annihilator and values coincide with kernel and image, respectively.

(b) Every monomorphism (resp. epimorphism) in $R\operatorname{Mod}$ is a monorelation (resp. epirelation).

(c) The relation u has a canonical factorisation $u = (nq^{\sharp}).i.(pm^{\sharp})$ (and a dual factorisation, by dashed arrows)

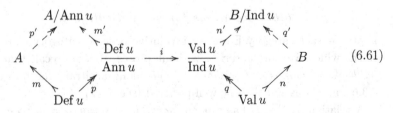

$$\tag{6.61}$$

where m, n are embeddings of submodules, p, q are projections on quotients and i is an isomorphism between *subquotients* of A and B (see 6.3.9).

This factorisation consists of an epirelation $(ipm^\sharp)$ followed by a monorelation $(s = nq^\sharp)$: it is a factorisation epi-mono in Rel Ab, and therefore *the image of u in this category should be defined as the subquotient* Val u/Ind u (rather than the subobject Val u).

(d) The relation u is a monorelation if and only if Def $u = A$ and Ann $u = 0$.

(d*) The relation u is an epirelation if and only if Val $u = B$ and Ind $u = 0$.

(e) The relation u is invertible in Rel $(R\,\mathsf{Mod})$ if and only if it is invertible in $R\,\mathsf{Mod}$, i.e. an isomorphism of modules, in which case $u^\sharp$ is the inverse isomorphism.

(f) Following our analysis in 6.3.9, a subquotient $S = M/N$ of a module A is presented by the following bicartesian square (where $N \subset M$)

$$
\begin{array}{ccc}
M & \overset{m}{\rightarrowtail} & A \\
{\scriptstyle p}\downarrow & \nearrow{\scriptstyle s} & \downarrow{\scriptstyle q} \\
S & \underset{n}{\rightarrowtail} & A/N
\end{array}
\qquad (6.62)
$$

This bicartesian square determines one monorelation

$$s = mp^\sharp = q^\sharp n \colon S \rightarrowtail A,$$

that sends the equivalence class $[x] \in M/N$ to all the elements of the lateral $[x] = x + N \subset A$. Because of the factorisation (6.61) of a relation, every monorelation $S' \rightarrowtail A$ is of this type, up to isomorphism.

The *subquotients* of the module A amount thus to the *subobjects of A in* Rel $(R\,\mathsf{Mod})$. This will be used in 6.6.6–6.6.8.

(g) All idempotents in Rel $(R\,\mathsf{Mod})$ split (see 2.8.2).

(h) Rel $(R\,\mathsf{Mod})$ lacks equalisers and coequalisers, provided R is not the trivial ring.

6.6.3 Relations for p-exact categories

Every p-exact category E has an involutive ordered category of relations Rel E, which we now *describe*, without proofs. Let us recall that E has *pullbacks of monos* and *pushouts of epis* along arbitrary morphisms (see 6.3.6), and is assumed to be well powered (see 6.3.2).

A relation $u\colon A \rightarrowtail B$ has a *quaternary factorisation* $u = nq^\sharp pm^\sharp$, and a *coquaternary factorisation* $u = q'^\sharp n' m'^\sharp p'$, each of them determined up to three coherent isomorphisms; they form two *bicartesian* squares (pullback

and pushout, see 6.3.6(b))

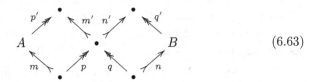

$$(6.63)$$

The composed relation vu is computed, on *quaternary* factorisations, by *pullbacks of monos* (along arbitrary arrows) and *pushout of epis along epis*, as in the following diagram; the relation $u \colon A \dashrightarrow B$ is drawn in the upper row, the relation $v \colon B \dashrightarrow C$ in the right-hand column and $vu \colon A \dashrightarrow C$ along the dashed diagonal

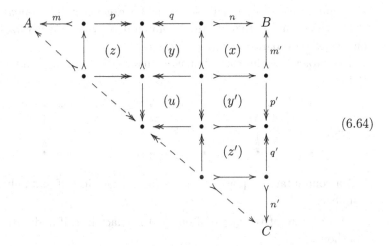

$$(6.64)$$

The square (x) is a pullback, (y) and (y') are commutative (by epi-mono factorisation), (z) and (z') are pullbacks and (u) is a pushout.

Dually, it can be computed on *coquaternary* factorisations, by means of *pushouts of epis* along arbitrary arrows and *pullbacks of monos along monos*.

The involution $u \mapsto u^\sharp$ is obvious. The order relation $u \leqslant v$ is expressed by the existence of a commutative diagram in E linking their quaternary factorisations (or equivalently the coquaternary ones)

$$(6.65)$$

In the construction of Rel E, a relation $A \nrightarrow B$ is defined as an equivalence class $[m, p, q, n]$ of quaternary diagrams of E as above, up to three coherent *isomorphisms* f, h, g (forming a commutative diagram as above); the composition is defined as in diagram (6.64); the proof of associativity is not easy (see one of the references [T2, BriP, G9]).

The category Rel Gp can be constructed by 'quaternary classes', as shown in [BriP] (and not in the dual way). The category of corelations Cor Set mentioned in 1.8.9 can be constructed by 'coquaternary classes', as essentially shown in [Da1]. This explains why, in both cases, the involution is regular.

6.6.4 Exact squares

A commutative square of E is said to be *exact* if it is *bicommutative* in Rel E, which means that it stays commutative when one reverses, by the involution, two parallel edges of the square.

To characterise this fact, first note that (by computing compositions in Rel E):

$$(6.66)$$

(i) a commutative square of monos (a) is exact in E if and only if it is a pullback,

(i*) a commutative square of epis (b) is exact in E if and only if it is a pushout,

(ii) a commutative *mixed* square (c) is exact in E if and only if it is a pullback, if and only if it is a pushout.

More generally, every preimage of a mono (by pullback) produces a bicommutative square, as well as every direct image of an epi (by pushout).

Finally, a commutative square in E is exact if and only if its canonical factorisation is composed of four exact squares of the previous types

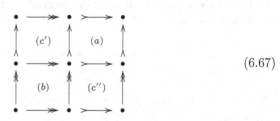

$$(6.67)$$

Every p-exact functor $F\colon \mathsf{E} \to \mathsf{E}'$ preserves preimages of monos and direct images of epis; therefore it has a unique extension to a functor that preserves involution and ordering

$$\operatorname{Rel} F\colon \operatorname{Rel}\mathsf{E} \to \operatorname{Rel}\mathsf{E}'. \tag{6.68}$$

*Every natural transformation $\varphi\colon F \to G\colon \mathsf{E} \to \mathsf{E}'$ determines a *lax transformation of order-preserving functors*

$$\operatorname{Rel}\varphi\colon \operatorname{Rel} F \to \operatorname{Rel} G\colon \operatorname{Rel}\mathsf{E} \to \operatorname{Rel}\mathsf{E}', \tag{6.69}$$

whose components are precisely those of φ (and belong to E'). For this notion the interested reader is referred to [G9].

Essentially, it means that a relation $u\colon A \nrightarrow B$ in $\operatorname{Rel}\mathsf{E}$ produces a 'lax square' in the ordered category $\operatorname{Rel}\mathsf{E}'$, as follows (writing $\operatorname{Rel} F$ as $\underline{F}$)

$$
\begin{array}{ccc}
F(A) & \xrightarrow{\ \varphi A\ } & G(A) \\
\underline{F}u \downarrow & \leqslant & \downarrow \underline{G}u \\
F(B) & \xrightarrow[\varphi B]{} & G(B)
\end{array}
\qquad \varphi B.\underline{F}u \leqslant \underline{G}u.\varphi A. * \tag{6.70}
$$

6.6.5 The abelian case

Restricting everything to abelian categories gives little advantage in this domain, except for the existence of 'two-arrow factorisations' of relations. Yet, one should not forget the fact that (co)quaternary factorisations are adapted to the study of subquotients, as we have seen in 6.6.2.

In fact, every relation has now (diverse) *binary factorisations* $u = gf^{\sharp}$: $A \nrightarrow B$. Such a factorisation will be said to be *strong* if the pair (f, g) is jointly mono (corresponding to a monomorphism with values in $A \oplus B$); this can be obtained from a quaternary factorisation of u, by the pullback of the central pair of epimorphisms:

$$\tag{6.71}$$

Dually, there is a *strong cobinary factorisation* $u = f'^{\sharp}g'$, where (f', g') is jointly epi. Two cobinary factorisations yield the same relation u if and only if they have the same pullback, which gives a *strong* binary factorisation of u.

Remarks. An abelian category E is a regular category. The present construction of Rel E is equivalent to the construction of Section 4.5, by equivalence classes of spans and pullbacks: see [G11], Section A4.

6.6.6 Induced relations and canonical isomorphisms

We end with some points on the study of induced relations and canonical isomorphisms between subquotients, at an introductory level.

Let us begin by working in a category of modules. We have seen in 6.6.2 that a subquotient can be represented as a subobject in Rel $(R\,\mathsf{Mod})$.

Given two subquotients $s\colon S \twoheadrightarrow A$ and $t\colon T \twoheadrightarrow B$, a relation $a\colon A \to B$ has an *induced relation*

$$
\begin{array}{ccc}
A & \xrightarrow{\;a\;} & B \\
{\scriptstyle s}\uparrow & /\!/ & \uparrow{\scriptstyle t} \\
S & \dashrightarrow{a'} & T
\end{array}
\qquad\qquad a' = t^{\sharp} a s\colon S \twoheadrightarrow T. \qquad\qquad (6.72)
$$

Generally, this procedure does not give a commutative square, and is *not* preserved by composition.

More particularly, for $a = \mathrm{id}\,A$, two subquotients $s\colon S \twoheadrightarrow A$ and $t\colon T \twoheadrightarrow A$ have an *induced relation* $t^{\sharp} s\colon S \twoheadrightarrow T$. If it is an isomorphism, it is called the *canonical isomorphism* from S to T.

The reader is warned that, as already remarked in Mac Lane [M2], *this does not give a transitive relation*, in general: if we have a third subquotient $u\colon U \twoheadrightarrow A$ and two canonical isomorphisms

$$
t^{\sharp} s\colon S \twoheadrightarrow T, \qquad\qquad u^{\sharp} t\colon T \twoheadrightarrow U,
$$

the composed isomorphism $(u^{\sharp} t)(t^{\sharp} s)$ *need not be the induced relation* $u^{\sharp} s\colon S \twoheadrightarrow U$. The term 'canonically isomorphic subquotients' should be reserved to situations where the corresponding relation is indeed an equivalence relation, as discussed below.

6.6.7 A case of incoherence

The following examples show some instances of *incoherence of induction on subquotients*:

- first, canonical isomorphisms need not be closed under composition,

- second, if we extend them in this sense the result need not be determined.

In other words, canonical isomorphisms – in general – do not form nor generate a coherent system of isomorphisms (as defined in 1.4.9(b)).

As in Mac Lane's book [M2], our examples of inconsistency are based on the lattice $\mathrm{Sub}A$ of subgroups of the abelian group $A = \mathbb{Z} \oplus \mathbb{Z}$, and more particularly on the (*non-distributive*) triple formed of the diagonal Δ and two of its complements, the subgroups A_1 and A_2

$$A_1 = \mathbb{Z} \oplus 0, \qquad A_2 = 0 \oplus \mathbb{Z},$$
$$A_i \vee \Delta = A, \qquad A_i \wedge \Delta = 0. \tag{6.73}$$

We consider the subquotients $m_i \colon A_i \to A$ and $s = p^\sharp \colon A/\Delta \to A$, in Rel Ab.

(a) The identity of A induces two canonical (Noether) isomorphisms

$$u_i = pm_i \colon A_i \to A/\Delta \qquad (A_i/(A_i \wedge \Delta) \to (A_i \vee \Delta)/\Delta),$$

and the canonical isomorphism $u_2^{-1} \colon A/\Delta \to A_2$.

Then, the composed isomorphism $w = u_2^{-1} u_1 \colon A_1 \to A_2$ *is not canonical.* Indeed:

$$w \colon A_1 \to A/\Delta \to A_2,$$
$$(x, 0) \mapsto [(x, 0)] = [(0, -x)] \mapsto (0, -x), \tag{6.74}$$

while the canonical relation $m_2^\sharp . m_1 \colon A_1 \to A_2$ has graph $\{(0, 0)\}$ (and is not even a homomorphism).

(b) Using the subgroup $\Delta' = \{(x, -x) \mid x \in \mathbb{Z}\}$ instead of the diagonal Δ, we get the *opposite* composed isomorphism from A_1 to A_2

$$A_1 \to A/\Delta' \to A_2 \qquad (x, 0) \mapsto [(x, 0)] = [(0, x)] \mapsto (0, x). \tag{6.75}$$

This shows that a composite $A_1 \to A_2$ of canonical isomorphisms between subquotients of $\mathbb{Z}^2$ *is not determined.*

A change of sign can be quite important, in Homological Algebra and Algebraic Topology. For instance, it is the case in the usual argument proving that 'even-dimensional spheres cannot be combed': if the sphere $\mathbb{S}^n$ has a non-null vector field, then its antipodal map $t \colon \mathbb{S}^n \to \mathbb{S}^n$ is homotopic to the identity, and the degree $(-1)^{n+1}$ of t must be 1, whence n must be odd.

The conclusion cannot be obtained if we only know the induced homomorphism $t_{*n} \colon H_n(\mathbb{S}^n) \to H_n(\mathbb{S}^n)$ up to sign change.

(c) We are lead to think that these cases of incoherence might depend on the non-distributivity of the lattice of subobjects involved in our subquotients. We show now that it is indeed the case.

6.6.8 *Coherence and distributivity*

We come back to the framework of p-exact categories, where all the previous arguments work: even the abelian setting would now be insufficient.

These informal comments are taken from the Introduction of [G9]; we refer to this book for the theory hinted at here.

Let us recall that a p-exact category E is said to be distributive if all its lattices of subobjects are distributive, or in other words if its transfer functor Sub: E → Mlc takes values in the full (p-exact) subcategory Dlc of distributive lattices and modular connections (see 6.3.3(c)). A nontrivial abelian category cannot be distributive: the lattice $Sub(A \oplus A)$ is not distributive, as soon as A is not a zero object. But there are distributive p-exact categories, like Dlc and other examples in 6.3.3.

Now, the coherence problem mentioned above is solved by Theorem 2.7.6 of [G9], whose basic part says the following.

Coherence Theorem of Homological Algebra. *In a p-exact category* E, *induced isomorphisms between subquotients are preserved by composition (and form a coherent system) if and only if* E *is distributive, if and only if its category of relations* Rel E *is orthodox (i.e. its idempotent endomorphisms are closed under composition).*

As a consequence, in a distributive p-exact category, all the diagrams of canonical isomorphisms between subquotients (induced by the identity of a given object) commute.

Now, every p-exact category (including the abelian ones!) has an associated distributive p-exact category, the distributive expansion Dst E (mentioned in 6.3.3(f) and constructed in [G9], 2.8.4.)

Combining the coherence theorem with the distributive expansion, it follows that the subquotients M/N of a given object A (in a p-exact, or abelian category) have a coherent system of isomorphisms (well-determined and closed under composition) *when we restrict to those subquotients whose numerator* (M) *and denominator* (N) *belong to a distributive sublattice* X *of the (modular) lattice of all subobjects.*

The crucial fact is that many important homological systems, and in particular most of those giving rise to spectral sequences (like the filtered complex, the double complex, Massey's exact couple and Eilenberg's exact system), actually 'live' in the distributive expansion of an abelian category, like Dst Ab, and can therefore be studied within *distributive p-exact categories.*

On the other hand, systems that lead to *non*-distributive sublattices of subobjects, like a *trifiltered* object or a *bifiltered* chain complex, *should be handled with great care*, to avoid the risk of using 'canonical isomorphisms' in an illegitimate way. Examples of such problems can be found in [G9], 1.4.5.

6.7 *Extending Homological Algebra

Kernels, cokernels and all related notions of exactness can be defined, more generally, in a category equipped with an ideal of (so called) *null morphisms* – that extends the ideal of the zero morphisms, in the pointed case.

Homological Algebra in this framework is studied in [G10], where the hints below are developed. The main goal was to set up a categorical framework adapted to the exact sequences and spectral sequences of unstable homotopy. Recently this setting has been used by A. Connes and C. Consani, to study 'homological algebra in characteristic one' [CoC].

The present section expands a subsection of the first edition (at the end of Section 6.1). It is mainly addressed to readers with some knowledge of homology theories, or interest in them. But in fact the hints described here are quite elementary, provided the reader is willing to follow this outline without being hindered by details, that can be filled in a second time.

6.7.1 Connected sequences of functors

Let there be given a short exact sequence (m, p) in $R\,\mathsf{Mod}$

$$A' \stackrel{m}{\rightarrowtail} A \stackrel{p}{\twoheadrightarrow} A'' \qquad m \sim \ker p, \quad p \sim \operatorname{cok} m, \qquad (6.76)$$

so that, up to isomorphism, A' is a submodule of A and $A'' = A/A'$.

Applying the functor $- \otimes_R B \colon R\,\mathsf{Mod} \to R\,\mathsf{Mod}$ (see 3.4.2, 3.4.6), we get an exact sequence

$$A' \otimes_R B \xrightarrow{m \otimes B} A \otimes_R B \xrightarrow{p \otimes B} A'' \otimes_R B \to 0 \qquad (6.77)$$

where the first morphism need not be mono: we know that the left adjoint $- \otimes_R B$ is just right exact.

This failure brings forth a sequence of functors

$$\operatorname{Tor}_n^R(-, B) \colon R\,\mathsf{Mod} \to R\,\mathsf{Mod} \qquad (n \geqslant 0), \qquad (6.78)$$

the *left satellites* of $\operatorname{Tor}_0^R(-, B) = - \otimes_R B$, called the Tor-*products*, or *torsion products* on the ring R. (The ring can be left understood.)

These functors produce, on the short exact sequence (m, p) in (6.76), a left-unbounded exact sequence of R-modules which extends at the left the sequence (6.77), and 'corrects' its failure (in a universal way)

$$
\begin{array}{c}
\dots \longrightarrow \operatorname{Tor}_2(A'', B) \\
\overset{\partial_2}{\swarrow} \\
\operatorname{Tor}_1(A', B) \rightarrow \operatorname{Tor}_1(A, B) \rightarrow \operatorname{Tor}_1(A'', B) \\
\overset{\partial_1}{\swarrow} \\
A' \otimes_R B \longrightarrow A \otimes_R B \longrightarrow A'' \otimes_R B \longrightarrow 0
\end{array}
\qquad (6.79)
$$

The functors $\mathrm{Tor}_n^R(-, B)$ are half-exact. The *connective* homomorphisms ∂_n are the components of natural transformations

$$\partial_n \colon \mathrm{Tor}_n P'' \to \mathrm{Tor}_{n-1} P' \colon \mathrm{Sh}(R\,\mathsf{Mod}) \to R\,\mathsf{Mod},$$
$$\partial_n \colon \mathrm{Tor}_n(A'', B) \to \mathrm{Tor}_{n-1}(A', B) \qquad (n \geqslant 1), \tag{6.80}$$

where $\mathrm{Sh}(R\,\mathsf{Mod})$ is the category of short exact sequences of R-modules (a full subcategory $R\,\mathsf{Mod}^3$, see 1.6.2(a)) and P' (resp. P'') is its projection on the first (resp. last) term of the sequence.

More generally, we are interested in *exact connected sequences of functors* (F_n, ∂_n), formed of functors and natural transformations

$$F_n \colon \mathsf{A} \to \mathsf{B} \qquad (n \geqslant 0),$$
$$\partial_n \colon F_n P'' \to F_{n-1} P' \colon \mathrm{Sh}(\mathsf{A}) \to \mathsf{B} \qquad (n \geqslant 1), \tag{6.81}$$

which produce, on every short exact sequence (6.76) in A, a left-unbounded exact sequence in B, as in (6.79).

This makes sense, provided that short exact sequences make sense in A, and exact sequences make sense in B. Classically, connected sequences of functors (and satellites, and derived functors) are studied in Cartan–Eilenberg [CE] for categories of modules, and in Grothendieck [Gt] for abelian categories.

But all this can be extended far beyond abelian categories, p-exact categories and pointed categories, to comprehend the homology theories of topological spaces, and other sequences of functors. We hint now at this extension, studied in [G10].

Before that, we only mention the dual aspect of right connected sequences of functors. The left exact functor $\mathrm{Hom}_R(B, -) \colon R\,\mathsf{Mod} \to R\,\mathsf{Mod}$ has a sequence of *right satellites*, the Ext-functors

$$\mathrm{Ext}_R^n(B, -) \colon R\,\mathsf{Mod} \to R\,\mathsf{Mod} \qquad (n \geqslant 0), \tag{6.82}$$

with $\mathrm{Ext}_R^0(B, A) = \mathrm{Hom}_R(B, A)$.

On the short exact sequence (6.76), these functors produce a right-unbounded exact sequence of R-modules, extending at the right the Hom-sequence

$$\begin{array}{c} 0 \to \mathrm{Hom}_R(B, A') \to \mathrm{Hom}_R(B, A) \to \mathrm{Hom}_R(B, A'') \\ \xrightarrow{ d_0 } \\ \mathrm{Ext}_R^1(B, A') \xrightarrow{} \dots \end{array} \tag{6.83}$$

6.7.2 Hints at relative singular homology

Singular homology is generally studied in the extended form of *relative singular homology*, defined on the category Top_2 of *pairs of topological spaces*, where the term 'pair' has a special, traditional meaning (cf. Eilenberg–Steenrod [EiS]).

An object of this category is a *pair* (X, A) formed of a space X and a *subspace* A. A morphism $f \colon (X, A) \to (Y, B)$ is a continuous mapping $f \colon X \to Y$ such that $f(A) \subset B$; composition is as in Top. The latter is embedded in Top_2, identifying the space X with the pair $(X, \emptyset)$.

The object (X, A) is usually read as 'X modulo A' and viewed as a surrogate of a quotient, more 'flexible' than collapsing the subspace A to a point.

Relative singular homology is a sequence of functors (for $n \geqslant 0$)

$$
\begin{aligned}
H_n \colon \mathsf{Top}_2 &\to \mathsf{Ab}, \qquad (X, A) \mapsto H_n(X, A), \\
(f \colon (X, A) \to (Y, B)) &\mapsto (f_{*n} \colon H_n(X, A) \to H_n(Y, B)),
\end{aligned}
\tag{6.84}
$$

which extend the absolute form: $H_n(X) = H_n(X, \emptyset)$.

$H_n(X, A)$ is called the n-th *group of singular homology of X modulo A*. (The definition is deferred to 6.7.5(f), not to break this outline. It can also be found in any book on singular homology.)

For subspaces $B \subset A \subset X$, we have a sequence of morphisms in Top_2

$$
(A, B) \xrightarrow{\ f\ } (X, B) \xrightarrow{\ g\ } (X, A)
\tag{6.85}
$$

given by the inclusion $A \subset X$ and $\mathrm{id}X$.

There is now a left-unbounded exact sequence in Ab, called *the exact homology sequence of the triple* (X, A, B)

$$
\begin{array}{ccccc}
& & \cdots & \longrightarrow & H_2(X, A) \\
& & & \nearrow \!\!\!\!\!\!\!\!\!\!\!\!\!\!\!\! \scriptstyle{\partial_2} & \\
H_1(A, B) & \rightarrow & H_1(X, B) & \to & H_1(X, A) \\
& & & \nearrow \!\!\!\!\!\!\!\!\!\!\!\!\!\!\!\! \scriptstyle{\partial_1} & \\
H_0(A, B) & \rightarrow & H_0(X, B) & \to & H_0(X, A) \to 0
\end{array}
\tag{6.86}
$$

In other words relative singular homology, equipped with the connective homomorphisms ∂_n, forms an *exact connected sequence of functors*

$$
H_n \colon \mathsf{Top}_2 \to \mathsf{Ab},
$$

provided we are able to interpret (6.85) as a short exact sequence in Top_2, and actually as *the* general short exact sequence of topological pairs (up to isomorphism).

Remarks. (a) If we take $B = \emptyset$ in (6.85), the 'special' short exact sequence

$$(A, \emptyset) \rightarrowtail (X, \emptyset) \twoheadrightarrow (X, A) \tag{6.87}$$

will say that the pair (X, A) is indeed the quotient of $X = (X, \emptyset)$ modulo $A = (A, \emptyset)$, in Top_2.

(b) What we said about singular homology holds much more generally: every homology theory defined on Top_2 and satisfying the first four axioms of Eilenberg–Steenrod (cf. [EiS], Section I.3) behaves in this way.

In fact, these axioms say that the theory takes any 'special' sequence (6.87) to a homology sequence (6.86), which is exact and 'natural'. As a consequence, the same holds for a general short exact sequence (6.85), by Theorem 10.2 of [EiS], Chapter I.

6.7.3 Categories with null maps

Now, the category Top_2 is not pointed: kernels, cokernels and exact sequences cannot be defined in the usual way.

Yet, it is natural to say that the object (X, A), viewed as a surrogate of a quotient X modulo A, is *null* if $X = A$, and that a map $f \colon (X, A) \to (Y, B)$ is *null* if it factorises through a null object, or equivalently if $f(X) \subset B$. Null morphisms form an *ideal* in Top_2: they are easily seen to be closed under composition with any morphism.

Kernels and cokernels *with respect to this ideal* do exist and are easily computed. Not to break the argument, we go on without proving our claims; this will be the subject of a list of easy exercises, in 6.7.5.

For a general morphism $f \colon (X, A) \to (Y, B)$, the kernel is the morphism

$$k \colon (f^{-1}(B), A) \to (X, A), \tag{6.88}$$

given by the inclusion $f^{-1}(B) \to X$; in fact the composite fk is null, in a universal way.

The cokernel of f is the morphism

$$c \colon (Y, B) \to (Y, f(X) \cup B), \tag{6.89}$$

given by the identity of Y; again, the composite cf is null, in a universal way.

Therefore the sequence of natural morphisms that we have considered above for $B \subset A \subset X$, in (6.85)

$$(A, B) \xrightarrow{\;f\;} (X, B) \xrightarrow{\;g\;} (X, A) \tag{6.90}$$

is indeed short exact in Top_2: $f = \ker g$ and $g = \operatorname{cok} f$, with respect to the given ideal. Moreover, it is easy to see that every short exact sequence of Top_2 is of this type, up to isomorphism.

Further properties show that Top_2 is a 'homological category', in the sense of [G10], a structure which allows one to develop a consistent part of homological algebra.

Essentially, this means that:

- kernels and cokernels exist, with respect to the given ideal,

- normal monos and normal epis are both stable under composition,

- a 'subquotient axiom' is satisfied.

A similar analysis holds for the category Set_2 *of pairs of sets* (where A is a subset of X), or Gp_2 *of pairs of groups* (where A is a subgroup of X), and in fact for all 'categories of pairs', under very general hypotheses. (Of course $f(X) \cup B$ is now interpreted as $f(X) \vee B$, a join of subobjects.)

The basic framework for (relative) homology and cohomology of groups is the homological category Gp_2, whose short exact sequences have the same form (6.90), for arbitrary subgroups $B \subset A \subset X$.

The categories $\mathsf{Set}_\bullet$, $\mathsf{Top}_\bullet$ and Ban are *pointed homological*, i.e. homological with respect to the ideal of zero morphisms.

6.7.4 Semiexact categories

We only make explicit the basic notion of a semiexact category, in the sense of [G10], a first step towards the notion of homological category.

We are interested in a category E equipped with a *closed ideal* $\mathcal{N}$, whose elements are called *null morphisms*. The condition that $\mathcal{N}$ be an ideal means that every composite with a null morphism is null. The 'closedness' of $\mathcal{N}$ means that every null morphism factorises through a null identity (see Exercise 6.7.5(b)).

Kernels and cokernels in E are defined by an (obvious) reformulation of the universal properties that define them in a pointed category (in 6.1.2).

The *kernel* of $f\colon A \to B$, written as $\ker f\colon \mathrm{Ker}\, f \rightarrowtail A$, is characterised up to isomorphism by the 'usual' universal property, written with respect to the ideal of null morphisms:

(i) $f.(\ker f)$ is null, and for every map h such that fh is null, there is a unique u such that $h = (\ker f)u$

$$\mathrm{Ker}\, f \xrightarrow{\ \ker f\ } A \xrightarrow{\ f\ } B \qquad (6.91)$$

The *cokernel* of f, written as $\operatorname{cok} f \colon B \twoheadrightarrow \operatorname{Cok} f$, is defined by the dual property:

(i*) $(\operatorname{cok} f).f$ is null, and for every map h such that hf is null, there is a unique u such that $h = u(\operatorname{cok} f)$

$$
\begin{array}{ccc}
A \xrightarrow{\ f\ } B \xrightarrow{\ \operatorname{cok} f\ } \operatorname{Cok} f \\
h \downarrow \quad \swarrow u \\
\bullet
\end{array}
\tag{6.92}
$$

It follows easily that $\ker f$ is mono and $\operatorname{cok} f$ is epi. A *normal mono* is, by definition, a kernel (of some morphism), and a *normal epi* is a cokernel; the arrows $\rightarrowtail$, $\twoheadrightarrow$ *are reserved here for these morphisms.*

Now, a *semiexact* category $(\mathsf{E}, \mathcal{N})$ is a category E equipped with a closed ideal $\mathcal{N}$, which has all kernels and cokernels with respect to the latter. We generally write it as E.

Extending the canonical factorisation in a p-exact category (in (6.20)), one proves that every morphism f has a unique *normal factorisation* $f = mgp$ ([G10], Section 1.1.5) through its *normal coimage* p (the cokernel of the kernel morphism) and its *normal image* m (the kernel of the cokernel morphism)

$$
\begin{array}{ccc}
\operatorname{Ker} f \xrightarrow{\ \ker f\ } A \xrightarrow{\ f\ } B \xrightarrow{\ \operatorname{cok} f\ } \operatorname{Cok} f \\
p \downarrow \qquad\qquad \uparrow m \\
\operatorname{Ncm} f \xrightarrow{\ g\ } \operatorname{Nim} f
\end{array}
\tag{6.93}
$$

$$
p = \operatorname{cok}(\ker f), \qquad m = \ker(\operatorname{cok} f).
$$

This factorisation is natural; f is said to be an *exact* morphism if this g is an isomorphism.

Also here, a sequence

$$
M \xrightarrow{\ m\ } A \xrightarrow{\ p\ } P
\tag{6.94}
$$

in E is said to be *short exact* if m is a kernel of p, and p a cokernel of m.

(Exact sequences are also defined in E, but are not need here, as the connected sequences of functors we are considering take values in abelian categories.)

A *p-semiexact category* is a pointed category which is semiexact with respect to the ideal of zero morphisms. Plainly, a p-exact category is the same as a p-semiexact category where every morphism is exact.

6.7.5 Exercises and complements

(a) Every category E has a trivial (closed) ideal $\mathcal{N} = $ E, and is semiexact for it.

(b) In a category E, assigning a closed ideal of morphisms is equivalent to assigning a class of objects *closed under retracts* (the *null* objects).

(c) Verify the formulas (6.88) and (6.89), for kernels and cokernels in Top_2.

(d) Verify the characterisation of short exact sequences (6.90), in Top_2.

(e) The following facts in Top_2 (or Set_2) are in sharp contrast with the behaviour of pointed categories.

A null morphism $(X, A) \to (Y, B)$ between two given objects needs neither exist (take $X \neq \emptyset = B$) nor be unique. A monomorphism (given by any injective map) need not have a null kernel; a null monomorphism need not be normal. The initial object $(\emptyset, \emptyset)$ and the terminal object $(\{*\}, \{*\})$ are distinct and both null, but do not determine the null morphisms; they are also null in the trivial semiexact structure of Exercise (a).

(f) To define the relative singular homology of a pair (X, A) in Top_2, we consider the complex $C_+(X)$ of singular chains of the space X (either in the simplicial form $\mathrm{Ch}_+(\mathrm{Smp}(X))$ of 5.5.4, or in the cubical form $\mathrm{Ch}_+(\mathrm{Cub}(X))$ of 5.5.5), and its subcomplex $C_+(A)$.

The embedding $C_+(A) \to C_+(X)$ gives a short exact sequence in the abelian category $\mathrm{Ch}_+(\mathsf{Ab})$

$$C_+(A) \;\rightarrowtail\; C_+(X) \;\twoheadrightarrow\; C_+(X, A) \tag{6.95}$$

where $C_+(X, A) = C_+(X)/C_+(A)$ is, by definition, the *complex of relative chains* of the pair (X, A).

The *relative homology groups* $H_n(X, A)$ are the homology groups of this complex.

The exact homology sequence of the pair (X, A) follows from a celebrated theorem on the homology groups of a short exact sequence of chain complexes (closely related to the Snake Lemma).

7
Hints at higher dimensional category theory

We sketch here the main two-dimensional categorical structures, namely 2-categories and double categories, together with their weak versions, namely bicategories and weak double categories. This chapter is written in a less elementary form than the previous ones, as a sequence of hints that might encourage a reader to investigate this topic.

The theory of 2-categories and bicategories (or weak 2-categories), briefly presented in Section 7.1, is the best known part of higher dimensional category theory. A general item is a 2-*cell* $\varphi\colon u \to v\colon X \to Y$, as in the left diagram below

$$
X \underset{v}{\overset{u}{\underset{\downarrow\varphi}{\rightrightarrows}}} Y
\qquad
\begin{array}{ccc}
X & \xrightarrow{\ f\ } & X' \\
u\downarrow & \varphi & \downarrow v \\
Y & \xrightarrow[g]{} & Y'
\end{array}
\qquad
\begin{array}{c}
X \\
u\left\downarrow{\overset{\varphi}{\to}}\right\downarrow v \\
Y
\end{array}
\qquad (7.1)
$$

Strict and weak double categories (denoted by letters $\mathbb{A}, \mathbb{B}, ...$) are an extension of the former, presented in Section 7.2 and used in Sections 7.5, 7.6 for applications in Homological Algebra.

Now a general item is a *double cell* $\varphi\colon (u \overset{f}{\underset{g}{}} v)$ among horizontal and vertical arrows, as in the central diagram above. The previous case is obtained by reducing the horizontal arrows to vertical identities, as in the right diagram above. (We use the *horizontal* direction as the main one, that remains strict in the weak case; this choice, and its influence on terminology, will be discussed in 7.2.5.)

One of our main reasons to study double categories, briefly recalled in 7.6.3, is that the general theory of two-dimensional adjunctions can only be formulated in this framework.

In fact, in the general case, *the left adjoint $F\colon \mathbb{C} \to \mathbb{D}$* (between weak double categories, or *also* between bicategories) *is colax while the right ad-*

274

joint $G: \mathbb{D} \to \mathbb{C}$ *is lax*, they should not be composed (this would destroy their comparisons), but viewed as *vertical* and *horizontal* arrows, respectively, in a suitable double category $\mathbb{D}\text{bl}$ introduced in [GP2]. The unit and counit of the adjunction are double cells of the latter

$$\tag{7.2}$$

and their coherence conditions are based – separately – on the comparisons of F and G.

Within 2-categories or bicategories, one can only formulate two particular cases: a colax-pseudo adjunction and a pseudo-lax adjunction, which do not cover the simple colax-lax adjunction presented in 7.6.3; furthermore, adjunctions of these different kinds can only be composed in the general setting.

Another reason is that many 2-categories and bicategories, starting with Rel Ab and the bicategories of spans or cospans, have few limits, while the corresponding (strict or weak) double categories have all horizontal limits, studied in [GP1].

Higher dimensional category theory naturally extends to finite and infinite dimension

- from (strict and weak) 2-categories to n-categories and ω-categories,

- from (strict and weak) double categories to cubical categories, or more generally to multiple categories.

We only refer to [Le] for the first family, to [G11] for the second.

7.1 From categories to 2-categories and bicategories

The prime example of a 2-category is Cat, when we add to its objects and arrows the natural transformations $\varphi: F \to G: \mathsf{C} \to \mathsf{D}$, viewed as *2-cells*, or two-dimensional arrows between the functors – as in the left diagram (7.1), above.

We briefly describe the strict notion of 2-category (from [Eh2]) and the corresponding weak notion of bicategory [Be2]. The elementary cases of a locally ordered 2-category and a locally preordered bicategory have been presented in 4.5.1 and 4.5.5, respectively, while dealing with relations on regular categories.

7.1.1 Sesquicategories

It is convenient to start from a preliminary notion, introduced in [St2]. A *sesquicategory* is a category C equipped with:

(a) for each pair of parallel arrows $f, g\colon X \to Y$, a set $C_2(f, g)$ of *2-cells*, or *homotopies*, written as $\varphi\colon f \to g$ (or $\varphi\colon f \to g\colon X \to Y$), so that each map f has an *identity endocell* $\mathrm{id}f\colon f \to f$,

(b) a *concatenation*, or *vertical composition* of 2-cells, written as $\psi\varphi$ or $\psi.\varphi$

$$X \quad \substack{\xrightarrow{f} \\ \xrightarrow{\downarrow\varphi} \\ \xrightarrow{\downarrow\psi} \\ \xrightarrow{h}} \quad Y \qquad \psi\varphi\colon f \to h\colon X \to Y, \qquad (7.3)$$

(c) a *whisker composition* of 2-cells and maps, or *reduced horizontal composition*, written as $k\varphi h$ (or $k{\circ}\varphi{\circ}h$, when useful)

$$X' \xrightarrow{h} X \quad \substack{\xrightarrow{f} \\ \xrightarrow{\downarrow\varphi} \\ \xrightarrow{g}} \quad Y \xrightarrow{k} Y' \qquad (7.4)$$

$$k\varphi h\colon kfh \to kgh\colon X' \to Y'.$$

These data must satisfy the following axioms, for *associativities, identities* and *distributivity of the whisker composition* (assuming that all compositions are legitimate):

$$\chi(\psi\varphi) = (\chi\psi)\varphi, \qquad\qquad k'(k\varphi h)h' = (k'k)\varphi(hh'),$$

$$\varphi.\mathrm{id}f = \varphi = \mathrm{id}\,g.\varphi, \qquad 1_Y{\circ}\varphi{\circ}1_X = \varphi, \qquad k{\circ}\mathrm{id}f{\circ}h = \mathrm{id}(kfh), \qquad (7.5)$$

$$k(\psi\varphi)h = (k\psi h).(k\varphi h).$$

Note that each set $C(X, Y)$ is a category, under vertical composition. We also use partial whisker compositions: $\varphi h = 1_Y{\circ}\varphi{\circ}h$ and $k\varphi = k{\circ}\varphi{\circ}1_X$.

Examples. (a) The category Top equipped with its homotopies $\varphi\colon f \to g$ does *not* become a sesquicategory, because the vertical composition of homotopies is not associative and has no identities: we have a more complicated 2-dimensional structure (see [G8] and its references).

(b) The category $Ch_+(Ab)$ of chain complexes of abelian groups, equipped with its homotopies $\varphi\colon f \to g$ (cf. 5.5.3), becomes a sesquicategory: the vertical composition and the whisker composition of homotopies are defined in the solution of 5.5.3(a) (see 8.5.8).

(c) This works, with the same computations, for chain complexes over any preadditive category A. If A is additive, we can define a cylinder endofunctor $I\colon Ch_+(A) \to Ch_+(A)$ that extends the definition in (5.63) and produces the homotopies of $Ch_+(A)$.

7.1.2 Two-categories

A *2-category* can be defined as a sesquicategory C which satisfies the following *reduced interchange property*:

$$X \;\overset{f}{\underset{g}{\Rightarrow}}\; Y \;\overset{h}{\underset{k}{\Rightarrow}}\; Z \qquad\qquad (\psi g).(h\varphi) = (k\varphi).(\psi f). \qquad (7.6)$$

The (complete) *horizontal composition* of 2-cells φ, ψ which are *horizontally consecutive*, as above, is defined using the previous identity:

$$\psi \circ \varphi = (\psi g).(h\varphi) = (k\varphi).(\psi f) \colon hf \to kg \colon X \to Z. \qquad (7.7)$$

(The whisker composition is now a particular case: $h\varphi = \mathrm{id}h \circ \varphi$ and $\psi f = \psi \circ \mathrm{id}f$.)

As proved in the exercises below, the horizontal composition of 2-cells is associative, has identities (the identity cells of identity arrows) and satisfies the *middle-four interchange property* with vertical composition (an extension of the reduced interchange property (7.6)):

$$X \;\overset{\scriptstyle\varphi}{\underset{\scriptstyle\psi}{\Rrightarrow}}\; Y \;\overset{\scriptstyle\sigma}{\underset{\scriptstyle\tau}{\Rrightarrow}}\; Z \qquad\qquad (\tau.\sigma) \circ (\psi.\varphi) = (\tau \circ \psi).(\sigma \circ \varphi). \qquad (7.8)$$

As a prime example of such a structure, Cat will also denote the *2-category* of small categories, their functors *and* their natural transformations, with the usual operations, defined in 1.5.3; we have seen in 1.5.4 that the axioms are satisfied.

In fact, the usual definition of a 2-category is based on the complete horizontal composition, although one usually works with the reduced one. Moreover, there are important sesquicategories where the reduced interchange property does not hold (and one does not define a full horizontal composition): for instance, the sesquicategories of chain complexes recalled above.

Two-dimensional limits are studied in [Kl3, St2, Gra].

Adjunctions, monads, equivalences and adjoint equivalences can be defined *inside* any 2-category. (For an adjunction one should use the 'algebraic' form (iii) of definition 3.1.2.)

An ordered category C, as defined in 1.8.2, is a 2-category where each category $\mathsf{C}(X,Y)$ is an ordered set; it is also called a *locally ordered 2-category*. Typical examples are Ord, AdjOrd and $\mathsf{Rel\,Set}$ (cf. 4.5.1).

Exercises. Let C be a sesquicategory satisfying (7.6).

(a) Prove the associativity of the horizontal composition (7.7).

(b) The identity cells of the identity arrows are identities for this composition.

(c) The middle-four interchange property (7.8) holds true.

7.1.3 Bicategories

A *bicategory* is a weak version of a 2-category, introduced by J. Bénabou [Be2] in 1967.

The general framework of objects, arrows, 2-cells and their operations is the same as in a 2-category. The vertical composition of 2-cells is still categorical, and the interchange property still holds. But the horizontal composition $f \otimes g$ of arrows (now written in diagrammatic order) is associative and has units *up to comparisons cells* (for $f \colon X \to Y$, $g \colon Y \to Z$, $h \colon Z \to T$)

$$\kappa(f, g, h) \colon f \otimes (g \otimes h) \to (f \otimes g) \otimes h \qquad (associator),$$
$$\lambda(f) \colon \mathrm{id}X \otimes f \to f, \qquad \rho(f) \colon f \otimes \mathrm{id}Y \to f \qquad (unitors),$$
(7.9)

that are invertible under vertical composition and natural with respect to 2-cells; moreover, they are to satisfy suitable coherence axioms.

In Section 7.2 we shall see the more general structure of a weak double category, from which the interested reader can deduce the naturality and coherence axioms of the previous comparisons, in a bicategory. Typical examples, the bicategories of spans and cospans, will be seen in the extended form of weak double categories, in 7.2.6.

A monoidal (resp. strict monoidal) category C is the same as a bicategory (resp. 2-category) on one formal object $*$. The arrows $X \colon * \to *$ are the objects of C, the cells $f \colon X \to Y$ are the maps of C, the vertical composition is that of C, the horizontal one is the tensor product $X \otimes Y$.

A bicategory C is said to be *locally preordered* when each category $C(X, Y)$ is a preordered set, as in the structure $\mathrm{Rel}'(C)$ of 'jointly monic relations' on a regular category, described in 4.5.5.

In this case the comparison cells (7.9) are uniquely determined, as instances of the equivalence relation associated to the preorder

$$f \otimes (g \otimes h) \sim (f \otimes g) \otimes h, \qquad \mathrm{id}X \otimes f \sim f \sim f \otimes \mathrm{id}Y,$$

and their coherence is automatically satisfied. A locally preordered bicategory C has an associated locally ordered 2-category $C/{\sim}$.

The coherence theorem for bicategories can be found in [MaP]. For higher dimensional extensions of bicategories one can see Leinster [Le] and references therein.

7.1.4 Two-dimensional functors and transformations

A 2-*functor* $F \colon C \to D$ between 2-categories sends objects to objects, arrows to arrows and cells to cells, strictly preserving the whole structure: (co)domains, units and compositions. (Lax versions can be found in [Be2, KlS]; here they will be defined for weak double categories, in 7.2.4.)

A 2-*natural transformation* $h: F \to G: \mathsf{C} \to \mathsf{D}$ between 2-functors is a natural transformation that behaves in a natural way also on every 2-cell $\varphi: f \to g: X \to Y$ in C

$$
\begin{array}{ccc}
FX & \xrightarrow{\ hX\ } & GX \\
{\scriptstyle F\varphi}\big\downarrow\raisebox{-2pt}{$\scriptstyle\to$} & & {\scriptstyle G\varphi}\big\downarrow\raisebox{-2pt}{$\scriptstyle\to$} \\
FY & \xrightarrow[\ hY\ ]{} & GY
\end{array}
\qquad\qquad hY.F\varphi = G\varphi.hX. \qquad (7.10)
$$

It is now easy to define a 2-adjunction (between 2-categories) and a 2-monad (on a 2-category), by 2-functors and 2-natural transformations satisfying the same identities as in the ordinary one-dimensional case. In Section 7.7 we shall use 2-monads on the 2-category Cat.

7.1.5 Two-dimensional universal arrows

Universal arrows of functors (defined in 2.7.1) have a strict and a weak extension to the two-dimensional case.

A 2-*universal arrow* from an object X of X to the 2-functor $U: \mathsf{A} \to \mathsf{X}$ is a pair $(A_0, h: X \to UA_0)$ which gives an isomorphism of categories (of arrows and cells, with vertical composition):

$$
\mathsf{A}(A_0, A) \to \mathsf{X}(X, UA), \qquad g \mapsto Ug.h. \qquad (7.11)
$$

This amounts to saying that the functor (7.11) is bijective on objects, full and faithful, that is:

(i) for every A in A and every $f: X \to UA$ in X there is a unique morphism $g: A_0 \to A$ in A such that $f = Ug.h$,

(ii) for every pair $g, g': A_0 \to A$ in A and every cell $\varphi: Ug.h \to Ug'.h$ in X, there is a unique cell $\psi: g \to g'$ in A such that $\varphi = U\psi.h$.

More generally, a *biuniversal arrow* from X to $U: \mathsf{A} \to \mathsf{X}$ is a pair $(A_0, h: X \to UA_0)$ such that the functor (7.11) is an *equivalence* of categories. This can be rephrased saying that the functor (7.11) is essentially surjective on objects, full and faithful (see 1.5.5).

In other words, we replace property (i) with a weaker version (and keep (ii) as it is):

(i') for every A in A and every $f: X \to UA$ in X there is *some* $g: A_0 \to A$ in A such that $f \cong Ug.h$ (*isomorphic objects* in the category $\mathsf{X}(X, UA)$).

In concrete situations one can often replace (i) with the property of *surjectivity* on objects, which is intermediate between (i) and (i'):

(i'') for every A in A and every $f : X \to UA$ in X there is *some* $g : A_0 \to A$ such that $f = Ug.h$.

Of course, the solution of a 2-universal problem is *determined up to isomorphism*, while the solution of a biuniversal one is *determined up to equivalence* (in a 2-category).

The *2-limit* (resp. *bilimit*) of a 2-functor is defined by a 2-universal (resp. biuniversal) property, replacing the ordinary one. Further information on 2-dimensional limits and related matter can be found in [St2].

Exercises and complements. (a) The conjunction of the properties (i), (ii) above is equivalent to a global universal property: for every cell $\varphi : f \to f' : X \to UA$ in X, there is precisely one cell $\psi : g \to g'$ in A such that $\varphi = U\psi.h$. (This implies that $f = Ug.h$ and $f' = Ug'.h$.)

(b) The 2-category Cat has (small) 2-limits.

(c) The term 'biproduct' (in the sense of product *and* coproduct, as in 2.1.9 and Section 6.4) clashes with the (more recent) term 'bilimit'. The clash is reduced by the fact that 2-products seem to be more important than the corresponding bilimits – essentially because they are 'flexible limits', see [BKPS].

7.1.6 Natural transformations and mates

Let us suppose that we have two adjunctions, between ordinary categories

$$F : X \rightleftarrows Y : G, \qquad \eta : 1 \to GF, \qquad \varepsilon : FG \to 1,$$
$$F' : X' \rightleftarrows Y' : G', \qquad \eta' : 1 \to G'F', \qquad \varepsilon' : F'G' \to 1, \qquad (7.12)$$

and two functors $H : X \to X'$, $K : Y \to Y'$.

As shown in [KlS], Section 2.2, there is a bijection between sets of natural transformations:

$$\mathrm{Nat}(F'H, KF) \to \mathrm{Nat}(HG, G'K), \qquad \varphi \leftrightarrow \psi,$$

$$\psi = G'K\varepsilon.G'\varphi G.\eta'HG : HG \to G'F'HG \to G'KFG \to G'K, \qquad (7.13)$$

$$\varphi = \varepsilon'KF.F'\psi F.F'H\eta : F'H \to F'HGF \to F'G'KF \to KF,$$

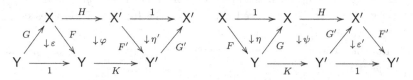

The natural transformations φ, ψ are said to be *mates under the given data*; or *under the given adjunctions* when H and K are identities.

Actually, this holds true for *internal* adjunctions in any 2-category.

An exercise. (a) The mappings $\varphi \mapsto \psi$ and $\psi \mapsto \varphi$ in (7.13) are inverse to each other.

7.1.7 The 2-category of adjoint functors

The category AdjCat of small categories and adjunctions

$$(F, G, \eta, \varepsilon) \colon \mathsf{C} \twoheadrightarrow \mathsf{D},$$

has been introduced in 3.2.1.

We show now that it has a natural 2-categorical structure, derived from that of Cat and extending the 2-categorical structure of AdjOrd given by its order (defined in (1.52)).

Given two adjunctions between the same categories

$$(F, G, \eta, \varepsilon) \colon \mathsf{C} \twoheadrightarrow \mathsf{D}, \qquad (F', G', \eta', \varepsilon') \colon \mathsf{C} \twoheadrightarrow \mathsf{D}, \tag{7.14}$$

a 2-cell of AdjCat

$$(\varphi, \psi) \colon (F, G, \eta, \varepsilon) \to (F', G', \eta', \varepsilon') \colon \mathsf{C} \twoheadrightarrow \mathsf{D}, \tag{7.15}$$

is defined as a pair of mate natural transformations (under the given adjunctions), each of them determining the other

$$\varphi \colon F \to F' \colon \mathsf{C} \to \mathsf{D}, \qquad \psi \colon G' \to G \colon \mathsf{D} \to \mathsf{C},$$

$$\psi = (G\varepsilon'.G\varphi G'.\eta G' \colon G' \to GFG' \to GF'G' \to G), \tag{7.16}$$

$$\varphi = (\varepsilon F'.F\psi F'.F\eta' \colon F \to FG'F' \to FGF' \to F').$$

The vertical composition of 2-cells in AdjCat comes from the vertical composition of natural transformations

$$(\varphi', \psi').(\varphi, \psi) = (\varphi'\varphi, \psi\psi'). \tag{7.17}$$

The whisker composition of (φ, ψ) with two adjunctions

$$(H, K, \rho, \sigma) \colon \mathsf{C}' \twoheadrightarrow \mathsf{C}, \qquad (H', K', \rho', \sigma') \colon \mathsf{D} \twoheadrightarrow \mathsf{D}',$$

is given by two whisker compositions of natural transformations

$$H'\varphi H \colon H'FH \to H'F'H \colon \mathsf{C}' \to \mathsf{D}',$$
$$K\psi K' \colon KG'K' \to KGK' \colon \mathsf{D}' \to \mathsf{C}'. \tag{7.18}$$

Finally the (contravariant) involution of (3.18) can be extended (covariantly) to 2-cells

$$(-)^{\mathrm{op}} \colon \mathsf{AdjCat} \to \mathsf{AdjCat}, \qquad \mathsf{C} \mapsto \mathsf{C}^{\mathrm{op}},$$

$$((F, G, \eta, \varepsilon) \colon \mathsf{C} \nrightarrow \mathsf{D}) \mapsto ((G^{\mathrm{op}}, F^{\mathrm{op}}, \varepsilon^{\mathrm{op}}, \eta^{\mathrm{op}}) \colon \mathsf{D}^{\mathrm{op}} \nrightarrow \mathsf{C}^{\mathrm{op}}),$$

$$((\varphi, \psi) \colon (F, G, \eta, \varepsilon) \to (F', G', \eta', \varepsilon')) \mapsto \tag{7.19}$$

$$((\psi^{\mathrm{op}}, \varphi^{\mathrm{op}}) \colon (G^{\mathrm{op}}, F^{\mathrm{op}}, \varepsilon^{\mathrm{op}}, \eta^{\mathrm{op}}) \to (G'^{\mathrm{op}}, F'^{\mathrm{op}}, \varepsilon'^{\mathrm{op}}, \eta'^{\mathrm{op}})).$$

In the same way (except for the involution) one can construct a 2-category AdjC of *internal adjunctions* in an arbitrary 2-category C.

7.2 Double categories

Strict double categories were introduced and studied by C. Ehresmann and Andrée C. Ehresmann [Eh1]–[Eh5], [BasE], [EE] since 1962; the weak notion much later, in [GP1]–[GP4]; many other references can be found in these articles.

The strict case extends the more usual notion of 2-category, while the weak one extends bicategories: see 7.2.5.

Applications of double categories in Computer Science can be found, for instance, in [BruMM]. Applications in Homological Algebra will be presented in Sections 7.5 and 7.6.

Double and multiple categories (strict, weak and partially lax) are the subject of a recent book [G11].

7.2.1 Definition

A (strict) *double category* $\mathbb{A}$ consists of the following structure.

(a) A set $\mathrm{Ob}\,\mathbb{A}$ of *objects* of $\mathbb{A}$.

(b) *Horizontal morphisms* $f \colon X \to X'$ between the previous objects; they form the category $\mathrm{Hor}_0\mathbb{A}$ *of the objects and horizontal maps* of $\mathbb{A}$, with composition written as gf and identities $1_X \colon X \to X$.

(c) *Vertical morphisms* $u \colon X \nrightarrow Y$ (often denoted by a dot-marked arrow) between the same objects; they form the category $\mathrm{Ver}_0\mathbb{A}$ *of the objects and vertical maps* of $\mathbb{A}$, with composition written as $v{\bullet}u$ (or $u \otimes v$, in diagrammatic order) and identities written as $e_X \colon X \nrightarrow X$ or $1^{\bullet}_X$.

(d) (*Double*) *cells* $a\colon (u\ {}^{f}_{g}\ v)$

$$
\begin{array}{ccc}
X & \xrightarrow{\ f\ } & X' \\
{\scriptstyle u}\big\downarrow\!\!\bullet & a & \bullet\!\!\big\downarrow{\scriptstyle v} \\
Y & \xrightarrow[\ g\ ]{} & Y'
\end{array}
\tag{7.20}
$$

with a *boundary* formed of two horizontal arrows f, g and two vertical arrows u, v (often marked with a dot).

Writing $a\colon (X\ {}^{X}_{g}\ v)$ or $a\colon (e\ {}^{1}_{g}\ v)$ we mean that $f = 1_X$ and $u = e_X$. The cell a is also written as $a\colon u \to v$ (with respect to its *horizontal* domain and codomain, which are *vertical* arrows) or as $a\colon f \twoheadrightarrow g$ (with respect to its *vertical* domain and codomain, which are *horizontal* arrows).

We refer now to the following diagrams of cells, where the first is called a *consistent matrix* $\left({}^{a\ b}_{c\ d}\right)$ of cells

$$
\begin{array}{ccccc}
X & \xrightarrow{\ f\ } & X' & \xrightarrow{\ f'\ } & X'' \\
{\scriptstyle u}\big\downarrow\!\!\bullet & a & \bullet\!\!\big\downarrow{\scriptstyle v} & b & \bullet\!\!\big\downarrow{\scriptstyle w} \\
Y & {\scriptstyle -g}\!\!\succ & Y' & {\scriptstyle -g'}\!\!\succ & Y'' \\
{\scriptstyle u'}\big\downarrow\!\!\bullet & c & \bullet\!\!\big\downarrow{\scriptstyle v'} & d & \bullet\!\!\big\downarrow{\scriptstyle w'} \\
Z & \xrightarrow[\ h\ ]{} & Z' & \xrightarrow[\ h'\ ]{} & Z''
\end{array}
\qquad
\begin{array}{ccc}
X & \xrightarrow{\ 1\ } & X \\
{\scriptstyle u}\big\downarrow\!\!\bullet & 1_u & \bullet\!\!\big\downarrow{\scriptstyle u} \\
Y & \xrightarrow[\ 1\ ]{} & Y
\end{array}
\qquad
\begin{array}{ccc}
X & \xrightarrow{\ f\ } & X' \\
{\scriptstyle e}\big\downarrow\!\!\bullet & e_f & \bullet\!\!\big\downarrow{\scriptstyle e} \\
X & \xrightarrow[\ f\ ]{} & X'
\end{array}
\tag{7.21}
$$

(e) Cells have a *horizontal composition*, consistent with the horizontal composition of arrows and written as $a|b$, or

$$
(a \mid b)\colon (u\ {}^{f'f}_{g'g}\ w).
$$

This composition gives the category $\mathrm{Hor}_1\mathbb{A}$ *of vertical arrows and cells* $a\colon u \to v$ of $\mathbb{A}$, with identities $1_u\colon (u\ {}^{1}_{1}\ u)$.

(f) Cells have also a *vertical composition*, consistent with the vertical composition of arrows and written as $a \otimes c$, or

$$
\left(\frac{a}{c}\right)\colon (u \otimes u'\ {}^{f}_{h}\ v \otimes v').
$$

This composition gives the category $\mathrm{Ver}_1\mathbb{A}$ *of horizontal arrows and cells* $a\colon f \twoheadrightarrow g$ of $\mathbb{A}$, with identities $e_f\colon (e\ {}^{f}_{f}\ e)$, or $1^{\bullet}_f$.

(g) The two compositions satisfy the *interchange laws* (for binary and 0-ary compositions), which means that we have, in diagram (7.21):

$$
\left(\frac{a \mid b}{c \mid d}\right) = \left(\frac{a}{c} \,\middle|\, \frac{b}{d}\right), \qquad \left(\frac{1_u}{1_{u'}}\right) = 1_{u \otimes u'},
$$

$$
(e_f \mid e_{f'}) = e_{f'f}, \qquad\qquad 1_{e_X} = e_{1_X}.
\tag{7.22}
$$

The first condition says that a consistent matrix $\begin{pmatrix} a & b \\ c & d \end{pmatrix}$ has a precise *pasting*; the last says that an object X has an *identity cell* $\square_X = 1_{e_X} = e_{1_X}$.

The expressions $(a \mid f')$ and $(f \mid b)$ will stand for $(a \mid e_{f'})$ and $(e_f \mid b)$, when this makes sense.

The double category $\mathbb{A}$ is said to be *flat* if every double cell $a \colon (u \, ^f_g \, v)$ is determined by its boundary – namely the arrows f, g, u, v. A standard example is the double category $\mathbb{RelSet}$ of sets, mappings and relations, described below in 7.2.7.

7.2.2 Weak double categories

More generally, in a *weak double category* $\mathbb{A}$ the horizontal composition behaves categorically (and we still have ordinary categories $\mathrm{Hor}_0\mathbb{A}$ and $\mathrm{Hor}_1\mathbb{A}$), while the composition of vertical arrows is categorical up to *comparison cells*; the latter are *special* (which means that their horizontal arrows are identities) and *horizontally invertible*.

- For $u \colon X \nrightarrow Y$ we have a *left unitor* and a *right unitor*

$$\lambda u \colon e_X \otimes u \to u, \qquad \rho u \colon u \otimes e_Y \to u, \tag{7.23}$$

which are natural for a cell $a \colon (u \, ^f_g \, v)$ in the sense that

$$(e_f \otimes a \mid \lambda v) = (\lambda u \mid a), \qquad (a \otimes e_g \mid \rho v) = (\rho u \mid a).$$

- For three consecutive vertical arrows $u \colon X \nrightarrow Y$, $v \colon Y \nrightarrow Z$ and $w \colon Z \nrightarrow T$ we have an *associator*

$$\kappa(u, v, w) \colon u \otimes (v \otimes w) \to (u \otimes v) \otimes w, \tag{7.24}$$

natural for cells $a \colon (u \, ^f_g \, u')$, $b \colon (v \, ^g_h \, v')$, $c \colon (w \, ^h_k \, w')$, in the sense that

$$\big(a \otimes (b \otimes c) \mid \kappa(u', v', w')\big) = \big(\kappa(u, v, w) \mid (a \otimes b) \otimes c\big).$$

Interchange between horizontal and vertical composition of cells holds strictly, as above.

Finally, *the comparison cells* λ, ρ, κ *are assumed to satisfy coherence axioms*, listed below.

A is said to be *unitary* if the unitors are identities, so that the vertical identities behave as strict units: a constraint which in concrete cases can often be easily met.

The terminology of the strict case is extended to the weak one, as far as possible.

7.2.3 Coherence conditions

In a weak double category the following diagrams of cells must commute under horizontal composition.

(a) *Coherence pentagon of the associator* κ, on four consecutive vertical arrows x, y, z, u

$$
\begin{array}{ccc}
& (x \otimes y) \otimes (z \otimes u) & \\
{}^{\kappa}\nearrow & & \searrow{}^{\kappa} \\
x \otimes (y \otimes (z \otimes u)) & & ((x \otimes y) \otimes z) \otimes u \\
{}^{1 \otimes \kappa}\searrow & & \nearrow{}^{\kappa \otimes 1} \\
& x \otimes ((y \otimes z) \otimes u) \xrightarrow[\kappa]{} (x \otimes (y \otimes z)) \otimes u &
\end{array}
\qquad (7.25)
$$

(b) *Coherence conditions for the unitors*, for $x \colon A \rightarrowtail B$ and $y \colon B \rightarrowtail C$

$$
\begin{array}{ccc}
x \otimes (e_B \otimes y) & \xrightarrow{\ \kappa\ } & (x \otimes e_B) \otimes y \\
{}_{1 \otimes \lambda}\searrow & & \swarrow{}_{\rho \otimes 1} \\
& x \otimes y &
\end{array}
\qquad (7.26)
$$

$$
\begin{array}{ccc}
e_A \otimes (x \otimes y) & \xrightarrow{\ \kappa\ } & (e_A \otimes x) \otimes y \\
{}_{\lambda}\searrow & & \swarrow{}_{\lambda \otimes 1} \\
& x \otimes y &
\end{array}
\qquad (7.27)
$$

$$
\begin{array}{ccc}
x \otimes (y \otimes e_C) & \xrightarrow{\ \kappa\ } & (x \otimes y) \otimes e_C \\
{}_{1 \otimes \rho}\searrow & & \swarrow{}_{\rho} \\
& x \otimes y &
\end{array}
\qquad (7.28)
$$

$$
\lambda(eA) = \rho(eA) \colon eA \otimes eA \to eA. \qquad (7.29)
$$

We write the conditions (b) in the form used by Mac Lane in his classical paper on coherence of monoidal categories [M3]. As proved by Kelly [Kl1], this set of axioms is redundant: properties (7.25) and (7.26) imply the other three.

This is also true in the present more general case. Listing all the previous conditions has nevertheless some interest: there are defective structures where the unitors are *not* invertible and the redundancy recalled above disappears.

7.2.4 Lax functors and horizontal transformations

A *lax (double) functor* $F: \mathbb{X} \to \mathbb{A}$ between weak double categories amounts to assigning:

(a) two functors $\mathrm{Hor}_0 F$ and $\mathrm{Hor}_1 F$, consistent with domain and codomain

$$
\begin{array}{ccc}
\mathrm{Hor}_1\mathbb{X} \xrightarrow{\mathrm{Hor}_1 F} \mathrm{Hor}_1\mathbb{A} & \quad & \mathrm{Hor}_1\mathbb{X} \xrightarrow{\mathrm{Hor}_1 F} \mathrm{Hor}_1\mathbb{A} \\
\Big\downarrow{\scriptstyle \mathrm{Dom}} \qquad \Big\downarrow{\scriptstyle \mathrm{Dom}} & \quad & \Big\downarrow{\scriptstyle \mathrm{Cod}} \qquad \Big\downarrow{\scriptstyle \mathrm{Cod}} \\
\mathrm{Hor}_0\mathbb{X} \xrightarrow[\mathrm{Hor}_0 F]{} \mathrm{Hor}_0\mathbb{A} & \quad & \mathrm{Hor}_0\mathbb{X} \xrightarrow[\mathrm{Hor}_0 F]{} \mathrm{Hor}_0\mathbb{A}
\end{array}
\qquad (7.30)
$$

(b) for any object X in $\mathbb{X}$ a special cell, the *identity comparison* of F

$$\underline{F}(X): e_{FX} \to Fe_X: FX \nrightarrow FX,$$

(c) for any vertical composite $u \otimes v: X \nrightarrow Y \nrightarrow Z$ in $\mathbb{X}$ a special cell, the *composition comparison*

$$\underline{F}(u,v): Fu \otimes Fv \to F(u \otimes v): FX \nrightarrow FZ.$$

Again, these comparisons must satisfy axioms of naturality and coherence with the comparisons of $\mathbb{X}$ and $\mathbb{A}$ (cf. [GP2]).

A *colax (double) functor* has comparisons in the opposite direction. A *pseudo (double) functor* is a lax functor whose comparison cells are horizontally invertible; it is made colax by the inverse cells.

A *horizontal transformation of lax functors* $h: F \to G: \mathbb{X} \to \mathbb{A}$ has the following components in $\mathbb{A}$

- a horizontal map $hX: FX \to GX$ and a cell $hu: (Fu \; \begin{smallmatrix} hX \\ hY \end{smallmatrix} \; Gu)$,

for each object X and each vertical map $u: X \nrightarrow Y$ in $\mathbb{X}$, under conditions of naturality and coherence with the comparisons of F and G (cf. [GP2]).

7.2.5 Dualities

A weak double category has a *horizontal opposite* $\mathbb{A}^h$ (reversing the horizontal direction) and a *vertical opposite* $\mathbb{A}^v$ (reversing the vertical direction).

A strict structure also has a *transpose* $\mathbb{A}^t$ (interchanging the horizontal and vertical issues).

The prefix 'co', as in *colimit, coequaliser* or *colax double functor*, refers to horizontal duality, the main one.

Let us note that a weak double category whose horizontal arrows are identities is the same as a *bicategory written in vertical* (as in the right diagram of (7.1)), with arrows and weak composition in the vertical direction and strict composition of cells in the horizontal one. This is why *the oplax functors of bicategories correspond here to colax double functors.*

(Transposing the theory of double categories, as is done in some papers, would avoid this conflict of terminology, but would produce other conflicts at a more basic level: for instance, colimits in Set would become 'op-limits' in RelSet and SpanSet.)

7.2.6 Spans and cospans

For a category C with (a fixed choice of) pullbacks there is a weak double category $\mathbb{S}\mathrm{pan}(C)$ *of spans* over C.

Objects, horizontal arrows and their composition come from C, so that $\mathrm{Hor}_0(\mathbb{S}\mathrm{pan}C) = C$.

A vertical arrow $u \colon X \rightarrowtail Y$ is a span $u = (u', u'')$, i.e. a diagram

$$X \leftarrow U \rightarrow Y$$

in C, or equivalently a functor $u \colon \vee \to C$ defined on the formal-span category $\bullet \leftarrow \bullet \rightarrow \bullet$. A vertical identity is a pair $e_X = (1_X, 1_X)$.

A cell $\sigma \colon (u \, {}_g^f \, v)$ is a natural transformation $u \to v$ of such functors and amounts to the commutative left diagram below

$$
\begin{array}{ccc}
X \xrightarrow{\ f\ } X' & \qquad & X \longrightarrow X' \longrightarrow X'' \\
\end{array}
\tag{7.31}
$$

We say that the cell σ is *represented* by its middle arrow $m\sigma \colon U \to V$, which determines it together with the boundary (the present structure is not flat).

The horizontal composition $\sigma | \tau$ of σ with a second cell $\tau \colon v \to w$ is a composition of natural transformations as in the right diagram above; it gives the category $\mathrm{Hor}_1(\mathbb{S}\mathrm{pan}C) = \mathsf{Cat}(\vee, C)$.

The vertical composition $u \otimes v$ of spans is computed by (chosen) pullbacks in C

$$X \nwarrow \quad \nearrow Y \nwarrow \quad \nearrow Z$$
$$U \quad V \qquad W = U \times_Y V. \qquad (7.32)$$
$$W$$

This composition is extended to double cells, in the obvious way. For the sake of simplicity *we make* $\mathrm{Span}(C)$ *unitary*, by adopting the 'unit constraint' for pullbacks (as in 2.4.8): the chosen pullback of an identity along any morphism is an identity. The associator κ is determined by the universal property of pullbacks.

Dually, for a category C with (a fixed choice of) pushouts there is a unitary weak double category $\mathbb{C}\mathrm{osp}(C)$ *of cospans* over C, that is horizontally dual to $\mathbb{S}\mathrm{pan}(C^{\mathrm{op}})$. We have now

$$\mathrm{Hor}_0(\mathbb{C}\mathrm{osp}C) = C, \qquad \mathrm{Hor}_1(\mathbb{C}\mathrm{osp}C) = \mathrm{Cat}(\wedge, C), \qquad (7.33)$$

where $\wedge = \vee^{\mathrm{op}}$ is the formal-cospan category $\bullet \to \bullet \leftarrow \bullet$.

A vertical arrow $u \colon X \rightarrowtail Y$ is a cospan $X \to U \leftarrow Y$ in C, and a functor $u \colon \wedge \to C$; a cell $\sigma \colon u \to v$ is a natural transformation. The vertical composition is computed by pushouts in C; again, we generally follow the 'unit constraint' for pushouts.

We write as $\mathrm{Span}(C)$ and $\mathrm{Cosp}(C)$ the bicategories obtained from the previous weak double categories, by restricting horizontal arrows to identities.

7.2.7 Other examples

We briefly recall other strict or weak double categories. The interested reader will find the details in [GP1], or [G11].

(a) The ordered category $\mathrm{Rel}\,\mathrm{Set}$ of sets and relations has been introduced in 1.8.3. We now form a flat double category $\mathbb{A} = \mathbb{R}\mathrm{el}\mathrm{Set}$, with $\mathrm{Hor}_0\mathbb{A} = \mathrm{Set}$, $\mathrm{Ver}_0\mathbb{A} = \mathrm{Rel}\,\mathrm{Set}$ and double cells of the following kind

$$\begin{array}{ccc} X & \xrightarrow{f} & X' \\ u \downarrow & \leqslant & \downarrow v \\ Y & \xrightarrow{g} & Y' \end{array} \qquad gu \leqslant vf \quad (\text{in } \mathrm{Rel}\,\mathrm{Set}). \qquad (7.34)$$

Double cells are determined by their boundary; their horizontal and vertical composition amount to composing mappings or relations, respectively.

Similarly, one can form a double category $\mathbb{Rel}(C)$ over any regular category, 'pasting' C with the ordered category $\mathrm{Rel}\,(C)$ constructed in Section 4.5. Or over any p-exact category, using the ordered category constructed in Section 6.6; this structure will be used in Section 7.6.

(b) The ordered category AdjOrd of ordered sets and Galois connections has been introduced in 1.7.4. We now form a flat double category $\mathbb{A} = \mathsf{AdjOrd}$, with $\mathrm{Hor}_0\mathbb{A} = \mathsf{Ord}$, $\mathrm{Ver}_0\mathbb{A} = \mathsf{AdjOrd}$ and double cells of the following kind

$$
\begin{array}{ccc}
X & \xrightarrow{\ f\ } & X' \\
u\downarrow & \downarrow & \downarrow v \\
Y & \xrightarrow{\ g\ } & Y'
\end{array}
\qquad
\begin{aligned}
v_\bullet f &\leqslant g u_\bullet : X \to Y' \quad (\text{in Ord}), \\[4pt]
f u^\bullet &\leqslant v^\bullet g : Y \to X' \quad (\text{in Ord}).
\end{aligned}
\tag{7.35}
$$

(The two conditions are easily seen to be equivalent.) Again, the horizontal and vertical composition amount to composing horizontal or vertical arrows.

We are also interested in the double subcategory $\mathsf{Adj}_0\mathsf{Ord}$ where we restrict the double cells to the *bicommutative* ones, which here means that $v_\bullet f = g u_\bullet$ and $f u^\bullet = v^\bullet g$.

(c) More generally there is a double category AdjCat of (small) *categories, functors and adjunctions.*

Here $\mathrm{Hor}_0(\mathsf{AdjCat}) = \mathsf{Cat}$ and $\mathrm{Ver}_0(\mathsf{AdjCat}) = \mathsf{AdjCat}$, the category of adjunctions introduced in 3.2.1. A vertical arrow is thus an adjunction, conventionally directed as the *left* adjoint

$$
\begin{aligned}
u &= (u_\bullet, u^\bullet, \eta, \varepsilon) \colon X \nrightarrow Y, \qquad (u_\bullet \colon X \to Y) \dashv (u^\bullet \colon Y \to X), \\
\eta &\colon 1_X \to u^\bullet u_\bullet, \qquad\qquad\qquad\quad \varepsilon \colon u_\bullet u^\bullet \to 1_Y.
\end{aligned}
\tag{7.36}
$$

A double cell $a = (a_\bullet, a^\bullet)$ is a pair of *mate* natural transformations (cf. 7.1.6)

$$
\begin{array}{ccc}
X & \xrightarrow{\ f\ } & X' \\
u\downarrow & \downarrow a & \downarrow v \\
Y & \xrightarrow{\ g\ } & Y'
\end{array}
\qquad
\begin{aligned}
a_\bullet &\colon v_\bullet f \to g u_\bullet \colon X \to Y', \\[4pt]
a^\bullet &\colon f u^\bullet \to v^\bullet g \colon Y \to X'.
\end{aligned}
\tag{7.37}
$$

Each of them determines the other via the units and counits of the adjunctions u and v

$$
\begin{aligned}
a^\bullet &= (f u^\bullet \to v^\bullet v_\bullet f u^\bullet \to v^\bullet g u_\bullet u^\bullet \to v^\bullet g), \\
a_\bullet &= (v_\bullet f \to v_\bullet f u^\bullet u_\bullet \to v_\bullet v^\bullet g u_\bullet \to g u_\bullet).
\end{aligned}
\tag{7.38}
$$

(d) This can be further extended to a double category $\mathsf{Adj}\mathsf{C}$, where C is any 2-category.

7.2.8 *Profunctors

Another important prototype is the weak double category Cat of (small) *categories, functors and profunctors*. In a general cell

$$
\begin{array}{ccc}
X & \xrightarrow{\ f\ } & X' \\
{\scriptstyle u}\downarrow & \overset{a}{\underset{\to}{}} & \downarrow{\scriptstyle v} \\
Y & \xrightarrow{\ g\ } & Y'
\end{array}
\tag{7.39}
$$

each object is a category, a horizontal arrow is a functor, a vertical arrow $u\colon X \nrightarrow Y$ is a *profunctor*, i.e. a functor $X^{\mathrm{op}} \times Y \to \mathsf{Set}$. The cell 'is' a natural transformation

$$
a\colon u \to v.(f^{\mathrm{op}} \times g)\colon X^{\mathrm{op}} \times Y \to \mathsf{Set}.
$$

The vertical composition of u with $u'\colon Y \nrightarrow Z$ is given by (choosing) a coend in Set (see 2.8.6)

$$
(u \otimes u')(x, z) = \int^{y} u(x, y) \times u'(y, z),
\tag{7.40}
$$

and has identities $e_X = \mathrm{Mor}\colon X^{\mathrm{op}} \times X \to \mathsf{Set}$.

Its comparison cells for unitarity and associativity derive from the universal property of coends. The horizontal composition of double cells is obvious, the vertical one is computed by coends.

All this can be interpreted in a more elementary way.

In a profunctor $u\colon X \nrightarrow Y$, we view the elements $\lambda \in u(x, y)$ as formal *new arrows* $\lambda\colon x \nrightarrow y$ from the objects of X to those of Y (also called *heteromorphisms*).

Together with the disjoint union of the objects and arrows of X and Y, we form thus a new category $X +_u Y$ known as the *gluing*, or *collage*, of X and Y along u. (It is the *cotabulator* of u in Cat [GP1, G11].)

The whisker composition between 'old' and 'new' arrows is determined by the action of u on the old ones

$$
\eta\lambda\xi = u(\xi, \eta)(\lambda),
\tag{7.41}
$$

for $\xi\colon x' \to x$ in X, $\lambda\colon x \nrightarrow y$ and $\eta\colon y \to y'$ in Y.

The profunctor u amounts thus to a category $U = X +_u Y$ containing the disjoint union $X + Y$ and new arrows from objects of X to objects of Y (possibly none).

More formally, the collage U is a *category over* $\mathbf{2}$, i.e. an object $U \to \mathbf{2}$ of the slice category $\mathsf{Cat}/\mathbf{2} = (\mathsf{Cat}\!\downarrow\mathbf{2})$ (see (3.29)), whose fibres over 0 and 1 are X and Y, respectively.

For the vertical composite $u \otimes u' \colon X \nrightarrow Z$, an element of $(u \otimes u')(x, z)$ is an equivalence class $\mu{\circ}\lambda \colon x \nrightarrow z$, where $\lambda \colon x \nrightarrow y$, $\mu \colon y \nrightarrow z$ are new arrows of u and v; the equivalence relation is generated by

$$
\mu\eta \circ \lambda \sim \mu \circ \eta\lambda, \qquad (7.42)
$$

where $\eta \colon y \to y'$ belongs to Y.

The cell a in (7.42) corresponds to a *functor over* $\mathbf{2}$, that restricts to f and g over 0 and 1

$$
a \colon X +_u Y \to X' +_v Y', \qquad (\lambda \colon x \nrightarrow y) \mapsto (a(\lambda) \colon fx \nrightarrow gy). \qquad (7.43)
$$

The horizontal composition of cells is the composition of these functors, so that $\mathrm{Hor}_1\mathbb{C}\mathsf{at} = \mathsf{Cat}/\mathbf{2}$. In a vertical composite $\alpha \otimes \beta \colon u \otimes u' \to v \otimes v'$ the equivalence class $\mu{\circ}\lambda \colon x \nrightarrow z$ of $X +_{u\otimes u'} Z$ is sent to the equivalence class $\beta(\mu){\circ}\alpha(\lambda) \colon \alpha(x) \nrightarrow \beta(z)$.

Finally, it will be useful to note that a functor $f \colon X \to Y$ has two associated profunctors:

$$
f_* = \mathrm{Mor}_Y(f^{\mathrm{op}} \times \mathrm{id}\,Y) \colon X^{\mathrm{op}} \times Y \to \mathsf{Set},
$$
$$
f^* = \mathrm{Mor}_Y(\mathrm{id}\,Y^{\mathrm{op}} \times f) \colon Y^{\mathrm{op}} \times X \to \mathsf{Set}, \qquad (7.44)
$$
$$
f_*(x, y) = Y(fx, y), \qquad f^*(y, x) = Y(y, fx).
$$

Profunctors were introduced by Bénabou [Be3] and Lawvere [Lw1], under the names of *distributors* and *bimodules*, respectively. We are using the opposite direction with respect to the original one, justified by the direction of the heteromorphisms.

Exercises and complements. (a) Characterise, in the weak double category $\mathbb{C}\mathsf{at}$, the *special* cells $a \colon u \to v$ (whose horizontal arrows are identities) and the *globular* cells $b \colon f \nrightarrow g$ (whose vertical arrows are identities)

$$(7.45)$$

(b) For a monoidal category C, the tensor product $T\colon \mathsf{C}^2 \to \mathsf{C}$ gives a *profunctor of heteromorphisms*

$$T_*\colon \mathsf{C}^2 \nrightarrow \mathsf{C}, \qquad T_*((A,B),C) = \mathsf{C}(A \otimes B, C). \qquad (7.46)$$

Describe the collage of T_*.

(c) In particular, the tensor product $T\colon R\,\mathsf{Mod}^2 \to R\,\mathsf{Mod}$ of modules on a commutative ring gives a profunctor T_* whose collage $R\,\mathsf{Bil}$ contains the R-bilinear maps. The universal property of $\eta\colon A \times B \to A \otimes_R B$ (in (3.37)) can be rewritten as a diagram in $R\,\mathsf{Bil}$.

7.2.9 Vertical involutions

As we have hinted at in 1.8.1, the (covariant) reversor of Ord and the (contravariant) involution of AdjOrd have the same action $X \mapsto X^{\mathrm{op}}$ on the objects, and can be combined to form a *vertical involution* of the double category AdjOrd: this is horizontally covariant (on monotone mappings) and vertically contravariant (on adjunctions).

More precisely, we have an involutive double functor defined on the vertical dual AdjOrd^v

$$(-)^{\mathrm{op}}\colon \mathsf{AdjOrd}^v \to \mathsf{AdjOrd}, \qquad X \mapsto X^{\mathrm{op}}, \qquad (7.47)$$

which acts as follows on the arrows and the (flat) double cells

$$
\begin{array}{ccc}
X \xrightarrow{\ f\ } X' & & Y^{\mathrm{op}} \xrightarrow{\ g^{\mathrm{op}}\ } Y'^{\mathrm{op}} \\
{\scriptstyle u}\big\downarrow \quad \downarrow \quad \big\downarrow{\scriptstyle v} & \mapsto & {\scriptstyle u^{\mathrm{op}}}\big\downarrow \quad \downarrow \quad \big\downarrow{\scriptstyle v^{\mathrm{op}}} \\
Y \xrightarrow[\ g\ ]{} Y' & & X^{\mathrm{op}} \xrightarrow[\ f^{\mathrm{op}}\ ]{} X'^{\mathrm{op}}
\end{array}
\qquad (7.48)
$$

On a strict double category $\mathbb{A}$, a *vertical involution* will thus be an involutive double functor $\mathbb{A}^v \to \mathbb{A}$.

This also exists in $\mathbb{Rel}\mathsf{Set}$. More generally, we have a vertical involution in $\mathbb{Rel}(\mathsf{C})$, where C is a regular or p-exact category.

For a weak double category $\mathbb{A}$, we must relax the previous procedure: one generally encounters a *weak vertical involution*, defined as an involutive *pseudo* double functor $\mathbb{A}^v \to \mathbb{A}$.

For instance, let us consider the weak double category $\mathbb{Span}(\mathsf{C})$ of 7.2.6, where the vertical arrows are spans composed with a choice of pullbacks. We cannot assume that this choice is symmetric, as already remarked in 2.3.2(c): the procedure of reversing a span is only consistent with composition up to isomorphism of spans (contravariantly, of course).

Similarly, the weak double category $\mathbb{C}$at of categories, functors and profunctors has a weak vertical involution $X \mapsto X^{\mathrm{op}}$, produced by opposite categories

$$
\begin{array}{ccc}
X \xrightarrow{\ f\ } X' & & Y^{\mathrm{op}} \xrightarrow{\ g^{\mathrm{op}}\ } Y'^{\mathrm{op}} \\
u \downarrow \quad \underset{\rightarrow}{a} \quad \downarrow v & \mapsto & u^{\mathrm{op}} \downarrow \quad \underset{\rightarrow}{a^{\mathrm{op}}} \quad \downarrow v^{\mathrm{op}} \\
Y \xrightarrow[\ g\ ]{} Y' & & X^{\mathrm{op}} \xrightarrow[\ f^{\mathrm{op}}\ ]{} X'^{\mathrm{op}}
\end{array}
\qquad (7.49)
$$

7.3 Enriched, internal and ordered categories

A category with small hom-sets is enriched on Set (its hom-sets live there, together with the composition mappings) and a small category is internal in Set (its sets of objects and morphisms live there, with the whole structure). These facts can be extended.

The classical references for categories enriched on a monoidal basis are the article [EiK], by Eilenberg and Kelly, and Kelly's book [Kl2]. Categories internal to a category were introduced by C. Ehresmann [Eh3]; see also [St3].

We end this section by reviewing other forms of enrichment, including enrichment on an ordered category – the subject of Section 7.4.

7.3.1 Categories enriched over a monoidal category

In an additive category C, like R Mod and Ban, or more generally in a preadditive one, each hom-set $C(X, Y)$ is equipped with a structure of abelian group, so that composition is bilinear over $\mathbb{Z}$.

We show now that, if C has small hom-sets, this amounts to saying that C is 'enriched' over Ab, with respect to the symmetric monoidal structure of its tensor product.

In general, let $V = (V, \otimes, E)$ be a symmetric monoidal category, whose comparisons will be left understood. A V-*enriched category*, or V-*category*, C consists of:

(a) a set Ob C, of objects,

(b) for every pair X, Y in Ob C, a V-object of *morphisms* $C(X, Y)$,

(c) for every $X \in$ Ob C, a V-morphism of *identity* $\mathrm{id}_X \colon E \to C(X, X)$,

(d) for every triple X, Y, Z in Ob C, a V-morphism of *composition*

$$
k \colon C(X, Y) \otimes C(Y, Z) \to C(X, Z).
$$

These data are to satisfy three axioms of unitarity and associativity. Namely, the following diagrams must commute, for all X, Y, Z, U in Ob C

$$
\begin{array}{ccc}
E \otimes \mathsf{C}(X,Y) & \xrightarrow{\;\mathrm{id} \otimes 1\;} & \mathsf{C}(X,X) \otimes \mathsf{C}(X,Y) \\
& \searrow{\scriptstyle 1} & \Big\downarrow{\scriptstyle k} \\
& & \mathsf{C}(X,Y)
\end{array} \tag{7.50}
$$

$$
\begin{array}{ccc}
\mathsf{C}(X,Y) \otimes E & \xrightarrow{\;1 \otimes \mathrm{id}\;} & \mathsf{C}(X,Y) \otimes \mathsf{C}(Y,Y) \\
& \searrow{\scriptstyle 1} & \Big\downarrow{\scriptstyle k} \\
& & \mathsf{C}(X,Y)
\end{array} \tag{7.51}
$$

$$
\begin{array}{ccc}
\mathsf{C}(X,Y) \otimes \mathsf{C}(Y,Z) \otimes \mathsf{C}(Z,U) & \xrightarrow{\;1 \otimes k\;} & \mathsf{C}(X,Y) \otimes \mathsf{C}(Y,U) \\
{\scriptstyle k \otimes 1}\Big\downarrow & & \Big\downarrow{\scriptstyle k} \\
\mathsf{C}(X,Z) \otimes \mathsf{C}(Z,U) & \xrightarrow[\;\;k\;\;]{} & \mathsf{C}(X,U)
\end{array} \tag{7.52}
$$

An enriched category has an *underlying category* $|\mathsf{C}|$, with the same objects and hom-sets constructed with the canonical forgetful functor $U = \mathsf{V}(E, -)\colon \mathsf{V} \to \mathsf{Set}$ (not necessarily faithful, see (3.34))

$$
|\mathsf{C}|(X,Y) = U(\mathsf{C}(X,Y)) = \mathsf{V}(E, \mathsf{C}(X,Y)). \tag{7.53}
$$

The identity of X in $|\mathsf{C}|$ is precisely $\mathrm{id}_X\colon E \to \mathsf{C}(X,X)$. Given two consecutive morphisms $f \in |\mathsf{C}|(X,Y)$, $g \in |\mathsf{C}|(Y,Z)$ of the underlying category, their composite gf is

$$
gf = k(f \otimes g)\colon E \to C(X,Y) \otimes C(Y,Z) \to C(X,Z). \tag{7.54}
$$

7.3.2 Enriched functors and transformations

A V-*functor* $F\colon \mathsf{C} \to \mathsf{D}$ between V-categories consists of:

(a) a mapping $F_0\colon \mathrm{Ob}\,\mathsf{C} \to \mathrm{Ob}\,\mathsf{D}$ whose action is written as $X \mapsto F(X)$,

(b) for every pair X, Y in Ob C, a V-morphism

$$
F_{XY}\colon \mathsf{C}(X,Y) \to \mathsf{D}(FX, FY)
$$

(which will be written as $f \mapsto F(f)$ *when* V *is a category of structured sets*).

The following diagrams are assumed to commute, for all X, Y, Z in Ob C

$$
\begin{array}{ccc}
E & \xrightarrow{\;\mathrm{id}\;} & \mathsf{C}(X,X) \\
& \searrow{\scriptstyle \mathrm{id}} & \Big\downarrow{\scriptstyle F} \\
& & \mathsf{D}(FX, FX)
\end{array} \tag{7.55}
$$

$$C(X,Y) \otimes C(Y,Z) \xrightarrow{\quad k \quad} C(X,Z)$$

$$F \otimes F \downarrow \qquad\qquad\qquad\qquad \downarrow F \qquad\qquad (7.56)$$

$$D(FX,FY) \otimes D(FY,FZ) \xrightarrow[k]{} D(FX,FZ)$$

A V-*transformation* $\varphi \colon F \to G \colon C \to D$ between V-functors consists of a family of V-morphisms

$$\varphi X \colon E \to D(FX, GX) \qquad\qquad (X \in \mathrm{Ob}\, C),$$

making the following diagrams commutative

$$E \otimes C(X,Y) \xrightarrow{\varphi X \otimes G} D(FX,GX) \otimes D(GX,GY) \xrightarrow{k} D(FX,GY)$$

$$\| \qquad\qquad\qquad\qquad\qquad\qquad\qquad \| \qquad\qquad (7.57)$$

$$C(X,Y) \otimes E \xrightarrow[F \otimes \varphi Y]{} D(FX,FY) \otimes D(FY,GY) \xrightarrow[k]{} D(FX,GY)$$

Appropriate compositions give the 2-category V-Cat of V-categories (see [Kl2]).

Again, a V-functor has an underlying functor and a V-transformation has an underlying natural transformation, so that we have a 2-functor

$$| - | \colon \text{V-Cat} \to \text{Cat}.$$

Complements. Profunctors between categories (defined in 7.2.8) also have an enriched version: a V-profunctor $F \colon C \nrightarrow D$ between V-categories [Bo2].

Their composition requires cocompleteness conditions on V, corresponding in Set to the existence of coends, preserved by the cartesian product.*

7.3.3 Exercises and complements

The reader can examine and (easily) complete the following examples.

They are based on the cartesian structure of the categories Set, Cat, Ord, **2**, and the symmetric monoidal structure of the categories Ab and Abm. In each case the enriched functors and their transformations are the obvious ones.

(a) A Set-enriched category is the same as an ordinary category with small hom-sets.

(b) A Cat-enriched category is the same as a 2-category with small hom-categories.

(c) An Ord-enriched category is an ordered category, as defined in 1.8.2, with small hom-sets. (A different notion of 'ordered category' is recalled in 7.3.6.)

(d) The arrow category **2** is equivalent to the subcategory of Set consisting of all sets of cardinal $\leqslant 1$. It has all limits, and an obvious cartesian structure. A **2**-enriched category is a preordered set, possibly large.

(e) Let C be a *preadditive category* (see 6.4.1) with small hom-sets: as we have already recalled at the beginning of this section, this amounts to saying that C is enriched over Ab.

Completing some remarks in 1.2.1, every object X of C has a (possibly large, unital) ring $\text{End}(X)$. A ring is essentially the same as a preadditive category on a single object, while a preadditive category can be thought to be a 'many-object generalisation' of a ring.

(f) Similarly an $\mathbb{N}$-linear category (see 6.4.1) with small hom-sets is enriched over Abm, the symmetric monoidal category of abelian monoids, with tensor product over $\mathbb{N}$ defined 'as' in Ab. An $\mathbb{N}$-linear category is a 'many-object generalisation' of a semiring.

(g) A symmetric monoidal closed category C is canonically enriched on itself, by its internal hom $\text{Hom}(X, Y)$, with $\text{id}_X \colon E \to \text{Hom}(X, X)$ corresponding to the unitor $\lambda X \colon E \otimes X \to X$.

Examining under this aspect the symmetric monoidal closed structure of Ban_1 reviewed in 3.4.7(k), it is important to note that the enrichment of Ban_1 on itself does not mean to provide the set $\text{Ban}_1(X, Y)$ with a structure of Banach space (!), but *to assign a Banach space* $\text{Hom}(X, Y)$ *whose unit ball is that set.*

The basic view of enriching an ordinary category by providing *its homsets* with a V-structure is only adequate when the canonical forgetful functor of V does give (what we view as) the underlying set of any V-object. The next example is even farer from this basic view.

7.3.4 Lawvere metric spaces

The strict monoidal category $\underline{\text{R}} = [0, \infty]^{\text{op}}$ was introduced by Lawvere [Lw1] to formalise 'generalised metric spaces' as enriched categories – a far-reaching idea, also at the basis of our approach to manifolds in the next section.

An object of $\underline{\text{R}}$ is an 'extended' positive real number $\lambda \in [0, \infty]$, an arrow $\lambda \to \mu$ is given by the order relation $\lambda \geqslant \mu$; the tensor product is given by the ordinary sum $\lambda + \lambda'$.

A category enriched over $\underline{\text{R}}$ is a set X with *hom-values*

$$\delta(x, y) = X(x, y),$$

satisfying the axioms:

$$\delta(x, y) \in [0, \infty],$$
$$\delta(x, x) = 0, \qquad \delta(x, y) + \delta(y, z) \geqslant \delta(x, z). \tag{7.58}$$

Note that 'composition' becomes the triangular inequality, and symmetry is not required (while it was for a pseudometric, in 5.1.7).

We call this structure X a *directed metric space*, or a *δ-metric space*. An enriched functor $f \colon X \to Y$ is a weak contraction, satisfying

$$\delta(x, x') \geqslant \delta(f(x), f(x')),$$

for all $x, x' \in X$. This forms the category $\delta\mathsf{Mtr} = \underline{\mathrm{R}}\mathsf{Cat}$.

All this leads to the important notion of Cauchy-completeness of enriched categories, see [Lw1].

Exercises and complements. (a) Compute the category underlying a δ-metric space X.

(b) We already mentioned in 5.7.6 that, in the context of Directed Algebraic Topology, the category $\delta\mathsf{Mtr}$ is studied as a category of directed spaces, where paths have a cost – which is infinite for the illegitimate ones ([G8], Chapter 6).

The reversor $R \colon \delta\mathsf{Mtr} \to \delta\mathsf{Mtr}$ obviously sends the space X to the *opposite δ-metric space*, with opposite δ-metric:

$$R(X) = X^{\mathrm{op}}, \qquad \delta^{\mathrm{op}}(x, y) = \delta(y, x).$$

7.3.5 Internal categories

(a) Let us begin by remarking that a small category C can be described as a diagram in Set having the form of a 3-truncated simplicial set (see Section 5.3)

$$C_0 \; \underset{e_0}{\overset{\partial_i}{\rightleftarrows}} \; C_1 \; \underset{e_i}{\overset{\partial_i}{\rightleftarrows}} \; C_2 \; \overset{\partial_i}{\underset{e_i}{\dashrightarrow}} \; C_3 \tag{7.59}$$

where

- $C_0 = \mathrm{Ob}\,\mathsf{C}$ is the (small) set of objects of C,
- $C_1 = \mathrm{Mor}\,\mathsf{C}$ is the set of morphisms,
- C_2 is the set of consecutive pairs of morphisms,
- C_3 is the set of consecutive triples of morphisms.

The faces ∂_i and the degeneracies e_i are defined as follows (on an object $X \in C_0$, an arrow $f \in C_1$, a pair $(f,g) \in C_2$ and a triple $(f,g,h) \in C_3$):

$$e_0(X) = \mathrm{id}_X, \qquad \partial_0(f) = \mathrm{Dom}\,(f), \qquad \partial_1(f) = \mathrm{Cod}\,(f),$$
$$e_0(f) = (f,1), \qquad e_1(f) = (1,f),$$
$$\partial_0(f,g) = f, \qquad \partial_1(f,g) = gf, \qquad \partial_2(f,g) = g,$$
$$e_0(f,g) = (f,g,1), \qquad e_1(f,g) = (f,1,g), \qquad e_2(f,g) = (1,f,g),$$
$$\partial_0(f,g,h) = (f,g), \qquad \partial_1(f,g,h) = (f,hg), \qquad \partial_2(f,g,h) = (gf,h),$$
$$\partial_3(f,g,h) = (g,h).$$

The simplicial identities hold, so that (7.59) is indeed a 3-truncated simplicial set:

$$\partial_i\partial_j = \partial_{j-1}\partial_i \;\text{ for } i < j, \qquad e_j e_i = e_i e_{j-1} \;\text{ for } i < j,$$
$$\partial_i e_j = e_{j-1}\partial_i \;\text{ for } i < j, \qquad \partial_i e_j = e_j \partial_{i-1} \;\text{ for } i > j+1, \qquad (7.60)$$
$$\partial_i e_j = \mathrm{id} \quad \text{for } i = j, j+1.$$

Finally, the following two squares are pullbacks (of sets)

$$
\begin{array}{ccc}
C_2 \xrightarrow{\ \partial_0\ } C_1 & \qquad & C_3 \xrightarrow{\ \partial_0\ } C_2 \\
{\scriptstyle\partial_2}\downarrow \quad\ \downarrow{\scriptstyle\partial_1} & & {\scriptstyle\partial_3}\downarrow \quad\ \downarrow{\scriptstyle\partial_2} \\
C_1 \xrightarrow[\ \partial_0\ ]{} C_0 & & C_2 \xrightarrow[\ \partial_0\ ]{} C_1
\end{array}
\qquad (7.61)
$$

The morphism $m = \partial_1 \colon C_2 \to C_1$ is called the *composition morphism* of C, or *partial multiplication*.

(b) Replacing Set with an arbitrary category X, an *internal category* C in X, or a *category object* in C, consists of a diagram (7.59) in X where $C_0 = \mathrm{Ob}\,C$ is called the *object of objects* of C, $C_1 = \mathrm{Mor}\,C$ is called the *object of morphisms* and C_2 is called the *object of consecutive pairs of morphisms*.

The following (redundant) axioms must be satisfied:

(i) the simplicial identities (7.60) hold, so that (7.59) is a 3-truncated simplicial object in X,

(ii) the square diagrams of (7.61) are pullbacks in X.

(Note that the category X is not required to have all pullbacks: this would exclude cases of interest.)

In a category with finite limits, an internal monoid with respect to the cartesian product, as defined in 3.4.8, is the same as an internal category where $\mathrm{Ob}\,C$ is the terminal object.

(c) An *internal functor* $F: \mathsf{C} \to \mathsf{D}$ between internal categories in X is defined as a morphism of 3-simplicial objects. In other words we have four arrows in X

$$F_i: C_i \to D_i \qquad\qquad (i = 0, ..., 3), \qquad\qquad (7.62)$$

that commute with faces and degeneracies. The components F_2, F_3 are determined by the pullback condition, so that for $\mathsf{X} = \mathsf{Set}$ we simply have

$$F_2(f, g) = (F_1 f, F_1 g), \qquad F_3(f, g, h) = (F_1 f, F_1 g, F_1 h).$$

The composition of internal functors is obvious.

(d) An *internal transformation* $\varphi: F \to G: \mathsf{C} \to \mathsf{D}$ between internal functors in X is given (or represented) by an X-morphism $\hat{\varphi}$ satisfying the following conditions

$$\hat{\varphi}: C_0 \to D_1, \qquad \partial_0.\hat{\varphi} = F_0, \qquad \partial_1.\hat{\varphi} = G_0,$$
$$m.\langle F_1, \hat{\varphi}\partial_1 \rangle = m.\langle \hat{\varphi}\partial_0, G_1 \rangle. \qquad (7.63)$$

In the last condition (of *naturality*):

- the morphism $m = \partial_1: D_2 \to D_1$ is the partial multiplication of D,

- the morphism $\langle F_1, \hat{\varphi}\partial_1 \rangle: C_1 \to D_2$ comes from the commutative square $\partial_1 F_1 = F_0 \partial_1 = \partial_0 \, \hat{\varphi} \, \partial_1$, and the pullback-property of D_2,

- the morphism $\langle \hat{\varphi}\partial_0, G_1 \rangle: C_1 \to D_2$ comes from the commutative square $\partial_1 \, \hat{\varphi} \, \partial_0 = G_0 \partial_0 = \partial_0 G_1$.

The interested reader can verify that there is a 2-category $\mathsf{Cat}(\mathsf{X})$ of internal categories, internal functors and internal transformations in X. The vertical composition of internal transformations and their whisker composition with internal functors are defined as follows

$$(\psi.\varphi)\hat{} = m.\langle \hat{\varphi}, \hat{\psi} \rangle: C_0 \to D_1, \qquad\qquad \text{for } \psi: G \to H: \mathsf{C} \to \mathsf{D},$$
$$(M\varphi L)\hat{} = M_1.\hat{\varphi}.L_0: C_0' \to D_1', \qquad \text{for } L: \mathsf{C}' \to \mathsf{C}, \ M: \mathsf{D} \to \mathsf{D}. \qquad (7.64)$$

7.3.6 Exercises and complements

(a) We have already seen that an internal category in Set is a small category.

(b) An internal category in Cat is a small double category, as defined in 7.2.1. Internal functors are the double functors; internal transformations are the horizontal ones (or the vertical ones, according to the way we are 'presenting' the double category).

This approach to double categories masks their transpose symmetry, that interchanges horizontal and vertical arrows. But it is adequate to weak double categories, which are 'pseudo-category objects' in Cat and do not have such a symmetry.

*(c) We say that a category C is *internally ordered* when the sets Ob C and Mor C are equipped with consistent order relations (respected by domain, codomain, identity and composition). If C is small, this the same as an internal category in Ord.

> An internally ordered category with a discrete order on the set of objects is the same as an ordered category in the present enriched sense: see 1.8.2.

This notion was extensively used by C. Ehresmann for his theory of 'pseudogroups' and 'inductive categories' in Differential Geometry. One should recall that an 'ordered category' in the sense of Ehresmann is also assumed to satisfy the following condition: the order induced on each hom-set $C(X, Y)$ is discrete (see [Eh5], p. 711).

The main example of such a structure is the category Set ordered by (weak) inclusion, of objects $(X \subset X')$ and morphisms: the mapping $f \colon X \to Y$ being *included* in $f' \colon X' \to Y'$ if

$$X \subset X', \quad Y \subset Y', \qquad f(x) = f'(x) \quad \text{for } x \in X.$$

(Indeed, if this holds with $X = X'$ and $Y = Y'$, then $f = f'$.)

This example, as many others used by Ehresmann, rests heavily on set-theory. One can replace this internally ordered category by the category $\mathcal{S}$ of sets and partial mappings (see 1.8.6), ordered in the present sense, where inclusion is only used between subsets of given sets.

Similarly, the ordered category $\mathcal{C}$ of partial continuous mappings defined on open subspaces (introduced in 1.8.8) can be used in the theory of topological manifolds, while its differential analogues can be used for differentiable manifolds. This was the subject of [G2], briefly reviewed in the next section.

7.3.7 Kinds of enrichment

Actually, 'categorical enrichment' is a wide subject: a category can be enriched on a *basis* of the following types (the arrows denote inclusion):

$$
\begin{array}{ccc}
\text{monoidal category} & \longrightarrow & \text{bicategory} \\
\uparrow & & \uparrow \\
\text{ordered monoid} & \longrightarrow & \text{ordered category}
\end{array}
\qquad (7.65)
$$

(a) *Enrichment on a monoidal category*, reviewed above, is the first case considered in category theory, and the more readily understood. The classical references are [EiK, Kl2].

(b) *Enrichment on an ordered category* is sketched below and used in the next section.

(c) *Enrichment on an ordered monoid* is a basic, particular case of the previous ones, examined in 7.3.8(c).

(d) *Enrichment on a bicategory* unifies all these cases, and is not dealt with here. It was introduced in [Bet, Wa], and also studied in many articles. In fact:

- a monoidal category is the same as a bicategory on a single object, as already remarked in 7.1.3,
- a preordered category is the same as a 2-category with at most one cell $f \to g$ between two given arrows, as remarked in 7.1.2.

In each of these cases one has enriched functors and enriched profunctors, correlated by a notion of Cauchy completeness: it was introduced by Lawvere [Lw1], for enrichment on monoidal categories.

7.3.8 Categories enriched on an ordered category

We present this topic in a form adapted to study formal manifolds, in the next section. The basis C is an ordered category.

A C-*enriched category* U, or C-*category*, consists of:

(a) a set $I = \mathrm{Ob}\, U$ of objects (the *indices*),

(b) for every $i \in I$, a C-object U_i (a *chart*),

(c) for all $i, j \in I$, a C-morphism $u_j^i : U_i \to U_j$ (a *transition morphism*),

that satisfy the following axioms

(i) $u_i^i \geqslant 1_{U_i}$ (*identity law*),

(ii) $u_k^j u_j^i \leqslant u_k^i$ (*composition law, or triangle inequality*).

With respect to enrichment on a monoidal category, the transition morphism u_j^i replaces the hom-object $U(i,j)$; axioms (i) and (ii) replace, respectively, the identity map $E \to U(i,i)$ and the composition map $U(i,j) \times U(j,k) \to U(i,k)$. The coherence axioms of 7.3.1 are here automatically satisfied.

This structure will also be written as $(U_i, u_j^i)_I$, or U_I.

A C-*functor* $f : U_I \to V_H$, where $V = ((V_h), (v_k^h))_H$ is an enriched category indexed by the set H, is a mapping $f : I \to H$ between the set of indices of U and V, such that we have, in C

$$ U_i = V_{fi}, \qquad u_j^i \leqslant v_{fj}^{fi} \qquad (i, j \in I). \tag{7.66} $$

Their composition is obvious.

Exercises and complements. (a) We have seen in 7.3.4 that the Lawvere (strict) monoidal category $\underline{R} = [0, \infty]^{op}$ produces the category $\delta\mathsf{Mtr} = \underline{R}\mathsf{Cat}$ of generalised metric spaces, as (small) categories enriched on R.

This can also be obtained in the present context, viewing $\underline{R}$ as an ordered category on one formal object $*$. The extended real numbers $\lambda \in [0, \infty]$ become the endomorphisms $\lambda\colon * \to *$, their composition is the ordinary sum $\lambda + \mu$, and the order is given by the relation $\lambda \geqslant \mu$.

It is now easy to see that a category enriched on $\underline{R}$, as a monoidal category or as an ordered category, is the same thing. Enriched functors also have the same meaning.

(b) Similarly, we have seen in 7.3.3(d) that a preordered set is an enriched category on the category **2**, with respect to its (strict) cartesian structure.

We can view **2** as an ordered category on one formal object $*$. There are two endomorphisms $0, 1\colon * \to *$, their composition is the ordinary multiplication, the order is the natural one: $0 < 1$. Again, enriched categories and enriched functors on **2** have the same meaning, in both approaches.

(c) In general, an ordered monoid M, with elements $\lambda, \mu...$ can be viewed as:

- a strict monoidal category with objects $\lambda, \mu...$, at most one morphism $\lambda \leqslant \mu$ between two objects (given by the ordering) and tensor product $\lambda\mu$ (given by the product of M),

- an ordered category on one formal object $*$, with morphisms $\lambda, \mu...\colon * \to *$, composition $\lambda\mu$ and order $\lambda \leqslant \mu$.

Enrichment on M has the same meaning, in both approaches.

7.3.9 Enriched profunctors

Let us come back to an ordered category C. With the same notation as in 7.3.8, an *enriched profunctor* $a = (a_h^i)\colon U \nrightarrow V$ between C-categories, or C-*profunctor*, is a family of morphisms $a_h^i\colon U_i \to V_h$ of C (for $i \in I$, $h \in H$) such that, for all $i, j \in I$ and $h, k \in H$:

(i) $a_h^j.u_j^i \leqslant a_h^i, \quad v_k^h.a_h^i \leqslant a_k^i$ (*profunctor laws*).

The composite of $a\colon U \nrightarrow V$ with a consecutive profunctor $b = (b_r^h)\colon V \nrightarrow W$ with values in $W = ((W_r), (w_s^r))_R$ 'should' be defined by joins in the ordered set $\mathsf{C}(U_i, W_r)$

$$(ba)_r^i = \vee_{h \in H}\, b_r^h\, a_h^i\colon U_i \to W_r, \qquad (7.67)$$

which do exist if each ordered hom-set of C is a complete lattice.

Everything works in this general form if the base category C *is a quantaloid* [Ro], that is an ordered category where all hom-sets $\mathsf{C}(U, V)$ are complete lattices, and composition distributes over arbitrary joins.

Then the composition of profunctors is well defined and associative, with identities $\operatorname{id} U$ defined as follows

$$(\operatorname{id} U)^i_j = u^i_j \colon U_i \to U_j. \tag{7.68}$$

We have thus the ordered category $\mathrm{Prf}(\mathsf{C})$ of C-enriched categories and profunctors. For two profunctors $a, b \colon U \nrightarrow V$, the order relation $a \leqslant b$ is defined by the ordering of C, as

$$a^i_h \leqslant b^i_h, \quad \text{for all indices } i, h. \tag{7.69}$$

The ordered categories $\underline{\mathrm{R}}$ and $\mathbf{2}$ considered in 7.3.8(a), (b) are indeed quantaloids.

On the other hand, the bases $\mathcal{S}$, $\mathcal{C}$, $\mathcal{C}^r$, $\mathcal{V}$ used in the next section (and many others used in [G2]) only have joins of 'compatible' parallel morphisms, which will allow us to compose 'compatible profunctors', not the general ones.

Exercises and complements. (a) Prove the associativity and unitarity of the composition of profunctors. *Hints:* note that condition (i) implies: $a^i_h.u^i_i = a^i_h = v^h_h.a^i_h$.

(b) A profunctor $a \colon U_I \nrightarrow V_H$ amounts to extending the enriched categories U and V to the disjoint union $I + H$, adding new transition morphisms a^i_h, and trivial transition morphisms from H-indices to I-indices

$$a^i_h \in \mathsf{C}(U_i, V_h), \qquad a^h_i = \min(\mathsf{C}(V_h, U_i)) \qquad (i \in I, h \in H).$$

(The sets I and H are assumed to be disjoint.) The profunctor laws (i) precisely say that the new transitions a^i_h are consistent with the old ones, while the trivial transitions a^h_i automatically are.

7.4 *Manifolds as enriched categories

This section is an addition to the first edition, and is written in an informal way, mostly without proofs. An interested reader is referred to the original article [G2], and a new book [G12].

We present manifolds (and other 'local structures') as categories enriched on ordered categories, as reviewed in 7.3.8. The basis of enrichment is not a quantaloid, but an 'e-category', like the categories of partial mappings $\mathcal{S}$ and $\mathcal{C}$ already considered in Section 1.8; because of this, the simple theory of enrichment reviewed at the end of the last section has to be modified.

The structure of e-category, introduced in [G2], has been later studied as a 'restriction category', under equivalent axioms, in [CoL] and subsequent articles.

Partial mappings are still denoted by dot-marked arrows.

7.4.1 The general idea

Local structures, for instance manifolds, fibre bundles, vector bundles and foliations, can be obtained by patching together a family (U_i) of suitable 'elementary spaces' by means of partial homeomorphisms $u^i_j \colon U_i \rightarrowtail U_j$ expressing the gluing conditions and forming a sort of 'intrinsic atlas', instead of the more usual system of charts, living in an external framework.

We can thus define an 'intrinsic manifold' as such an atlas, in a suitable category of elementary spaces: open euclidean spaces, or trivial bundles, or trivial vector bundles, and so on.

This uniform approach allows us to move from one basis to another: for instance, the elementary tangent bundle of an open euclidean space can be automatically extended to the tangent bundle of any differentiable manifold. The same holds for tensor calculus.

Technically, we handle these structures as 'symmetric enriched categories' over a suitable basis, generally an ordered category of partial mappings. The morphisms of these 'generalised manifolds' are obtained as 'compatible profunctors' between enriched categories, which can be composed because of the existence of 'compatible joins' in the basis.

This approach to 'local structures' is related to Ehresmann's one, based on pseudogroups of transformations [Eh5]. On the other hand, our setting derives from Lawvere's unconventional presentation of generalised metric spaces as enriched categories (see 7.3.4) and is another realisation of his claim, that many interesting mathematical structures not only can be organised in categories, but *are* themselves categories, enriched over a suitable basis – a source of research since the 1970's.

7.4.2 Classical manifolds

Smooth manifolds are based on mappings of class C^∞ defined on some open euclidean space, i.e. an open subspace of some space $\mathbb{R}^n$ with euclidean topology.

A C^∞-manifold is usually defined as a topological space X equipped with a C^∞-*atlas of charts*, indexed by a set I

$$u^i \colon U_i \to X_i. \tag{7.70}$$

A *chart* is a homeomorphism between an open euclidean space U_i (in the sense specified above) and an open subspace X_i of X. We assume that these open subsets cover X and that every *transition map* (between open euclidean spaces)

$$u^i_j = u_j u^i \colon U_i \dashrightarrow U_j \qquad (i, j \in I), \tag{7.71}$$

is of class C^∞. Here $u_j \colon X_j \to U_j$ is the inverse of u^j, and the 'composite' $u_j u^i \colon U_i \dashrightarrow U_j$ (an abuse of notation) is *partially defined* on the open subset $u_i(X_i \cap X_j)$, possibly empty.

The space X is thus locally euclidean, with a locally constant dimension.

(One generally requires the space X to be Hausdorff paracompact, but these conditions can be added when we want them.)

If Y is also a C^∞-manifold, with charts

$$v^h \colon V_h \to Y_h \qquad (h \in H),$$

a C^∞-*mapping* $f \colon X \to Y$ is a continuous mapping such that all the *partial* continuous mappings

$$f^i_h = v_h f u^i \colon U_i \dashrightarrow V_h \qquad (i \in I, \, h \in H), \tag{7.72}$$

are of class C^∞. Again, as in (7.71), there is an abuse of notation: we are 'composing' three arrows

$$u^i \colon U_i \to X_i, \qquad f \colon X \to Y, \qquad v_h \colon Y_h \to V_h,$$

which are not consecutive. This could be fixed with Ehresmann's *pseudo-product*, but we prefer to work with ordinary composition in categories of partial mappings.

(a) We write as $\mathcal{C}$ the category of *topological spaces and partial continuous mappings defined on open subspaces* (already mentioned in 1.8.8).

An object is a topological space, and a morphism $f \colon X \dashrightarrow Y$ is a partial continuous mapping defined on an open subspace Def $f \subset X$; the composite $gf \colon X \dashrightarrow Z$ (with a consecutive morphism $g \colon Y \dashrightarrow Z$) is defined letting $(gf)(x) = g(f(x))$, on the open subset $f^{-1}(\mathrm{Def}\, g)$.

(b) We write as $\mathcal{C}^\infty$ the subcategory of $\mathcal{C}$ of *open euclidean spaces and partial C^∞-mappings defined on open subspaces*.

Now, we replace the homeomorphism $u^i \colon U_i \to X_i$ with the topological embedding $u^i \colon U_i \to X$; the latter has a (partially defined) *partial inverse* $u_i \colon X \dashrightarrow U_i$ in $\mathcal{C}$, characterised by the conditions

$$u^i = u^i u_i u^i, \qquad u_i = u_i u^i u_i, \tag{7.73}$$

as in the definition of a partial inverse in semigroup theory.

We can now replace the 'illegitimate compositions' of (7.71) and (7.72) with legitimate ones, in $\mathcal{C}$

$$u_j u^i \colon U_i \twoheadrightarrow X \twoheadrightarrow U_j, \qquad v_h f u^i \colon U_i \twoheadrightarrow X \twoheadrightarrow Y \twoheadrightarrow V_h, \qquad (7.74)$$

and we can require that these composites (whose domain and codomain are open euclidean spaces) belong to the subcategory $\mathcal{C}^\infty$.

(c) More generally, for each $r \in \mathbb{N} \cup \{\infty, \omega\}$, we write as $\mathcal{C}^r$ the subcategory of $\mathcal{C}$ of *open euclidean spaces and partial C^r-mappings defined on open subspaces.* (C^0 means continuous and C^ω means analytic.)

C^r-manifolds are dealt with as above. Topological manifolds correspond to the case $r = 0$; note that, in this case, the transition maps (7.71) automatically belong to $\mathcal{C}^0$.

7.4.3 Intrinsic manifolds on e-categories

We want to define a C^r-manifold in an intrinsic way, *inside the category* $\mathcal{C}^r$, as a collection (U_i) of objects, equipped with a family $(u_j^i \colon U_i \twoheadrightarrow U_j)_{ij}$ of *transition morphisms* – a system of instructions specifying how the different charts U_i should be glued together. The gluing will be realised in an external category, namely in $\mathcal{C}$.

The morphisms $u_j^i \colon U_i \twoheadrightarrow U_j$ should satisfy some axioms, which will be expressed exploiting the fact that the category $\mathcal{C}^r$ has a canonical order: for two partial mappings $f, g \colon X \twoheadrightarrow Y$ the relation $f \leqslant g$ means that f is a restriction of g (with $\mathrm{Def}\, f \subset \mathrm{Def}\, g$). The order is consistent with composition: $\mathcal{C}^r$ is an ordered category. The same holds for $\mathcal{C}$.

Now, we define an (intrinsic) *manifold* on the ordered category $\mathcal{C}^r$ as a diagram $U = ((U_i), (u_j^i))_I$ of $\mathcal{C}^r$, consisting of

(a) a set $I = \mathrm{Ob}\, U$ of objects (the *indices*),

(b) for every $i \in I$, a C-object U_i (a *chart*),

(c) for all $i, j \in I$, a C-morphism $u_j^i \colon U_i \to U_j$ (a *transition morphism*), satisfying three axioms (for $i, j, k \in I$):

(i) $u_i^i = 1_{U_i}$ (*identity law*),

(ii) $u_k^j u_j^i \leqslant u_k^i$ (*composition law, or triangle inequality*),

(iii) $u_j^i = u_j^i u_i^j u_j^i$ (*symmetry law*).

U is thus a small category enriched on the ordered category $\mathcal{C}^r$ (see 7.3.8), with an additional symmetry condition that forces all transition maps to be partial homeomorphisms. (Note that (i) is equivalent to $u_i^i \geqslant 1_{U_i}$, in all the bases we are considering here.)

Plainly, if we start from the usual charts $u^i \colon U_i \to X$ and define their transition maps u^i_j as above (in (7.74)), these axioms are satisfied. Conversely, if we start from a family (u^i_j) satisfying the conditions above, one can reconstruct the space $X = \mathrm{gl}\, U$ as the *gluing* of the diagram U, a quotient of the disjoint union of all U_i, modulo the equivalence relation produced by all transition maps.

More precisely, the pair $(X, (u^i \colon U_i \to X))$ is the *lax colimit* of the diagram (u^i_j), in the ordered category $\mathcal{C}$ (see [G12], Section 3.6).

The diagram U will also be written as $((U_i), (u^i_j))_I$, or as U_I. The family (u^i_j) is called an *intrinsic atlas*, or a *gluing atlas*, of the manifold.

7.4.4 Projectors and supports

In the ordered categories $\mathcal{C}$ and $\mathcal{C}^r$ (and in all the others used in this analysis of local structures), a prominent role is played by the endomorphisms $e \colon X \twoheadrightarrow X$ which are restriction of identities, called *projectors* of X. They are idempotent endomorphisms and commute, forming a meet semilattice (i.e. an ordered set with finite meets)

$$\mathrm{Prj}(X) = \{e \colon X \twoheadrightarrow X \mid e \leqslant \mathrm{id}X\},$$

$$e \wedge e' = ee' = e'e. \tag{7.75}$$

In fact, these projectors determine the order: for parallel morphisms $f, g \colon X \twoheadrightarrow Y$, the relation $f \leqslant g$ is equivalent to the existence of $e \in \mathrm{Prj}(X)$ such that $f = ge$. We can always take as e the *support* $\underline{e}(f)$ of f, namely the partial identity on $\mathrm{Def}\, f$, or equivalently the least $e \in \mathrm{Prj}(X)$ such that $fe = f$.

The projectors satisfy suitable axioms, and supply the categories $\mathcal{C}$ and $\mathcal{C}^r$ with the structure of an *e-cohesive category*, or *e-category*, one of the main ingredients of our analysis.

More precisely, $\mathcal{C}$ and $\mathcal{C}^r$ are *totally cohesive e-categories*, which means that every family of 'linked' morphisms $f_i \colon X \twoheadrightarrow Y$ has a join $\vee f_i \colon X \twoheadrightarrow Y$, and composition distributes over these joins.

Being linked can be simply read as 'upper bounded', but the important fact is that this property is characterised by supports. Namely, for $f, g \colon X \twoheadrightarrow Y$, we say that f and g are *linked*, or *compatible* (written as $f \,!\, g$), if

$$f\underline{e}(g) = g\underline{e}(f), \tag{7.76}$$

which means that they coincide wherever they are both defined.

If the morphisms $(f_i)_{i \in I}$ are pairwise linked, the join $f = \vee f_i$ is the obvious mapping defined on $\cup \operatorname{Def} f_i$ (whose graph is the union of the graphs of all f_i).

(A more general notion of 'cohesive structure' $(\leqslant, !)$ on a category can be found in [G2, G12], not necessarily determined by supports.)

7.4.5 Morphisms of manifolds

We can now define the category $\operatorname{Mf} \mathcal{C}^r$ of C^r-manifolds, extending the formula (7.72), where a morphism $f \colon U \to V$ is determined by the family of its components

$$f_h^i = v_h f u^i \colon U_i \rightarrowtail V_h,$$

on the charts of domain and codomain. A morphism in $\operatorname{Mf} \mathcal{C}^r$ will be a 'linked profunctor', that is an enriched profunctor between enriched categories, satisfying a compatibility condition.

More precisely, an (enriched) *profunctor*

$$a = (a_h^i)_{I,H} \colon (U_i, u_j^i)_I \to (V_h, v_k^h)_H, \tag{7.77}$$

is a family of morphisms $a_h^i \colon U_i \rightarrowtail V_h$ in $\mathcal{C}^r$ such that, for all $i, j \in I$ and $h, k \in H$

(i) $a_h^j u_j^i \leqslant a_h^i, \qquad v_k^h a_h^i \leqslant a_k^i$ (*profunctor laws*).

It will be said to be *linked*, or *compatible*, if it has a *resolution* $e_{ih} \in \operatorname{Prj}(U_i)$ $(i \in I, h \in H)$, defined by the property:

(ii) $a_k^i e_{ih} = v_k^h a_h^i$ (*right linking law*),

which is meant to ensure:

- that linked profunctors can be composed (although the basis $\mathcal{C}^r$ is not a quantaloid),

- that the gluing of a linked profunctor gives a *single-valued* partial mapping $a \colon \operatorname{gl} U \rightarrowtail \operatorname{gl} V$.

The resolution can be expressed by supports, taking $e_{ih} = \underline{e}(a_h^i)$.

The usual matrix composition of profunctors works because $\mathcal{C}^r$ is totally cohesive: composing $a \colon U \to V$ with a consecutive linked profunctor $b \colon (V_h, v_k^h)_H \to (W_r, w_s^r)_R$, the component of $c = ba \colon U \to W$ is computed as the linked join of each composite $b_r^h a_h^i \colon U_i \rightarrowtail W_r$ through V_h

$$c_r^i \colon U_i \rightarrowtail W_r, \qquad c_r^i = \vee_{h \in H} b_r^h a_h^i. \tag{7.78}$$

For a reader acquainted with the theory of enriched categories, let us note that the property of Cauchy completion of enriched categories, which is crucial in other contexts, plays a limited role here: replacing an intrinsic atlas by a complete one simply produces an isomorphic object (within profunctors!).

Furthermore, this can only be done when the basis of enrichment is a small category: it is the case for differentiable manifolds, but not for fibre bundles.

7.4.6 The interest of an intrinsic approach

The formalisation sketched above allows one to move between different contexts.

For instance, the tangent bundle of an open n-dimensional euclidean space U is the trivial vector bundle $TU = U \times \mathbb{R}^n$. The present machinery automatically extends this obvious setting to the tangent functor of differentiable manifolds

$$T \colon \mathrm{Mf}\, \mathcal{C}^r \to \mathrm{Mf}\, \mathcal{V} \qquad (r > 0), \tag{7.79}$$

with values in the category of vector bundles, presented as manifolds on an ordered category $\mathcal{V}$ of trivial vector bundles and suitable fibrewise linear partial mappings. The same procedure works for tensor calculus.

Moreover, the embedding of $\mathcal{C}^r$ into the category $\mathcal{C}$ of topological spaces and partial continuous mappings defined on open subspaces gives the topological realisation of C^r-manifolds

$$\mathrm{Mf}\, \mathcal{C}^r \to \mathcal{C}, \tag{7.80}$$

taking into account that the second category is 'gluing complete' (each manifold has a gluing space), and therefore $\mathcal{C}$ is equivalent to $\mathrm{Mf}\, \mathcal{C}$.

7.5 *Double categories of lattices in Homological Algebra

Continuing Section 6.2 we show, *in an informal way*, how double categories of lattices can be of help in Homological Algebra.

For the sake of simplicity, we still work in categories of modules, but everything can be extended to Puppe-exact categories (which include the abelian ones).

7.5.1 A double category of modular lattices

The transfer functor for subobjects of $R\,\mathrm{Mod}$

$$\mathrm{Sub}_R \colon R\,\mathrm{Mod} \to \mathrm{Mlc}, \tag{7.81}$$

has been introduced in 6.2.1, and shown to be exact in 6.2.6.

Now let $F \colon R\,\mathrm{Mod} \to S\,\mathrm{Mod}$ be an exact functor. We have seen (in 6.3.8) that it preserves subobjects, with their meets and joins.

For every R-module A we have thus a *homomorphism of lattices*, which belongs to the category Mlh of 1.2.6 (*not* to Mlc)

$$(\mathrm{Sub}_F)_A \colon \mathrm{Sub}_R(A) \to \mathrm{Sub}_S(FA), \qquad X \mapsto F(X). \qquad (7.82)$$

Furthermore F preserves direct and inverse images along morphisms: for a homomorphism $f \colon A \to B$ in $R\,\mathsf{Mod}$ we have two commutative squares (of monotone mappings)

$$ (7.83) $$

$$F(f_*X) = (Ff)_*(FX), \qquad F(f^*Y) = (Ff)^*(FY),$$

because F preserves epi-mono factorisations and pullbacks. This diagram is a cell in the double category $\mathbb{A}\mathrm{dj}_0\mathbb{O}\mathrm{rd}$ of ordered sets, monotone mappings (as horizontal arrows), Galois connections (as vertical arrows) and bicommutative cells, defined in 7.2.7.

To express this interaction between lattice-homomorphisms and modular connections we shall use – more precisely – a double subcategory $\mathbb{M}\mathrm{lhc} \subset \mathbb{A}\mathrm{dj}_0\mathbb{O}\mathrm{rd}$ introduced below (in 7.5.2) and consisting of: modular lattices, homomorphisms, modular connections and bicommutative cells.

The transfer functor (7.81) can then be seen as a *vertical functor*

$$\mathrm{Sub}_R \colon R\,\mathsf{Mod} \to \mathbb{M}\mathrm{lhc}, \qquad\qquad (7.84)$$

i.e. a double functor defined on the obvious 'vertical' double category $\mathbb{V}(R\,\mathsf{Mod})$ of modules, identities and homomorphisms.

For every exact functor $F \colon R\,\mathsf{Mod} \to S\,\mathsf{Mod}$ we have a *horizontal transformation of vertical functors*

$$\mathrm{Sub}_F \colon \mathrm{Sub}_R \to \mathrm{Sub}_S.F \colon R\,\mathsf{Mod} \to \mathbb{M}\mathrm{lhc}, \qquad (7.85)$$

where the vertical functor $\mathrm{Sub}_S.F$ is – more precisely – the composite

$$\mathrm{Sub}_S.\mathbb{V}F \colon \mathbb{V}(R\,\mathsf{Mod}) \to \mathbb{V}(S\,\mathsf{Mod}) \to \mathbb{M}\mathrm{lhc}.$$

7.5.2 Double categories of lattices

The double categories $\mathbb{A}\mathrm{dj}\mathbb{O}\mathrm{rd}$ and $\mathbb{A}\mathrm{dj}_0\mathbb{O}\mathrm{rd}$ introduced in 7.2.7 have double subcategories that are of interest here.

(a) First we have two *cellwise-full* double subcategories of $\mathbb{A}\mathrm{dj}\mathbb{O}\mathrm{rd}$:

- AdjLt, the double category of *lattices, homomorphisms and adjunctions*, which amalgamates the category Lth of lattices and homomorphisms with the category Ltc of lattices and Galois connections (see 6.4.5(c)),

- AdjMl ⊂ AdjLt, the double category of *modular lattices, homomorphisms and modular connections*, which amalgamates the category Mlh of modular lattices and homomorphisms with the category Mlc of modular lattices and modular connections.

Both have double cells as in (7.35)

$$
\begin{array}{ccc}
X & \xrightarrow{\ f\ } & X' \\
u \uparrow\bullet & \downarrow & \bullet\downarrow v \\
Y & \xrightarrow[\ g\ ]{} & Y'
\end{array}
\qquad
\begin{array}{l}
v_\bullet f \leqslant g u_\bullet \quad (\text{in Ord}), \\[2mm]
(\Leftrightarrow \quad f u^\bullet \leqslant v^\bullet g).
\end{array}
\tag{7.86}
$$

(b) But we are more interested in their double subcategories contained in $\mathrm{Adj}_0\mathrm{Ord}$ (see 7.2.7), namely

$$
\begin{aligned}
\mathbb{L}\text{thc} &= \mathrm{Adj}_0\text{Lt} \subset \text{AdjLt}, \\
\mathbb{M}\text{lhc} &= \mathrm{Adj}_0\text{Ml} \subset \text{AdjMl}.
\end{aligned}
\tag{7.87}
$$

These have the same objects and arrows as in (a), but only bicommutative cells:

$$
v_\bullet f = g u_\bullet, \qquad\qquad f u^\bullet = v^\bullet g. \tag{7.88}
$$

We have already seen above, in 7.5.1, that $\mathbb{M}$lhc comes out naturally when we want to formalise direct and inverse images of subobjects for modules (or abelian groups).

The double category $\mathbb{L}$thc plays a similar role in the much more general context of 'semiexact' and 'homological' categories, as one can see in [G10].

7.5.3 Vertical limits and colimits for adjoints

It is interesting to note (or it may be, for some reader) that the kernels and cokernels of $\text{Mlc} = \text{Ver}_0\mathbb{M}\text{lhc}$ become vertical double (co)limits in $\mathbb{M}$lhc. These vertical (co)limits also exist in $\mathbb{L}\text{thc} = \mathrm{Adj}_0\text{Lt}$.

To begin with, the zero-object $\{*\}$ of the category Mlc (and Ltc) becomes a vertical zero object in $\mathbb{M}$lhc (and $\mathbb{L}$thc) (as a vertical double limit and colimit).

In fact, every horizontal arrow f has two bicommutative cells

$$
f \twoheadrightarrow 1_{\{*\}} \twoheadrightarrow f,
$$

which are uniquely determined

$$
\begin{array}{ccc}
X & \xrightarrow{\ f\ } & X' \\
t\downarrow & & \downarrow t \qquad\qquad t^{\bullet}(*) = 1, \\
\{*\} & \longrightarrow & \{*\} \\
s\uparrow & & \uparrow s \qquad\qquad s_{\bullet}(*) = 0. \\
X & \xrightarrow{\ f\ } & X'
\end{array}
\tag{7.89}
$$

We have also seen that the kernel and cokernel of a covariant connection $u = (u_{\bullet}, u^{\bullet})\colon X \nrightarrow Y$ in the category Mlc (and Ltc), are computed as follows

$$
\begin{aligned}
m &\colon \downarrow(u^{\bullet}0) \nrightarrow X, & m_{\bullet}(x) &= x, & m^{\bullet}(x) &= x \wedge u^{\bullet}0, \\
p &\colon Y \nrightarrow \uparrow(u_{\bullet}1), & p_{\bullet}(y) &= y \vee u_{\bullet}1, & p^{\bullet}(y) &= y.
\end{aligned}
\tag{7.90}
$$

These become a vertical double kernel and a vertical double cokernel in Mlhc (or $\mathbb{L}\mathsf{thc}$).

Indeed, given the central bicommutative cell $a\colon (u \overset{f}{\underset{g}{\,}} v)$ below (with $gu_{\bullet} = v_{\bullet}f$, $fu^{\bullet} = v^{\bullet}g$), the vertical kernels $m = \ker u$ and $n = \ker v$ determine the upper bicommutative cell of the following diagram, where the homomorphism f' is the restriction of f (and preserves the new maxima because $f(u^{\bullet}0) = v^{\bullet}g(0) = v^{\bullet}(0)$)

$$
\begin{array}{ccc}
\downarrow(u^{\bullet}0) & \xrightarrow{\ f'\ } & \downarrow(v^{\bullet}0) \\
m\downarrow & & \downarrow n \\
X & - f \to & X' \\
u\downarrow & & \downarrow v \\
Y & - g \to & Y' \\
p\downarrow & & \downarrow q \\
\uparrow(u_{\bullet}1) & \xrightarrow[\ g'\ ]{} & \uparrow(v_{\bullet}1)
\end{array}
\tag{7.91}
$$

This cell is the kernel of a in $\mathsf{Ver}_1\mathsf{Mlhc}$ (or $\mathsf{Ver}_1\mathbb{L}\mathsf{thc}$). Dually the cokernel of a is the lower cell above, where $p = \operatorname{cok} u$, $q = \operatorname{cok} v$ and the homomorphism g' is the restriction of g.

The double category Mlhc is *vertically Puppe-exact*, in an obvious sense:

(i) the category $\mathsf{Ver}_0\mathsf{Mlhc} = \mathsf{Mlc}$ is p-exact,

(ii) the category $\mathsf{Ver}_1\mathsf{Mlhc}$ is p-exact,

(iii) the faces and degeneracy functors between them are exact.

In fact we already know (i) and have now verified that $Ver_1 M llhc$ has kernels and cokernels, preserved by faces and degeneracy. Property (ii) follows now easily: the induced (bicommutative) cell from $Cok(ker\,a)$ to $Ker(cok\,a)$ is vertically invertible because its vertical arrows are.

Finally we recall that the category Ltc has a semiadditive structure, with finite biproducts (see 6.4.5). These 'limits-colimits' become vertical biproducts in $Lthc$. (One can prove that the double category $Lthc$ is *vertically homological*, extending as above the notion of homological category defined in [G10] and extending the previous proof.)

7.6 *Double categories of relations in Homological Algebra

Double categories of relations are also of interest. In particular, an adjunction between abelian (or p-exact) categories has a natural extension to a colax-lax adjunction between their double categories of relations.

Double adjunctions are studied in [GP2], where the interested reader can find the complete definitions; the crucial point – as recalled in the Introduction to this chapter – is that colax and lax double functors must be kept 'in separate directions', say vertical and horizontal, because composing them would destroy their comparisons.

E, E' are always p-exact categories. This is the natural framework for the present issues; but the reader can think of abelian categories, or even of categories of modules.

7.6.1 Double categories of relations on p-exact categories

We have seen in 6.6.3 that the p-exact category E can be embedded in the involutive ordered category $Rel\,E$ of its relations, constructed by quaternary factorisations.

Proceeding as in 7.2.7(a) for $RelSet$, we now form a flat double category $A = RelE$ with

$$Hor_0 A = E, \qquad Ver_0 A = Rel\,E,$$

and double cells of the following kind

$$
\begin{array}{ccc}
A & \xrightarrow{\ f\ } & B \\
u\downarrow & \leqslant & \downarrow v \\
A' & \xrightarrow[\ g\ ]{} & B'
\end{array}
\qquad gu \leqslant vf \quad (\Leftrightarrow fu^\sharp \leqslant v^\sharp g). \qquad (7.92)
$$

This amounts to a commutative diagram of E based on quaternary factorisations, where the three dashed arrows are uniquely determined, by induction on subobjects and quotients

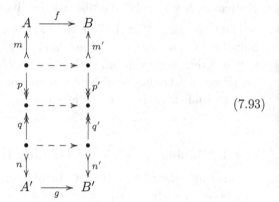

$$(7.93)$$

Equivalently, one can use a dual diagram for coquaternary factorisations. An exact functor $F\colon \mathsf{E} \to \mathsf{E}'$ can be uniquely extended to a double functor

$$\mathbb{Rel}(F)\colon \mathbb{Rel}\mathsf{E} \to \mathbb{Rel}\mathsf{E}'$$

that preserves involution and order. But we are interested in more general situations.

7.6.2 Left exact functors

Let us take a *left exact* functor $S\colon \mathsf{E} \to \mathsf{E}'$, i.e. a functor which preserves kernels (whence the zero-object, monomorphisms and their pullbacks).

Since the functor S preserves preimages of monomorphisms, which become bicommutative squares in $\mathbb{Rel}\mathsf{E}'$ (by 6.6.4), we can extend it to relations using – equivalently – the quaternary or coquaternary factorisation:

$$S'(nq^\sharp pm^\sharp) = S(n)(Sq)^\sharp(Sp)(Sm^\sharp),$$
$$S'(q'^\sharp n' m'^\sharp p') = (Sq')^\sharp(Sn')(Sm')^\sharp(Sp').$$

$$(7.94)$$

Note that S need *not* preserve epis, so that the right-hand parts of these formulas are *not* (co)quaternary factorisations, generally.

We obtain thus a *lax* double functor (see 7.2.4)

$$S' = \mathbb{Rel}S\colon \mathbb{Rel}\mathsf{E} \to \mathbb{Rel}\mathsf{E}'.$$

$$(7.95)$$

The fact that S' preserves cells follows trivially from the fact that the cell (7.92) amounts to a commutative diagram (7.93) in E, and factorises in four cells of $\mathbb{Rel}\mathsf{E}$, whose images are cells of $\mathbb{Rel}\mathsf{E}'$.

The comparison special cells for vertical composition

$$S'v.S'u \leqslant S'(v.u)$$

are an easy consequence of the composition $v.u$ as computed in (6.64)

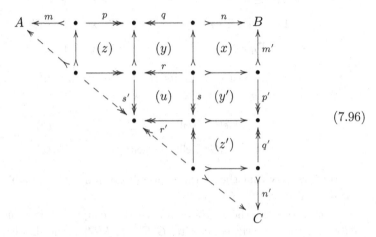

(7.96)

Indeed the functor S preserves the pullbacks of monomorphisms (x), (z), (z'), that are bicommutative in the category of relations, and the commutative squares (y), (y'). Furthermore it carries the pushout (u) to a commutative square of morphisms $Ss'.Sr = Sr'.Ss$ in E'; this gives

$$(Ss)(Sr)^\sharp \leqslant (Sr')^\sharp(Ss),$$

in Rel E'. It follows immediately that $S'v.S'u \leqslant S'(v.u)$.

Again, a natural transformation

$$\varphi\colon S \to T\colon \mathsf{E} \to \mathsf{E}'$$

of left exact functors gives a horizontal transformation of lax double functors

$$\mathbb{R}\mathrm{el}\varphi\colon \mathbb{R}\mathrm{el}S \to \mathbb{R}\mathrm{el}T\colon \mathbb{R}\mathrm{el}\mathsf{E} \to \mathbb{R}\mathrm{el}\mathsf{E}',$$

with the same components as φ (the coherence with the comparison cells of $\mathbb{R}\mathrm{el}S$ and $\mathbb{R}\mathrm{el}T$ is here automatic, since our double categories are flat).

Dually, a *right exact* functor $S\colon \mathsf{E} \to \mathsf{E}'$ (which preserves cokernels) can be extended to a *colax* double functor $S' = \mathbb{R}\mathrm{el}S\colon \mathbb{R}\mathrm{el}\mathsf{E}' \to \mathbb{R}\mathrm{el}\mathsf{E}$, by the same definition as above, in (7.94).

Of course if S is exact we get the double functor already considered in 7.6.1.

7.6.3 Adjoints between categories of relations

Let us start from an arbitrary adjunction $F \dashv G$ between p-exact categories (or just categories of modules)

$$F: \mathsf{E} \to \mathsf{E}', \qquad\qquad G: \mathsf{E}' \to \mathsf{E},$$
$$\eta: 1 \to GF: \mathsf{E} \to \mathsf{E}, \qquad \varepsilon: FG \to 1: \mathsf{E}' \to \mathsf{E}', \qquad (7.97)$$
$$\varepsilon F.F\eta = 1_F, \qquad\qquad G\varepsilon.\eta G = 1_G.$$

F preserves the existing colimits and G the existing limits; we have thus a colax and a lax extension, respectively

$$F' = \mathrm{Rel}F: \mathrm{Rel}\mathsf{E} \to \mathrm{Rel}\mathsf{E}' \qquad (colax),$$
$$G' = \mathrm{Rel}G: \mathrm{Rel}\mathsf{E}' \to \mathrm{Rel}\mathsf{E} \qquad (lax). \qquad (7.98)$$

We show now that they form a colax-lax adjunction of double categories (as defined in [GP2], Section 3).

As remarked in the Introduction to this chapter, this is not a trivial extension, because the composites $G'F'$ and $F'G'$ are neither lax nor colax, generally.

The extended adjunction consists of two double cells η, ε in a suitable double category $\mathbb{D}\mathsf{bl}$ of double categories, with lax and colax double functors as horizontal or vertical arrows, respectively

that satisfy the *triangle identities*

$$\eta \otimes \varepsilon = 1_{F'}, \qquad \varepsilon \mid \eta = e_{G'}.$$

The unit has components

$$\eta A: A \to GFA, \qquad \eta u: u \to G'F'u,$$

where ηA is the original component in (7.97) while ηu is defined below, for every vertical arrow u of $\mathrm{Rel}\mathsf{E}$.

Their coherence conditions (cf. [GP2], Section 2) are based, separately, on the comparisons of F' and G' (but here they hold automatically, because the double categories $\mathrm{Rel}\mathsf{E}$ and $\mathrm{Rel}\mathsf{E}'$ are flat).

The double cell ηu of $\mathbb{R}\mathsf{el}\mathbb{E}$ is constructed as follows

$$ \text{(7.100)} $$

The right-hand inequality comes from the colax-property of G' and the definition of F'

$$ (GFn)(GFq)^\sharp(GFp)(GFm^\sharp) \ \leqslant \ G'((Fn)(Fq)^\sharp(Fp)(Fm^\sharp)) $$
$$ = G'F'u. $$
$$ \text{(7.101)} $$

The components of the counit are defined dually.

7.6.4 The abelian case

A left exact functor S between abelian categories preserves arbitrary pullbacks; its extension $S' = \mathbb{R}\mathsf{el}S$ can be equivalently computed over quaternary factorisations, or coquaternary factorisations, or *strong* binary factorisations, or *arbitrary* cobinary factorisations. Again, *we get a* lax *double functor*, as in the more general p-exact case considered above.

Dually for right exact functors.

As an example, starting from the following adjunction $F \dashv G$ of abelian groups

$$ F = - \otimes A\colon \mathsf{Ab} \to \mathsf{Ab}, \qquad G = \mathrm{Hom}(A, -)\colon \mathsf{Ab} \to \mathsf{Ab}, \qquad \text{(7.102)} $$

with $A = \mathbb{Z}/2$, both the extended functors fail to be strict (because F and G are not exact).

To show that the lax functor $G' = \mathbb{R}\mathsf{el}(G)$ is not strict we take the canonical projection $p\colon \mathbb{Z} \to \mathbb{Z}/2$; then $pp^\sharp = 1\colon \mathbb{Z}/2 \to \mathbb{Z}/2$ but $G(p) = 0\colon 0 \to \mathbb{Z}/2$, whence $(Gp)(Gp)^\sharp \neq G'(pp^\sharp) = 1$. This also shows that we cannot compute G' over the (non-strong) binary factorisation $1 = pp^\sharp$.

Similarly, the colax extension $F' = \mathbb{R}\mathsf{el}(F)$ is not strict. The monomorphism $m = 2.-\colon \mathbb{Z} \to \mathbb{Z}$ gives $m^\sharp m = 1_\mathbb{Z}$, but $F(m) = 0\colon \mathbb{Z}/2 \to \mathbb{Z}/2$, whence $(Fm)^\sharp(Fm) \neq F'(m^\sharp m) = 1$.

7.7 *Factorisation systems as pseudo algebras for 2-monads

Factorisation systems, both the general ones and the proper ones (see Section 2.5), have an interesting, non-obvious relationship with 2-dimensional monads and their pseudo algebras (or weak algebras).

We only give some hints at this topic; the interested reader can see the papers cited below. General pseudo algebras are dealt with in Street's article [St1].

7.7.1 A biuniversal property

The starting point is the fact that the category C^2, equipped with the factorisation system (E, M) introduced in 2.5.1(b), is the *weakly free category with factorisation system* generated by C.

This means that, with respect to the diagonal embedding

$$\eta_C \colon C \to C^2,$$
$$\eta_C(X) = 1_X, \qquad \eta_C(f \colon X \to Y) = (f, f) \colon 1_X \to 1_Y, \tag{7.103}$$

every functor $F \colon C \to D$ with values in a category with an assigned factorisation system (E', M') can be extended to a functor $G \colon C^2 \to D$ that preserves the systems (i.e. $G(E) \subset E'$, $G(M) \subset M'$), and is determined up to a unique functorial isomorphism.

More precisely, (C^2, η_C) is a biuniversal arrow from C to the forgetful 2-functor $\mathsf{fsCat} \to \mathsf{Cat}$, defined on the obvious 2-category of categories equipped with a factorisation system (see 7.1.5).

7.7.2 A two-dimensional monad

We have seen, in 3.4.9, the structure of the arrow category $\mathbf{2}$ as an internal comonoid in Cat

$$\mathbf{1} \xleftarrow{\ e\ } \mathbf{2} \xrightarrow{\ d\ } \mathbf{2} \times \mathbf{2} \qquad d = (1, 1), \tag{7.104}$$

with the diagonal d as comultiplication. By contravariant exponentials, the category Cat inherits a monad, with unit $\eta_C \colon C \to C^2$ as above

$$P \colon \mathsf{Cat} \to \mathsf{Cat}, \qquad P(C) = C^2,$$
$$\eta_C \colon C \to C^2, \qquad \mu_C \colon C^{2 \times 2} \to C^2, \tag{7.105}$$
$$\mu_C(q \colon \mathbf{2} \times \mathbf{2} \to C) = qd \colon \mathbf{2} \to C.$$

The multiplication simply turns a commutative square q of C into its diagonal qd.

(This monad is often called 'the squaring monad' on Cat, which is somewhat misleading: of course C^2 is not the cartesian power C^2. A better name might be 'the elementary-path monad', viewing $\mathbf{2}$ as the standard directed interval of Cat, and C^2 as the category of directed elementary paths in C, as in [G8].)

(P, η, μ) is actually a 2-monad on the 2-category Cat (as defined in 7.1.4) and allows us to consider its 'pseudo algebras' (satisfying the axioms of algebras up to coherent invertible cells). It is known, since some hints in Coppey [Cop] and a full proof in Korostenski–Tholen [KrT], that the factorisation systems of C are in bijective correspondence with the pseudo-isomorphism classes of unital pseudo algebras of P.

The idea of the proof can be described as follows: a pseudo algebra $(\mathsf{C}, F : \mathsf{C}^2 \to \mathsf{C})$ assigns to each morphism $f : X \to Y$ an object $F(f)$ (to be viewed as the *image* of f); the factorisation of

$$\eta_\mathsf{C}(f) = (f, f) : 1_X \to 1_Y$$

in the factorisation system of C^2 (see 2.5.1(b))

$$
\begin{array}{ccccc}
X & \xrightarrow{\ 1\ } & X & \xrightarrow{\ f\ } & Y \\
{\scriptstyle 1}\big\downarrow & & {\scriptstyle f}\big\downarrow & & \big\downarrow{\scriptstyle 1} \\
X & \xrightarrow[\ f\ ]{} & Y & \xrightarrow[\ 1\ ]{} & Y
\end{array}
\qquad (7.106)
$$

is transformed by F into a factorisation $X \to F(f) \to Y$ of f in C.

7.7.3 The case of proper factorisation systems

Proper factorisation systems on C can also be viewed as pseudo algebras for a 2-monad.

One starts from an interesting, unexpected quotient $\mathrm{Fr}(\mathsf{C}) = \mathsf{C}^2/R$, introduced by Freyd [Fr2] for an outstanding goal: to embed the stable homotopy category of spaces in an abelian category.

Namely, two parallel morphisms

$$f = (f', f'') : x \to y, \qquad g = (g', g'') : x \to y,$$

of C^2 (i.e. commutative squares of C, written as in 2.5.1(b)) are R-equivalent whenever their diagonals $\overline{f}$, $\overline{g}$ coincide

$$yf' = f''x = yg' = g''x. \qquad (7.107)$$

This quotient was later studied by the present author, in a general context, to study weak limits and proper factorisation systems [G4, G6].

The first paper shows that Fr(C) has an induced proper factorisation system (easily described in the strict form of 2.5.3(f)), and is the *weakly free category with proper factorisation system* generated by C.

The second studies the induced 2-monad on the 2-functor Fr: Cat → Cat, and proves that there is a canonical bijection between proper factorisation systems in C and pseudo-isomorphism classes of unital pseudo algebras of this 2-monad.

Similar, simpler relations hold in the *strict* case: *strict factorisation systems are strictly monadic on* Cat, *as well as the proper strict factorisation systems.*

8

Solutions of the exercises

Easy exercises and exercises marked with * may be left to the reader.

8.1 Exercises of Chapter 1

8.1.1 Solutions of 1.2.5

(a) The proof is straightforward, and an easy, useful exercise.

If the ordered set X is a lattice, as defined in 1.2.4, properties (L.1–5) are easily verified. For instance, both $x \vee (y \vee z)$ and $(x \vee y) \vee z$ coincide with $\sup\{x, y, z\}$.

Conversely, suppose that the algebraic structure $(X, \vee, \wedge)$ satisfies (L.1–5). By (L.4), the condition $x \vee y = y$ implies $x \wedge y = x \wedge (x \vee y) = x$, while $x \wedge y = x$ implies $x \vee y = y$.

We define $x \leqslant y$ by these two equivalent conditions, which plainly give an order relation. Then $x \vee y$ is indeed the least upper bound of x and y, since:

- $x \vee (x \vee y) = (x \vee x) \vee y = x \vee y, \qquad y \vee (x \vee y) = y \vee (y \vee x) = y \vee x,$

- if $x \leqslant z$ and $y \leqslant z$, then $z \vee (x \vee y) = (z \vee x) \vee y = z \vee y = z.$

Similarly $x \wedge y$ is indeed the greatest lower bound of x and y.

(b) A self-evident argument can be found in 1.7.9(e).

8.1.2 Solutions of 1.3.4

(a) In pOrd and Ord every mono is injective, with the same argument as in Set, based on the (ordered) singleton.

To prove that epis are surjective in pOrd one can use a two-point set with chaotic preorder. In Ord the conclusion is the same, but the proof is more complicated: for a monotone mapping $f \colon X \to Y$ and a point $y_0 \in Y \setminus f(X)$, one can construct an ordered set Y' where y_0 is duplicated.

We have thus two distinct morphisms $u, v \colon Y \to Y'$ such that $uf = vf$, and f is not epi.

(b) In Hsd monomorphisms coincide with the injective continuous mappings, with the same proof as in Top, based on the topological singleton.

The cancellation property of maps with a dense image is a well-known issue, based on the fact that the diagonal of a Hausdorff space Y is closed in the product $Y \times Y$. (Our argument for the surjectivity of epimorphisms in Top is based on the indiscrete two-point space and does not apply here.)

(c) For monomorphisms in Gp one can use the same argument as in Ab. For epimorphisms one cannot, because the image of a group homomorphism need not be normal in the codomain.

(d) Use the scalar field K ($\mathbb{R}$ or $\mathbb{C}$) as a Banach space; the norm is the absolute value.

8.1.3 Solutions of 1.3.5

(a) If $a \in A \subset X$, we can define a retraction $p \colon X \to A$ letting $p(x) = x$ for $x \in A$, and $p(x) = a$ otherwise. The empty set is only a retract of itself.

(b) If the subgroup $A \subset X$ has a retraction $p \colon X \to A$ in Ab, the subgroup $B = \operatorname{Ker} p$ meets our conditions. In fact, if $x \in A \cap B$, then $x = p(x) = 0$; if $x \in X$, then $x = p(x) + (x - p(x)) \in A + B$.

Conversely, if there exists a subgroup $B \subset X$ such that $A \cap B = 0$ and $A + B = X$, each element $x \in X$ can be uniquely written as $x = a + b$, with $a \in A$ and $b \in B$. This gives a canonical isomorphism $X \cong A \oplus B$, and a retraction $p \colon X \to A$.

(c) There is a simple retraction, the normalisation mapping N

$$N \colon X \to \mathbb{S}^n, \qquad N(x) = x / \|x\| \quad (x \neq 0), \tag{8.1}$$

defined by means of the euclidean norm $\|x\| = d(x, 0)$ and the linear structure of the vector space $\mathbb{R}^{n+1}$ on the real field.

8.1.4 Solutions of 1.5.2

(a) Every scalar $\lambda \in R$ gives a natural transformation $\varphi \colon U \to U$ of the forgetful functor $U \colon R\,\mathsf{Mod} \to \mathsf{Set}$, whose component on the left module A is the scalar multiplication by λ, as a mapping between the underlying sets

$$\varphi_A \colon |A| \to |A|, \qquad \varphi_A(x) = \lambda x. \tag{8.2}$$

Conversely, let $\varphi \colon U \to U$ be a natural transformation. The mapping $\varphi_R \colon |R| \to |R|$ determines the scalar $\lambda = \varphi_R(1)$. For every left R-module

A, the naturality of φ on the homomorphism $f\colon R \to A$ that sends 1 to an element $a \in A$ gives the relation:

$$\varphi_A(a) = \varphi_A(f(1)) = f(\varphi_R(1)) = f(\lambda) = \lambda a.$$

(b) We have the same characterisation as in (a), because all the mappings φ_A in (8.2) are homomorphisms of abelian groups.

(c) If λ belongs to the centre of R, these mappings φ_A are homomorphisms of left R-modules. Conversely, if $\varphi_R\colon R \to R$ is in $R\,\mathsf{Mod}$, then $\lambda = \varphi_R(1)$ commutes with any $\mu \in R$, since $\lambda\mu = \varphi_R(\mu) = \mu\varphi_R(1) = \mu\lambda$.

(d) There are four obvious natural transformations, whose components $\varphi X\colon \mathcal{P}^*X \to \mathcal{P}^*X$ are

$$A \mapsto A, \qquad A \mapsto X \setminus A, \qquad A \mapsto X, \qquad A \mapsto \emptyset. \tag{8.3}$$

To show that there are no others requires more work. One considers the component of an arbitrary natural transformation $\varphi\colon \mathcal{P}^* \to \mathcal{P}^*$ on the set $X_0 = \{0,1\}$; its value on the subset $\{1\}$ has four possibilities

$$\{1\} \mapsto \{1\}, \qquad \{1\} \mapsto \{0\}, \qquad \{1\} \mapsto X_0, \qquad \{1\} \mapsto \emptyset. \tag{8.4}$$

Now, for a general set X and a subset $A \subset X$, let $f\colon X \to X_0$ be the characteristic function of A: $f(x) = 1$ if and only if $x \in A$. Then $A = f^{-1}\{1\}$, and the naturality of φ on the mapping f gives

$$\varphi_X(A) = \varphi_X(f^{-1}\{1\}) = f^{-1}\varphi_{X_0}(\{1\}). \tag{8.5}$$

Therefore the transformation φ is determined by $\varphi_{X_0}(\{1\})$, and the four cases listed in (8.4) give the four natural transformations (8.3).

All this will be a straightforward consequence of the Yoneda Lemma: see Exercise 1.6.5(e).

(e) Direct images of subsets need not preserve the total subset, nor the complement of a subset. There are two obvious natural transformations $\varphi\colon \mathcal{P}_* \to \mathcal{P}_*$, namely the identity and $\varphi_X(A) = \emptyset$. Proving that there are no others requires some work, which can be interesting or not, according to one's taste.

(g) A natural transformation $\varphi\colon f \to g\colon X \to Y$ between homomorphisms of monoids has one component $\varphi \in Y$ (corresponding to the single object $*$ of X), which has to satisfy the naturality condition for every $x \in X$

$$\begin{array}{ccc} * & \xrightarrow{\ \varphi\ } & * \\ {\scriptstyle fx}\big\downarrow & & \big\downarrow{\scriptstyle gx} \\ * & \xrightarrow[\ \varphi\ ]{} & * \end{array} \qquad\qquad f(x).\varphi = \varphi.g(x) \text{ in } Y. \tag{8.6}$$

The natural transformation φ is invertible if and only if the element φ is invertible in Y, in which case it gives an inner automorphism $h = \varphi^{-1}. - .\varphi \colon Y \to Y$ such that $hf = g$.

8.1.5 Solutions of 1.5.4

(d) The claim follows from the naturality of $\psi \colon H \to K$ on the general component $\varphi X \colon FX \to GX$

$$
\begin{array}{ccc}
HFX & \xrightarrow{\ \psi FX\ } & KFX \\
{\scriptstyle H(\varphi X)}\Big\downarrow & & \Big\downarrow{\scriptstyle K(\varphi X)} \\
HGX & \xrightarrow[\ \psi GX\ ]{} & KGX
\end{array}
\tag{8.7}
$$

8.1.6 Solutions of 1.5.6

For (a), (c), (d), (f) and (g) we simply use the characterisation 1.5.5(iii) of equivalences of categories.

(e) It is sufficient to prove that an equivalence F between two skeletal categories is an isomorphism. In fact F reflects isomorphisms, because it is full and faithful (see 1.4.7(c)), and is essentially surjective on objects: it follows that it is bijective on objects and morphisms.

8.1.7 Solutions of 1.6.2

(c) A natural transformation $\varphi \colon F \to G \colon \mathsf{C} \to \mathsf{D}$ gives a functor

$$\Phi \colon \mathsf{C} \times \mathbf{2} \to \mathsf{D},$$

$$\Phi(-,0) = F, \qquad \Phi(-,1) = G, \tag{8.8}$$

$$\Phi(f \colon X \to X', \iota \colon 0 \to 1) = \varphi(f) \colon F(X) \to G(X'),$$

where $\iota \colon 0 \to 1$ is the non-identity arrow of $\mathbf{2}$ and $\varphi(f)$ is defined in (1.22). Similarly, we have a functor

$$\Phi' \colon \mathsf{C} \to \mathsf{D}^{\mathbf{2}}, \qquad \Phi'(X) = \varphi X \colon F(X) \to G(X),$$

$$\Phi'(f \colon X \to X') = (Ff, Gf) \colon \varphi X \to \varphi X'. \tag{8.9}$$

Both procedures are invertible.

(d) A (left) *action* of the group G on the set X means a mapping

$$\varphi \colon X \times G \to X, \qquad \varphi(x,g) = gx \qquad (x \in X, g \in G), \tag{8.10}$$

that satisfies the following axioms (for all $x \in X$ and $g, h \in G$):

(i) $h(gx) = (hg)x$ (*compatibility*),

(ii) $1_G x = x$ (*unitarity*).

A functor $H\colon G \to \mathsf{Set}$ amounts to a set $X = H(*)$ equipped with an action of G, where $gx = H(g)(x)$.

Similarly, a natural transformation $\varphi\colon H \to H'\colon G \to \mathsf{Set}$ amounts to a mapping $\varphi\colon X \to X'$ consistent with the actions of G, in the sense that $\varphi(gx) = g\varphi(x)$, for $x \in X$ and $g \in G$.

An action of G on the topological space X is defined in the same way, requiring that all the mappings $\varphi(-, g)\colon X \to X$ be continuous.

8.1.8 Solutions of 1.6.5

(a) These functors are represented by the singleton space $\{*\}$, by the left R-module R and by the polynomial ring $\mathbb{Z}[X]$, respectively. All verifications are obvious or easy.

(c) The contravariant functor $\mathcal{P}^*\colon \mathsf{Set}^{\mathrm{op}} \to \mathsf{Set}$ is represented by the cardinal set $X_0 = \{0, 1\}$, by the natural isomorphism

$$\chi\colon \mathcal{P}X \to \mathsf{Set}^{\mathrm{op}}(X_0, X) = \mathsf{Set}(X, X_0),$$
$$A \mapsto (\chi_A\colon X \to X_0), \tag{8.11}$$

that sends a subset $A \subset X$ to its characteristic function: $\chi_A(x) = 1$ if and only if $x \in A$.

*In other words, the cardinal $2 = \{0, 1\}$ is the *classifier of subobjects*, in the topos Set.*

(d) Take a representable functor $\mathsf{C}(X_0, -)\colon \mathsf{C} \to \mathsf{Set}$. If $f\colon X \to Y$ is mono in C, the mapping

$$\mathsf{C}(X_0, f)\colon \mathsf{C}(X_0, X) \to \mathsf{C}(X_0, Y), \qquad u \mapsto fu,$$

is obviously injective. The dual case is a consequence.

The morphism f is mono if and only if, for every X_0 in C, the mapping $\mathsf{C}(X_0, f)$ is injective.

8.1.9 Solutions of 1.6.7

(a) We have seen in 1.6.5(c) that the contravariant functor of subsets

$$\mathcal{P}^*\colon \mathsf{Set}^{\mathrm{op}} \to \mathsf{Set}$$

is represented by the cardinal set $X_0 = \{0, 1\}$. We give now a topological version of this fact. We put on X_0 the *Sierpinski* topology, whose open subsets are $\emptyset$, $\{1\}$ and X_0.

Now a mapping $\chi\colon X \to X_0$ is continuous if and only if the subset $\chi^{-1}(\{1\})$ is open in X. We have thus a natural isomorphism

$$\chi\colon U\mathcal{O}(X) \to \mathsf{Top}(X, X_0), \qquad A \mapsto \chi_A, \tag{8.12}$$

that sends an open subset $A \subset X$ to its characteristic map $\chi_A\colon X \to X_0$.

The Sierpinsky space is the ordinal **2** with Alexandrov topology: see 5.1.5.
*In **Top** this space, equipped with the map $t\colon \{*\} \to X_0$, $* \mapsto 1$, is the classifier of open subspaces.*

(b) Take two distinct maps $f, g\colon X \to Y$, so that $f(x) \neq g(x)$ for some $x \in X$. If Y is T_0, there is some open set V of Y that meets $\{f(x), g(x)\}$ in a single point, and then $f^*(V) \neq g^*(V)$.

8.1.10 Solutions of 1.7.2

(a) The letters x and y denote arbitrary elements of X and Y, respectively.

(i) $\Rightarrow$ (ii). Assuming (i), $f(x) \leqslant f(x)$ gives $x \leqslant g(f(x))$; moreover, if $x \leqslant g(y)$ then $f(x) \leqslant y$.

(ii) $\Rightarrow$ (iii). Assuming (ii), f is monotone. From $f(x) \in \{y \in Y \mid x \leqslant g(y)\}$ we get $x \leqslant g(f(x))$; moreover

$$f(g(y)) = \min\{y' \in Y \mid g(y) \leqslant g(y')\} \leqslant y.$$

(iii) $\Rightarrow$ (i). We assume (iii). From $f(x) \leqslant y$ we get $x \leqslant g(f(x)) \leqslant g(y)$. From $x \leqslant g(y)$ we have $f(x) \leqslant f(g(y)) \leqslant y$.

(b) If $f \dashv g \dashv f$, conditions 1.7.1(iii) give $gf = 1$ and $fg = 1$.

(c) The right adjoint to the embedding $i\colon \mathbb{Z} \to \mathbb{R}$ is the integral-part function, or *floor function*

$$[-]\colon \mathbb{R} \to \mathbb{Z}, \qquad [x] = \max\{k \in \mathbb{Z} \mid k \leqslant x\}, \tag{8.13}$$

and can be thought of as the 'highest lower inverse' of i.

The left adjoint is the *ceiling function*

$$\min\{k \in \mathbb{Z} \mid k \geqslant x\} = -[-x], \tag{8.14}$$

related (here) to the right adjoint by the anti-isomorphism $x \mapsto (-x)$ of the real and integral lines.

(d) An irrational number has no 'best' rational approximation, lower or upper. For instance, the subset $\{q \in \mathbb{Q} \mid q \leqslant \sqrt{2}\}$ has no maximum, while $\{q \in \mathbb{Q} \mid q \geqslant \sqrt{2}\}$ has no minimum.

(e) Let $x \leqslant x'$ in X. Trivially $f(x') \leqslant f(x')$; we deduce that $x \leqslant x' \leqslant gf(x')$, and then $f(x) \leqslant f(x')$. The monotony of g is proved in the same way.

8.1.11 Solutions of 1.7.9

(a) Plainly, $A \subset A' \subset X$ implies $A\bot \supset A'\bot$ and $A \subset \bot(A\bot)$.

(e) The condition $x = \vee A = \min(U(A))$ means that $x \in U(A) \cap LU(A)$.

On the other hand, the condition $x = \wedge(U(A)) = \max(LU(A))$ means that $x \in LU(A) \cap ULU(A)$.

But we have seen that $ULU(A) = U(A)$, in Exercise (d), and the previous conditions are equivalent.

8.1.12 Solutions of 1.8.4

(b) Given three consecutive relations $u \colon X \rightarrow Y$, $v \colon Y \rightarrow Z$ and $w \colon Z \rightarrow T$, the ternary composites $w(vu)$ and $(wv)u$ are both computed as the set of all pairs $(x, t) \in X \times T$ for which:

- there are $y \in Y$ and $z \in Z$ such that: $(x, y) \in u$, $(y, z) \in v$, $(z, t) \in w$.

The rest is obvious.

(c) For the set $X = \{0, 1\} \subset \mathbb{N}$, the endo-relation $e \colon X \rightarrow X$ corresponding to the natural order is strictly smaller than $ee^{\sharp}e$, which also contains $(1, 0)$.

8.2 Exercises of Chapter 2

8.2.1 Solutions of 2.1.4

(d) If $m \colon A \to X$ and $p \colon X \to A$ are such that $pm = \mathrm{id}A$, it is easy to see that m is the equaliser of mp and $\mathrm{id}X$.

8.2.2 Solutions of 2.1.6

(e) The solution described in (2.9) (as a subset of the cartesian product) always works in Set$_\bullet$. We want to prove that, for a finite set of indices, it also works in Top$_\bullet$ (as a subspace of the cartesian product).

It is sufficient to consider the binary case, and we let

$$(Z, z_0) = (X, x_0) + (Y, y_0),$$

the categorical sum in Top$_\bullet$ constructed as the quotient of the space $X + Y$ which collapses x_0 and y_0 to the point z_0. The spaces X and Y are embedded in Z, as subspaces.

We have a canonical injective map

$$u \colon (Z, z_0) \to (X, x_0) \times (Y, y_0),$$

that sends $x \in X$ to (x, y_0) and $y \in Y$ to (x_0, y).

To prove that u is a homeomorphism onto its image Z', take an open subset W of Z and let $U = W \cap X$ (open in X) and $V = W \cap Y$ (open in Y). Then $W = U \cup V$ and there are two cases:

(i) if $z_0 \in W$, then $x_0 \in U$ and $y_0 \in V$,

(ii) if $z_0 \notin W$, then $x_0 \notin U$ and $y_0 \notin V$.

In case (i) we let $P = U \times V$, which is open in the product space $X \times Y$, so that $u(W) = P \cap Z'$ is open in Z'. In case (ii) we let $P = (U \times Y) \cup (X \times V)$, and we conclude again that $u(W) = P \cap Z'$ is open in Z'.

8.2.3 Solutions of 2.1.9

(a) Using the description by components and cocomponents (as in 2.1.1 and 2.1.5), the morphisms

$$f_1 = (1,0) \colon X \to X \times Y, \qquad f_2 = (0,1) \colon Y \to X \times Y,$$

give a morphism $f = [f_1, f_2] \colon X + Y \to X \times Y$.

8.2.4 Solutions of 2.2.5

(d) The product in $\mathcal{S}$ of a family (X_i) of sets is easily constructed 'along the equivalence' $R \colon \mathsf{Set}_\bullet \rightleftarrows \mathcal{S} : S$

$$R(\Pi_i S(X_i)) = (\Pi_i (X_i \cup \{\overline{x}_i\})) \setminus \{(\overline{x}_i)\}, \tag{8.15}$$

first adding a base point $\overline{x}_i \notin X_i$ to each factor, then taking their cartesian product in $\mathsf{Set}_\bullet$ and discarding the base point of the product.

Thus the cardinal of the product in $\mathcal{S}$ of two finite sets of m and n elements is $(m+1)(n+1) - 1 = mn + m + n$.

*(f) The proof (also written in [M4], Section V.2, Proposition 3) is quite simple. Assuming that C is not a preordered set, there is some hom-set $\mathsf{C}(A, B)$ with at least two distinct arrows $f \neq g$. Take the (small) cartesian power $C = \Pi_{j \in J} B$, where J is the cardinal of $\mathrm{Mor}\,\mathsf{C}$; then the cardinal of $\mathsf{C}(A, C) \cong \Pi_{j \in J} \mathsf{C}(A, B)$ is at least $2^J > J$, a contradiction.

8.2.5 Solutions of 2.2.6

(a) In Ban_1 a binary product $X \times Y$ is the vector space $X \oplus Y$ with the l_∞-norm

$$\|(x,y)\|_\infty = \|x\| \vee \|y\|,$$

while a binary sum $X + Y$ is the same vector space $X \oplus Y$ with the l_1-norm

$$\|(x, y)\|_1 = \|x\| + \|y\|.$$

These objects are distinct in Ban_1, but they are isomorphic in Ban, where the *biproduct* $X \oplus Y$ (see 2.1.9) can be given any l_p-norm, for $1 \leqslant p \leqslant \infty$ (all of them being Lipschitz-equivalent).

Ban_1 and Ban are finitely complete: equalisers are the obvious closed linear subspaces, with the restricted norm.

(c) An arbitrary sum $X = \Sigma X_i$ in Set is still a sum in $\mathsf{Rel\,Set}$, with the same injections $u_i \colon X_i \to X$. Since $\mathsf{Rel\,Set}$ is an involutive category, X is also the product of all X_i, with projections given by the opposite *relations* $u_i{}^\sharp \colon X \nrightarrow X_i$. The zero object is $\emptyset$.

(d) In $\mathsf{Rel\,Ab}$ no object A can be a terminal object, because every non-trivial abelian group B has at least two distinct relations $B \nrightarrow A$, namely $0 \times A$ and $B \times A$. The dual claim is a consequence of the involution.

8.2.6 Solutions of 2.2.9

(a) Let $(L, (u_i \colon L \to X_i)_{i \in S})$ be a cone of the functor $X \colon S \to \mathsf{C}$, and suppose that

$$(FL, (Fu_i \colon FL \to FX_i)_{i \in S})$$

is a limit of $FX \colon S \to \mathsf{D}$. Then $(L, (u_i))$ is the unique cone of X taken by F to $(FL, (Fu_i))$, and is a limit of X in C.

(b) By hypothesis the composed functor FX has a limit $(L', (v_i))$ in D; moreover the limit of X exists and is taken by F to $(L', (v_i))$.

(c) Suppose that F creates unary products and let $f \colon X' \to X$ be a morphism in C with $Ff = 1_{FX}$.

The functor $X \colon 1 \to \mathsf{C}$ has thus two cones, namely $(X, 1_X)$ and (X', f), taken by F to $(FX, 1_{FX})$, the unary product of FX in D. It follows that $X = X'$ and $f = 1_X$. The converse is obvious.

(d) Let $(L, (u_i \colon L_1 \to X_i)_{i \in S})$ be a limit of X, preserved by F.

For an arbitrary cone of X

$$(M, (f_i \colon M \to X_i)_{i \in S}),$$

we have a unique morphism $f \colon M \to L$ such that $u_i f = f_i$ (for all $i \in \mathrm{Ob}\,S$). If F takes $(M, (f_i))$ to $(FL, (Fu_i))$, then

$$Fu_i . Ff = Ff_i = Fu_i \qquad (i \in S).$$

This gives $Ff = 1_{FL}$. As F reflects identities, $f = 1_L$ and the given cone $(M, (f_i))$ coincides with the limit cone $(L, (u_i))$.

8.2.7 Solutions of 2.3.6

(a) Finite products exist, by Exercises 2.3.2(b) and 2.1.2(e).

For a pair of maps $f, g \colon A \rightrightarrows B$, we form the morphisms $(f, g) \colon A \to B^2$ and $\Delta = (1, 1) \colon B \to B^2$ (the diagonal of B^2). Then we take their pullback E

$$Z \xrightarrow{h} A \underset{g}{\overset{f}{\rightrightarrows}} B \qquad (8.16)$$

Now $fm = u = gm \colon E \to B$. Similarly, a morphism $h \colon Z \to A$ gives the same composite $fh = gh = h' \colon Z \to B$ if and only if the pair (h, h') forms a commutative square with the arrows (f, g) and Δ: this means that $m \colon E \to A$ is the equaliser of f, g.

(b) For a general map $h \colon X \to Y$ in Hsd, let $C = (h(X))^-$ be the closure of its image in Y, and $u \colon C \to Y$ the inclusion map.

We form *in* Top the cokernel pair Z of the embedding u. For the sake of simplicity, we let Y' be a disjoint copy of the space Y, with a point y' corresponding to each point $y \in Y$ and $C' \subset Y'$ corresponding to $C \subset Y$.

Then we define Z as the quotient $Z = (Y \cup Y')/R$, that identifies $[y] = [y']$ when $y \in C$ (no other identifications are made)

$$C \xrightarrowtail{u} Y \underset{g}{\overset{f}{\rightrightarrows}} Z \qquad\qquad f(y) = [y], \quad g(y) = [y']. \qquad (8.17)$$

We prove that the space Z is also Hausdorff, as a consequence of the fact that C is closed in Y. We let $C = f(C) = g(C') \subset Z$. Taking two points $z_1 \neq z_2$ in Z, we have four cases (up to symmetry).

(i) Both points belong to C, and we let $z_i = [y_i]$, with $y_i \in Y$. Then the points y_i have disjoint neighbourhoods U_i in Y; the R-saturated neighbourhoods $V_i = U_i \cup U_i'$ are still disjoint, and give disjoint neighbourhoods of the two points z_i in Z.

(ii) We have $z_1 \notin C$ and $z_2 \in C$. We let $z_1 = y_1 \in Y \setminus C$ and $z_2 = [y_2]$ with $y_2 \in C \subset Y$. The points y_i have disjoint neighbourhoods U_i in Y and we can suppose that U_1 is contained in the open subset $Y \setminus C$. The R-saturated neighbourhoods $V_1 = U_1$ and $V_2 = U_2 \cup U_2'$ are disjoint, and give disjoint neighbourhoods of the two points z_i in Z.

(iii) Both points belong to $Z \setminus C$, and $z_i = y_i \in Y \setminus C$. Then the points y_i have disjoint neighbourhoods U_i in $Y \setminus C$, which are already saturated for R.

(iv) Both points belong to $Z \setminus C$, with $z_1 = y_1 \in Y \setminus C$ and $z_2 = y_2 \in Y' \setminus C'$. Then $f(Y \setminus C)$ and $g(Y \setminus C)$ are disjoint open neighbourhoods of our points.

Now (f, g) is also the cokernel pair of u in the (full) subcategory Hsd.

Assuming that $h \colon X \to Y$ is epi in Hsd, also u is epi (because h factorises as uh'), and the morphisms f, g are equal (by Exercise 2.3.5(a)). This implies that $C = Y$, i.e. h has a dense image in Y.

8.2.8 Solutions of 2.4.5

(b) We have the commutative solid diagram at the left

$$\text{(8.18)}$$

where p, q are strong epis and m is mono. As p is strong, there exists a morphism $w \colon X \to Y$ such that $wp = u$ and $mw = (vq)$. As q is strong, there is also a morphism $w' \colon X' \to Y$ which makes the right diagram above commute.

(c) We have the commutative solid diagram at the left, where gf is a strong epimorphism; we already know that g is epi

$$\text{(8.19)}$$

The right diagram above gives a morphism $w \colon X' \to Y$ such that $wgf = uf$ and $mw = v$; this ensures that w also commutes in the left diagram.

(d) Let $p \colon A \twoheadrightarrow X$ be a strong epi, and a monomorphism. The following commutative square

$$\text{(8.20)}$$

proves that there is a morphism $w \colon X \to A$ such that $wp = 1$ and $pw = 1$.

8.2.9 Solutions of 2.4.7

After proving (a), the exercises (b)–(d) are easy consequences, left to the reader. The points (e) and (f) are already referred to.

(a) We construct a contravariant Galois connection

$$c\colon M(A) \rightleftarrows P(A) : e,$$

between two preordered sets, generally large.

$M(A)$ is the set of monomorphisms with values in A, with the canonical preorder $m' \prec m$: there exists a morphism v such that $m' = mv$ (as in the following diagram)

$$\tag{8.21}$$

$P(A)$ is the set of pairs of parallel morphisms $f, g\colon A \rightrightarrows \bullet$, with preorder $(f, g) \prec (f', g')$ meaning that there exists a morphism u such that $f = uf'$ and $g = ug'$ (as in the diagram above).

The antitone mapping $c\colon M(A) \to P(A)$ sends a monomorphism m to its cokernel pair. The antitone mapping $e\colon P(A) \to M(A)$ sends a pair of morphisms (f, g) to its equaliser.

Now we do have a contravariant Galois connection between preordered sets:

$$\operatorname{id} M(A) \prec ec, \qquad \operatorname{id} P(A) \prec ce.$$

It follows that

$$e(f, g) \sim ece(f, g), \qquad c(m) \sim cec(m), \tag{8.22}$$

which proves our claims, (i) and (ii).

Remarks. The argument can be presented in many equivalent forms. For instance, one can restrict $M(A)$ to the ordered set Sub$A$ of subobjects, or expand it to the whole set of morphisms with values in A, also preordered by factorisation. On the other side, one can restrict $P(A)$ to its subset of jointly epi pairs of arrows, and further restrict it to a skeletal ordered subset.

8.2.10 Solutions of 2.4.8

(a*) In the solid commutative diagram at the left, the left square is a pushout and p is a strong epi

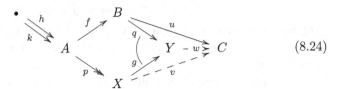

$$(8.23)$$

In the right rectangle, there is thus a morphism $w\colon X \to C$ that makes the diagram commute. This gives a commutative square $uf = wp$, and the pushout-property at Y yields a morphism $w'\colon Y \to C$ (in the left diagram) such that $w'q = u$ and $w'g = w$. It is automatically true that $mw' = v$, as already remarked in the definition of strong epis.

For the second point we draw a diagram where the square is a pushout and p is the coequaliser of the pair (h, k)

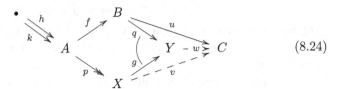

$$(8.24)$$

We want to prove that q is the coequaliser of the pair (fh, fk).

Plainly $qfh = gph = gpk = qfk$. Supposing we have a map $u\colon B \to C$ such that $ufh = ufk$, then uf coequalises h and k, and there is a unique $v\colon X \to C$ such that $uf = vp$. The pushout-property gives a morphism $w\colon Y \to C$ such that $wq = u$ and $wg = v$, and we have proved that u factorises through q.

8.2.11 Solutions of 2.6.3

(a) In fact

$$(X, (u_K\colon K \subset X)) = \mathrm{Colim}((K), (u_{KK'}\colon K \subset K')) \qquad (8.25)$$

where K varies in $\mathcal{P}_f(X)$ and the family of inclusions $u_{KK'}\colon K \subset K'$ is indexed by the pairs $K \subset K'$ of $\mathcal{P}_f(X)$.

(c) Plainly, the join $A' + A''$ of two finite subgroups of the abelian group A is finite. But the only finite subgroup of $\mathbb{Z}$ is the trivial one.

8.2.12 Solutions of 2.8.2

(a) Properties (iii) and (iv) are trivially equivalent.

(i) $\Rightarrow$ (ii) Obvious.

(ii) $\Rightarrow$ (iv) First, $em = mpm = m$. Second, take any morphism $f\colon Y \to X$ such that $ef = f$; then f factorises as $f = m(pf)$, through the monomorphism $m\colon A \to X$.

(iv) $\Rightarrow$ (i) If $m\colon A \to X$ is the equaliser of the pair $1, e\colon X \to X$, the morphism $e\colon X \to X$ factorises as $e = mp$ for a unique $p\colon X \to A$, and $pm = 1_A$ because $mpm = em = m$, and m is mono.

The rest comes from duality, since (i) and (ii) are selfdual.

*(e) One can prove that the idempotent endo-relation $e\colon \{0,1\} \nrightarrow \{0,1\}$ given by the natural order does not split. Therefore the pair (e, id) has no equaliser nor coequaliser.

8.3 Exercises of Chapter 3

8.3.1 Solutions of 3.1.4

(e) For an ordered set X, the unit $\eta X\colon X \to D(\pi_0 X)$ of the adjunction $\pi_0 \dashv D$ is the canonical projection on the quotient set $\pi_0 X$, equipped with the discrete order.

Its universal property is easily checked: for every set Y, a monotone mapping $f\colon X \to DY$ with values in the discrete ordered set DY induces a mapping $g\colon \pi_0 X \to Y$ that satisfies the condition $f = g.\eta X$ (more precisely $f = D(g).\eta X$), and is determined by the latter.

(f) For a small category C, the unit $\eta \mathsf{C}\colon \mathsf{C} \to J(\mathrm{po}\mathsf{C})$ is the identity on objects and the canonical projection on morphisms.

The distinct embeddings $f, g\colon \mathbf{1} \rightrightarrows \mathbf{2}$ of the singleton ordered set in the ordinal $\mathbf{2}$ have coequaliser $\mathbf{1}$ in pOrd, and a countable category in Cat, already met in Exercise 2.1.8(f).

The obvious embeddings $f, g\colon \mathbf{2} \rightrightarrows \mathsf{C}$ of the arrow category $\mathbf{2}$ in the category $\bullet \rightrightarrows \bullet$ have equaliser the discrete category 2 in Cat, and $\mathrm{po}(2)$ is a discrete ordered set. Applying the functor $\mathrm{po}\colon \mathsf{Cat} \to \mathsf{pOrd}$, we get $\mathrm{po}(f) = \mathrm{po}(g) = \mathrm{id}\colon \mathbf{2} \rightrightarrows \mathbf{2}$, whose equaliser is the ordinal $\mathbf{2}$.

(g) We can compose two previous adjunctions, in (f) and (e)

$$\mathrm{po}\colon \mathsf{Cat} \rightleftarrows \mathsf{pOrd}\colon J, \qquad \pi_0\colon \mathsf{pOrd} \rightleftarrows \mathsf{Set}\colon D.$$

In this way we get the set $\pi_0(\mathsf{C})$ of connected components of a small category C, applying the functor $\pi_0\colon \mathsf{pOrd} \to \mathsf{Set}$ to the preordered set $\mathrm{po}(\mathsf{C})$.

A more explicit description has already been given at the end of 3.1.4.

8.3.2 Solutions of 3.2.6

(a) The counit $\varepsilon A\colon tA \to A$ is the embedding of the torsion subgroup of an abelian group, namely the subgroup of all elements of finite order.

(b) We recall that an abelian group C is torsion-free if its only element of finite period is the identity, i.e. $tC = 0$. The unit of the reflection of the inclusion tfAb $\subset$ Ab is the canonical projection $\eta A\colon A \to A/tA$.

(c) See Exercises 1.7.2(c), (d).

(d) We extend the construction of the floor-function $[x]$, in 1.7.2(c). The right adjoint g of the embedding $A \subset \mathbb{R}$ is computed as in (1.46)

$$g(x) = \max A_x, \qquad A_x = \{a \in A \mid a \leqslant x\} \qquad (x \in \mathbb{R}), \qquad (8.26)$$

and exists if and only if all these maxima exist.

Now, if A is lower unbounded in $\mathbb{R}$, for every $x \in \mathbb{R}$ the set A_x is non-empty, and obviously upper bounded. Therefore $\sup A_x$ exists, and belongs to the closed subset A_x.

8.3.3 Solutions of 3.2.8

(d) One may prefer to write $- \times A \dashv A \times -$, so that the adjunction in 'form (i)' is even more evident: the natural isomorphism

$$\varphi_{XY}\colon \operatorname{Rel}\operatorname{Set}(X \times A, Y) \to \operatorname{Rel}\operatorname{Set}(X, A \times Y), \qquad (8.27)$$

sends a subset $\alpha \subset (X \times A) \times Y$ to the corresponding subset of $X \times (A \times Y)$.

8.3.4 Solutions of 3.3.2

(b) The equivalence comes from the following weakly inverse functors

$$H\colon \operatorname{gRng} \rightleftarrows \operatorname{Rng}/\mathbb{Z} \,:K,$$
$$H(R) = (R^+, p\colon R^+ \to \mathbb{Z}), \qquad K(R, p\colon R \to \mathbb{Z}) = \operatorname{Ker} p. \qquad (8.28)$$

Starting from a 'general ring' R, the unital ring R^+ is obtained by adding a unit (without considering whether R has a unit or not): it has additive part $\mathbb{Z} \oplus R$, multiplication

$$(h, r)(k, s) = (hk, hs + kr + rs),$$

and unit $(1, 0)$.

The morphism $p\colon R^+ \to \mathbb{Z}$ is the first projection; then $\operatorname{Ker} p \cong R$.

(The reader may know that in Algebra it is often preferred to define R^+ by 'adding a unit when it does not already exist'; this 'ad hoc' procedure is idempotent, but is not a functor gRng $\to$ Rng.)

The other way round, starting from a copointed ring $(R, p\colon R \to \mathbb{Z})$, we get an isomorphic copointed ring $(\operatorname{Ker} p)^+$, with a natural isomorphism

$$R \to (\operatorname{Ker} p)^+, \qquad r \mapsto (p(r), r - p(r).1_R).$$

(In fact there is a short exact sequence $\operatorname{Ker} p \rightarrowtail R \twoheadrightarrow \mathbb{Z}$ of general rings, which splits in Ab because $\mathbb{Z}$ is a free abelian group.)

8.3.5 Solutions of 3.4.2

(a) In a category C with finite products, the associator κ of the cartesian product

$$A \times B \times C \xrightarrow{\gamma'} A \times (B \times C)$$
$$\Big\downarrow \kappa(A,B,C) \qquad (8.29)$$
$$\gamma'' \searrow \qquad (A \times B) \times C$$

issues from two canonical isomorphisms γ', γ'': the latter was used in Exercise 2.1.2(e), the former is proved in the same way. (In Set, the associator simply takes $(a, (b, c))$ to $((a, b), c)$; similarly in Ord, Cat, Top, etc.)

The left (resp. right) unitor

$$\lambda(A)\colon \top \times A \to A \qquad (\text{resp. } \rho(A)\colon A \times \top \to A), \qquad (8.30)$$

derives from the fact that the product $\top \times A$ (resp. $A \times \top$) can be realised as A, with projections

$$A \to \top, \quad A = A \qquad (\text{resp. } A = A, \quad A \to \top). \qquad (8.31)$$

(f) $\mathcal{S}$ has an obvious symmetric monoidal structure, given by the cartesian product $X \times Y$ *of sets* and the cartesian product $f \times g$ of partial mappings

$$f \times g\colon X \times Y \nrightarrow X' \times Y',$$
$$(f \times g)(x, y) = (f(x), g(y)) \qquad (x \in \operatorname{Def} f,\ y \in \operatorname{Def} g). \qquad (8.32)$$

The unit is the singleton. (Let us recall that the categorical product in $\mathcal{S}$ is different: see 2.2.5(d).)

This is indeed the monoidal structure of $\mathcal{S}$ corresponding to the smash product of pointed sets, because the equivalence $R\colon \mathsf{Set}_{\bullet} \to \mathcal{S}$ (in 1.8.6) preserves our tensor products

$$R((X, x_0) \wedge (Y, y_0)) = (X \times Y) \setminus ((X \times \{y_0\}) \cup (\{x_0\} \times Y))$$
$$= (X \setminus \{x_0\}) \times (Y \setminus \{y_0\}) = R(X, x_0) \times R(Y, y_0). \tag{8.33}$$

The forgetful functor $\mathcal{S} \to \mathsf{Set}$ associated to the monoidal structure is represented by the singleton

$$\mathcal{S}(\{*\}, X) = X \cup \{0_X\}, \tag{8.34}$$

and adds to X a single element: the empty partial map $0_X\colon \{*\} \nrightarrow X$.

8.3.6 Solutions of 3.4.7

(a) The exponential law of Cat

$$\varphi_{XY}\colon \mathsf{Cat}(X \times A, Y) \to \mathsf{Cat}(X, \mathsf{Cat}(A, Y)),$$
$$(F\colon X \times A \to Y) \mapsto (G\colon X \to \mathsf{Cat}(A, Y)), \tag{8.35}$$

is an obvious extension of that of Set. We send the functor F to the functor

$$G\colon X \to Y^A, \qquad G(x) = F(x, -)\colon A \to Y,$$
$$G(f\colon x \to x')\colon F(X, -) \to F(X', -), \qquad G(f)_a = F(f, 1_a),$$

where $x \in \mathrm{Ob}\, X$, $f \in \mathrm{Mor}\, X$ and $a \in \mathrm{Ob}\, A$.

Conversely, given $G\colon X \to Y^A$, the value of $F(f\colon x \to x',\ u\colon a \to a')$ is defined as the value of the natural transformation $Gf\colon Gx \to Gx'\colon A \to Y$ on the morphism $u\colon a \to a'$ (cf. (1.22)).

The exponential law of Ord is obvious (and a particular case of the previous one).

(b) In a pointed category any object X is isomorphic to $X \times \top = X \times \bot$. If X is exponentiable, $X \times -$ preserves the initial object and X is also initial.

In our hypotheses, every object is initial and terminal, and there are some of them: C is equivalent to $\mathbf{1}$, by 1.5.6(g).

(d) The symmetric monoidal structure $(\times, \{*\})$ of $\mathcal{S}$ has internal hom

$$Y^A = \mathcal{S}(A, Y), \tag{8.36}$$

as one can easily verify, directly. Obviously, this corresponds to the internal hom of $\mathsf{Set}_{\bullet}$; indeed, the equivalence $R\colon \mathsf{Set}_{\bullet} \to \mathcal{S}$ (in 1.8.6) gives

$$\mathsf{Set}_{\bullet}((A, a_0), (Y, y_0)) = \mathcal{S}(R(A, a_0), R(Y, y_0)).$$

(h) This is already proved in Exercise 3.2.8(d).

*(i) The mapping

$$\varphi \colon R \times S \times R \times S \to R \otimes S, \qquad \varphi(x, y, x', y') = xx' \otimes yy',$$

is $\mathbb{Z}$-linear in each variable. It induces a homomorphism of abelian groups

$$(R \otimes_{\mathbb{Z}} S) \otimes_{\mathbb{Z}} (R \otimes_{\mathbb{Z}} S) \to R \otimes_{\mathbb{Z}} S,$$

which gives the bilinear multiplication $(R \otimes_{\mathbb{Z}} S) \times (R \otimes_{\mathbb{Z}} S) \to R \otimes_{\mathbb{Z}} S$.

*(j) A pair of homomorphisms $f_i \colon R_i \to S$ of commutative rings has a unique 'extension' to a group-homomorphism

$$f \colon R_1 \otimes R_2 \to S, \qquad f(x \otimes y) = f_1(x).f_2(y),$$

produced by the $\mathbb{Z}$-bilinear mapping

$$\varphi \colon R_1 \times R_2 \to S, \qquad \varphi(x, y) = f_1(x).f_2(y).$$

Moreover, f preserves the product, because the ring S is commutative.

8.3.7 Solutions of 3.6.3

(a) The unitarity axioms follow from the triangular identities of the adjunction:

$$\mu.T\eta = (G\varepsilon F).(GF\eta) = G(\varepsilon F.F\eta) = \operatorname{id} GF,$$

$$\mu.\eta T = (G\varepsilon F).(\eta GF) = (G\varepsilon.\eta G)F = \operatorname{id} GF.$$

The associativity axiom follows from the self-interchange property (1.27) of $\varepsilon \colon FG \to 1 \colon \mathsf{A} \to \mathsf{A}$, namely $\varepsilon.FG\varepsilon = \varepsilon.\varepsilon FG$

$$\mu.T\mu = G(\varepsilon.FG\varepsilon)F = G(\varepsilon.\varepsilon FG)F = \mu.\mu T.$$

8.3.8 Solutions of 3.6.5

(a) The object $F^T(X) = (TX, \mu X \colon T^2 X \to TX)$ is indeed a T-algebra, by the monad axioms

$$\mu X.\eta TX = 1_{TX}, \qquad \mu X.T\mu X = \mu X.\mu TX.$$

The rest is obvious.

(b) The transformation ε^T is obviously natural. The triangular identities hold

$$\varepsilon^T F^T(X).F^T \eta^T(X) = \varepsilon^T(TX, \mu X).T(\eta X)$$

$$= \mu X.T(\eta X) = 1_{TX} = \operatorname{id} F^T(X),$$

$$G^T \varepsilon^T(X, a).\eta^T G^T(X, a) = a.\eta X = 1_X = \operatorname{id} G^T(X, a).$$

(c) In fact

$$\mu^T(X) = G^T \varepsilon^T F^T(X) = G^T \varepsilon^T(TX, \mu X) = \mu X.$$

8.3.9 Solutions of 3.6.8

All proofs being straightforward, we only write down the main points. Exercise (b) is entirely left to the reader.

Three consecutive arrows of X_T are given, namely

$$f^\sharp \colon X \rightarrowtail Y, \qquad g^\sharp \colon Y \rightarrowtail Z, \qquad h^\sharp \colon Z \rightarrowtail W.$$

(a) In X_T, the composites $h^\sharp(g^\sharp f^\sharp)$ and $(h^\sharp g^\sharp)f^\sharp$ are respectively represented by:

$$(\mu W.Th).(\mu Z.Tg.f) = \mu W.\mu TW.T^2 h.Tg.f,$$

$$\mu W.T(\mu W.Th.g).f = \mu W.T\mu W.T^2 h.Tg.f.$$

They coincide, by the associativity axiom of the monad: $\mu.\mu T = \mu.T\mu$.

(c) First, $H \colon \mathsf{X}_T \to \mathsf{A}$ preserves the composition $g^\sharp f^\sharp$ (using the self-interchange property of ε)

$$H(g^\sharp f^\sharp) = H((\mu Z.Tg.f)^\sharp) = \varepsilon FZ.FG\varepsilon FZ.FTg.Ff$$
$$= \varepsilon FZ.\varepsilon FGFZ.FTg.Ff,$$

$$H(g^\sharp)H(f^\sharp) = (\varepsilon FZ.Fg)(\varepsilon FY.Ff) = \varepsilon FZ.Fg.\varepsilon FY.Ff$$
$$= \varepsilon FZ.\varepsilon FTZ.FTg.Ff.$$

Moreover H is full and faithful, because the adjunction $F \dashv G$ gives bijective mappings

$$\varepsilon FY.F(-) \colon \mathsf{X}(X, GFY) \to \mathsf{A}(FX, FY) \qquad \text{(for } X, Y \text{ in } \mathsf{X}\text{)}.$$

Its codomain-restriction $H' \colon \mathsf{X}_T \to \mathsf{B}$ is essentially surjective on the objects, by definition, and an equivalence of categories.

(d) Computing on the morphism $f \colon X \to Y$ in X, the morphism $f^\sharp \colon X \rightarrowtail Y$ in X_T, and the object X in X, we have:

$$HF_T(f) = H(\eta Y.f)^\sharp = \varepsilon FY.F\eta Y.F(f) = F(f),$$

$$GH(f^\sharp) = G\varepsilon FY.GFf = \mu Y.Tf = G_T(f^\sharp),$$

$$H\varepsilon_T(X) = H(1_{TX})^\sharp = \varepsilon FX.F(1_{TX}) = \varepsilon HX.$$

8.3.10 Solutions of 3.7.1

(b) Letting $g(y) = \Sigma_{z \in Z}\, g_z(y).z$, the composite $g^{\sharp} f^{\sharp} \colon X \nrightarrow Z$ is computed as

$$gf(x) = \Sigma_{yz}\, f_y(x).g_z(y).z \qquad (y \in Y,\, z \in Z).$$

(c), (d) A consequence of (b) and Exercise 3.6.8(c).

8.3.11 Solutions of 3.7.2

(a), (b) The left adjoint to the forgetful functor $G = |-| \colon \mathsf{VSlt} \to \mathsf{Set}$ is the power-set functor $\mathcal{P} \colon \mathsf{Set} \to \mathsf{VSlt}$, with the following unit and counit, on a set X and an upper complete semilattice L

$$\begin{aligned}
\eta X &\colon X \to |\mathcal{P}X|, & \eta X(x) &= \{x\} & (x \in X),\\
\varepsilon L &\colon \mathcal{P}(|L|) \to L, & (\varepsilon L)(A) &= \vee A & (A \subset L).
\end{aligned} \tag{8.37}$$

The associated monad (T, η, μ) (which one can also guess from the start) is

$$\begin{aligned}
T &= G\mathcal{P} \colon \mathsf{Set} \to \mathsf{Set}, & \eta X(x) &= \{x\} & (x \in X),\\
\mu &= G\varepsilon\mathcal{P} \colon T^2 \to T, & (\mu X)(\mathcal{A}) &= \cup\mathcal{A} & (\mathcal{A} \subset \mathcal{P}X).
\end{aligned} \tag{8.38}$$

(Note that T is the covariant power-set functor $\mathcal{P}_* \colon \mathsf{Set} \to \mathsf{Set}$.)
The comparison $K \colon \mathsf{VSlt} \to \mathsf{Set}^T$ defined in (3.68) gives

$$\begin{aligned}
K(L) &= (|L|, G\varepsilon L \colon T(|L|) \to |L|), & (G\varepsilon L)(A) &= \vee A,\\
K(h \colon L \to M) &= |h| \colon (|L|, G\varepsilon L) \to (|M|, G\varepsilon M),
\end{aligned} \tag{8.39}$$

for $A \subset L$ and $h \colon L \to M$ in VSlt. We want to prove that K is an isomorphism of categories. K is obviously faithful and easily seen to be full; we still have to prove that it is bijective on the objects.

K takes an upper complete semilattice L to the underlying set $|L|$, equipped with the structure

$$s \colon T(|L|) \to |L|, \qquad s(A) = \vee A \qquad (A \subset L),$$

that satisfies the axioms of T-algebra:

$$s(\{x\}) = x, \qquad s(Ts(\mathcal{A})) = s(\cup\mathcal{A}) \qquad (x \in X,\, \mathcal{A} \subset \mathcal{P}X). \tag{8.40}$$

Conversely, given a T-algebra $(X, s \colon TX \to X)$ (satisfying the axioms (8.40)), we have to prove that the set X has a (unique) structure of upper complete semilattice L for which $K(L) = (X, s)$.

We define $x \leqslant y$ in X if $y = s\{x, y\}$.

This relation is obviously reflexive and anti-symmetric; it is also transitive, because $x \leqslant y \leqslant z$ implies

$$s\{x,z\} = s\{s\{x\}, s\{y,z\}\} = s\{x,y,z\}$$
$$= s\{s\{x,y\}, s\{z\}\} = s\{y,z\} = z.$$

X is now an ordered set, and every subset $A \in TX$ gives an element $x = s(A)$ which is the least upper bound of A in X. In fact $a \in A$ implies $a \leqslant x$

$$s\{x,a\} = s\{s(A), s\{a\}\} = s\{A \cup \{a\}\} = s(A) = x,$$

and, assuming that $a \leqslant y$ for every $a \in A$, we get $x \leqslant y$:

$$s\{x,y\} = s\{s(A), y\} = s\{A \cup \{y\}\} = s\{\textstyle\bigcup_{a \in A}\{a,y\}\} = s\{s\{y\}\} = y.$$

(c) In the Kleisli category Set_T an object is a set. A morphism $u\colon X \rightarrow Y$ is represented by an ordinary mapping $f\colon X \to TY = |\mathcal{P}Y|$, and amounts to a relation $X \rightarrow Y$.

The composite in Set_T of $u\colon X \rightarrow Y$ with a consecutive relation $v\colon Y \rightarrow Z$ (represented by the mapping $g\colon Y \to TZ$) is represented by the following mapping $X \to TZ$

$$\mu Z.Tg.f\colon X \to TY \to T^2Z \to TZ,$$
$$x \mapsto \textstyle\bigcup\{g(y) \mid y \in f(x)\},$$

and coincides with the composite vu in $\mathsf{Rel\,Set}$.

(d) This is an easy variation of Exercise (b). We replace VSlt with the category vSlt of join semilattices (having all finite joins) and their homomorphisms. The left adjoint to the forgetful functor $G = |-|\colon \mathsf{vSlt} \to \mathsf{Set}$ is the functor $\mathcal{P}_f\colon \mathsf{Set} \to \mathsf{vSlt}$ of finite subsets.

We get a monad $T'\colon \mathsf{Set} \to \mathsf{Set}$, with $T'(X) = |\mathcal{P}_f(X)|$, and vSlt is monadic over Set. The Kleisli category Set_T can be identified with the wide subcategory of $\mathsf{Rel\,Set}$ formed by the relations $u\colon X \rightarrow Y$ where every element of the domain has a finite set of values in the codomain. (It is not an involutive subcategory.)

8.4 Exercises of Chapter 4

8.4.1 Solutions of 4.1.3

(e) For a morphism $f\colon X \to UB$, every $x \in X$ gives a unique morphism $g_x\colon A \to B$ such that $Ug_x.\eta_0(*) = f(x) \in UB$.

These arrows are the co-components of a morphism g of A such that $Ug.\eta = f$

$$g = [g_x] : \Sigma_{x \in X} A \to B,$$

$$(Ug)(\eta(x)) = U(gu_x)(\eta_0(*)) = (Ug_x)(\eta_0(*)) = f(x).$$

Conversely, if $Ug.\eta = f$, each co-component $g_x : A \to B$ satisfies the relation $(Ug_x)(\eta_0(*)) = f(x)$, and is uniquely determined by the universal property of η_0.

(f) We have already considered the left adjoints D: Set $\to$ Top (Exercise 3.1.4(d)) and D: Set $\to$ Cat (Exercise 3.1.4(g)).

The discrete space DX on the set X is indeed the topological sum $\Sigma_{x \in X}\{*\}$ of a family of singleton spaces. Similarly, the discrete category DX on the set X of objects is the categorical sum $\Sigma_{x \in X} 1$ of discrete categories on the singleton.

8.4.2 Solutions of 4.1.4

(a) The free monoid on the singleton is obviously the additive monoid $\mathbb{N}$, generated by 1. It follows that $F(X)$ 'is' the categorical sum $\Sigma_{x \in X} \mathbb{N}$ in Mon.

In fact $F(X)$ can be built in a similar way as the free group in 4.1.3(c), by equivalence classes $[w]$ of words $w = x_1^{k_1} x_2^{k_2} \dots x_p^{k_p}$, in the alphabet X with *natural* exponents. Rewriting the class of the word $x^2 xyz^2 x$ as $xxxyzzx$ we find again the simpler construction that we have used above.

8.4.3 Solutions of 4.1.5

(b) The elements of $\mathbb{Z}M$ are the formal linear combinations $\Sigma_{x \in M} \lambda_x.x$ of elements of M, with quasi-null integral coefficients. The multiplication of $\mathbb{Z}M$ is defined as the bilinear extension of the multiplication of M

$$(\Sigma_y \lambda_y\, y).(\Sigma_z \mu_z\, z) = \Sigma_{y,z}\, \lambda_y \mu_z (yz) = \Sigma_{x \in M} (\Sigma_{yz=x} \lambda_y \mu_z)\, x. \quad (8.41)$$

(c) The left adjoint takes a monoid M to an R-algebra whose linear part is the free R-module RM over the set M. The multiplication in RM is defined as in (8.41), with (quasi-null) scalars in R.

(d) For an abelian monoid M, the abelian group $F(M)$ can be obtained as the quotient of the monoid $M \times M$ modulo the congruence where $(x, y) \sim (x', y')$ when

$$x + y' + z = x' + y + z, \quad \text{for some } z \in M. \quad (8.42)$$

The unit-component $\eta M\colon M \to UF(M)$ sends x to the equivalence class $[x, 0]$; it is injective if and only if M satisfies the cancellation law, when the congruence means: $x + y' = x' + y$.

(e) If M is the multiplicative monoid $\mathbb{Z}^*$, we get $F(M) = \mathbb{Q}^*$, the multiplicative group of non-zero rational numbers, built as fractions

$$x/y = [x, y] \qquad (\text{for } x, y \in \mathbb{Z}^*),$$

with the usual identification: $x/y = x'/y'$ when $xy' = x'y$.

8.4.4 Solutions of 4.2.5

(b) Let $u, v\colon R \rightrightarrows A$ be the kernel pair of $f\colon A \to B$ in C, viewed as a subobject $(u, v)\colon R \rightarrowtail A \times A$. The properties of reflexivity and symmetry being obvious, we rewrite diagram (4.9) (for transitivity) as in the left diagram below

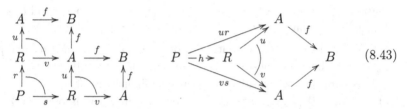

$$(8.43)$$

Now, $f(ur) = fvr = fus = f(vs)$. Applying this in the right diagram, there is a unique $h\colon P \to R$ such that $ur = uh$ and $vs = vh$.

8.4.5 Solutions of 4.3.3

(a) Suppose that $|f|$ is invertible. Let $g\colon B \to A$ be its inverse mapping in Set; then the mapping $g^n\colon A^n \to B^n$ is inverse to $f^n\colon A^n \to B^n$ (with some abuse of notation). For every n-ary operator $\omega \in \Omega(n)$ the condition (4.11) on f gives

$$g.\omega_B = g.\omega_B.f^n g^n = gf.\omega_B.g^n = \omega_B.g^n,$$

and g is a homomorphism, inverse to f. The converse is obvious.

(c) There is one 0-ary operation, under no axioms.

(f) Use Exercise 1.2.5(a).

(g) Add to the signature of lattices a unary operator $(-)'$ for the complements. Add to the previous axioms the distributivity axiom (D) (in 1.2.6) and the complement axiom

  (C) $x \wedge x' = 0, \qquad x \vee x' = 1.$

(h) Any field K can be embedded in the field $K(X)$ of rational functions on K, in one indeterminate. If K is a terminal object (in Fld or Fld$_p$), the embedding $K \to K(X)$ has a retraction $K(X) \to K$, which is injective (as any field homomorphism), and therefore an isomorphism.

Then the embedding $K \to K(X)$ is invertible, a contradiction.

8.5 Exercises of Chapter 5

8.5.1 Solutions of 5.1.2

(a) Let V be an open neighbourhood of $h_0(a_0)$ in Y. As A is locally compact, there is a compact neighbourhood K of a_0 in A such that $h_0(K) \subset V$. The distinguished open set $W(K,V) \times K$ of $Y^A \times A$ is a neighbourhood of (h_0, a_0), which the mapping ε_Y takes into V.

(b) We fix $x_0 \in X$ and prove the continuity of $h = g(x_0)$ at $a_0 \in A$.

Let V be an open neighbourhood of $h(a_0) = f(x_0, a_0)$ in Y. As the mapping $f \colon X \times A \to Y$ is continuous, there is a neighbourhood $U \times K$ of (x_0, a_0) such that $f(U \times K) \subset V$, and $h(K) = f(\{x_0\} \times K) \subset V$.

(c) We have to verify the continuity of $g \colon X \to Y^A$ at an arbitrary point $x_0 \in X$. (The uniqueness of g is obvious, set-theoretically.)

It is sufficient to consider an open neighbourhood of $g(x_0) = f(x_0, -)$ of the form $W(K,V)$, where K is compact in A, V is open in Y, and $f(\{x_0\} \times K) \subset V$.

For any $a \in K$ there is a basic open neighbourhood $U_a \times A_a$ of (x_0, a) in $X \times A$ such that $f(U_a \times A_a) \subset V$. All A_a form an open cover of K in A, and we can extract a finite family (A_{a_i}) which covers K.

Then the intersection $U = \cap U_{a_i}$ is an open neighbourhood of x_0, and $f(U \times K) \subset V$, because $(x, a) \in U \times K$ implies that a belongs to some A_{a_i} and $x \in U_{a_i}$. But $f(U \times K) \subset V$ means that $g(U) \subset W(K,V)$, and the proof is complete.

8.5.2 Solutions of 5.1.4

(c) For a topological space X, the space $T_0(X)$ is the quotient of X modulo the equivalence relation that identifies the pairs of points which have the same family of neighbourhoods.

8.5.3 Solutions of 5.1.6

(a) A topological space X is T_0 if and only if any two points x, y with the same closure always coincide, which means that the specialisation preorder of WX is antisymmetric. As the composite WA is the identity of pOrd, we are done.

(b) The square diagram of the right adjoints obviously commutes.

(c) A topological space X is T_1 if and only if the preordered set WX is discrete.

(d) The closed subsets of $A(\mathbb{R}, \leqslant)$ comprise the empty subset and every lower-unbounded interval

$$] - \infty, a[, \qquad\qquad] - \infty, a], \qquad\qquad \mathbb{R}.$$

To prove that there are no other closed subsets, let C be closed in the Alexandrov topology and non-empty. If it is not upper bounded, then $C = \mathbb{R}$. If C is upper bounded, and $a = \sup C$, then

$$] - \infty, a[\; \subset \; C \; \subset \;] - \infty, a],$$

and C is one of these intervals.

The non-empty open sets are the upper-unbounded intervals.

8.5.4 Solutions of 5.1.7

(d) The discrete pseudometric on the set X has $d(x, x') = \infty$ for $x \neq x'$; the indiscrete pseudometric has $d(x, x') = 0$ for all x, x'.

(e) By (d), all limits and colimits in psMtr_1 are constructed as in Set.

For a family (X_i) of pseudometric spaces, the product in psMtr_1 is the cartesian product $\prod_i X_i$ of the underlying sets, with the l_∞-pseudometric

$$d((x_i), (y_i)) = \sup_i d_{X_i}(x_i, y_i). \tag{8.44}$$

For a pair of parallel weak contractions $f, g \colon X \rightrightarrows Y$, the equaliser is the equaliser in Set, with the restricted pseudometric.

The sum of the family (X_i) is the disjoint union of the underlying sets

$$\Sigma_i X_i = \bigcup_i X_i \times \{i\},$$

with the greatest pseudometric that makes each injection of X_i a weak contraction

$$d((x, i), (y, i)) = d_{X_i}(x, y),$$
$$d((x, i), (y, j)) = \infty \qquad (i \neq j). \tag{8.45}$$

Finally, the coequaliser of $f, g \colon X \rightrightarrows Y$ is the coequaliser $p \colon Y \to Y/R$ in Set, with the *induced pseudometric*, namely the greatest that makes p a weak contraction.

This is computed as

$$d([y], [y']) = \inf \Sigma_i \, d_Y(y_{2i-1}, y_{2i}), \qquad (8.46)$$

where the greatest lower bound is taken over all finite sequences $y_1, y_2, ...,$ y_{2n} of points of Y such that

$$y \, R \, y_1, \qquad y_2 \, R \, y_3, \quad ..., . \qquad y_{2n} \, R \, y',$$

and $i = 1, 2, ..., n$.

Verifying that this is indeed a pseudometric, and the universal property, is straightforward.

8.5.5 Solutions of 5.2.4

(a) Using the $\mathbb{R}$-linear structure of $\mathbb{R}^n$, the origin is a deformation retract of $\mathbb{R}^n$

$$p \colon \mathbb{R}^n \to \{0\}, \qquad \varphi \colon ip \to \mathrm{id}\, \mathbb{R}^n, \qquad \varphi(x, t) = tx,$$

for $(x, t) \in \mathbb{R}^n \times \mathbb{I}$.

(b) We can use, as a retraction, the normalisation mapping N of 1.3.5(c)

$$N \colon X \to \mathbb{S}^n, \qquad N(x) = x/\|x\| \qquad (x \neq 0), \qquad (8.47)$$

with the homotopy

$$\varphi(x, t) = tx + (1 - t)N(x) \qquad ((x, t) \in X \times \mathbb{I}).$$

The latter moves along the segment from $N(x)$ to x, staying in the pierced space X.

8.5.6 Solutions of 5.2.7

(a) It is convenient to realise the cylinder $I(\mathbb{S}^n)$ as the product

$$\mathbb{S}^n \times [-1, 1] = \{(t_0, t_1, ..., t_{n+1}) \mid \Sigma_{i \leqslant n} t_i^2 = 1, \, t_{n+1}^2 \leqslant 1\}. \qquad (8.48)$$

The following picture shows the cylinder $\mathbb{S}^1 \times [-1, 1]$ and the standard sphere $\mathbb{S}^2$

$$(8.49)$$

Projecting the cylinder $I(\mathbb{S}^n)$ onto $\mathbb{S}^{n+1}$, orthogonally to the last axis t_{n+1}, we have a surjective continuous mapping

$$p: I(\mathbb{S}^n) \to \mathbb{S}^{n+1},$$
$$p(t_0, t_1, ..., t_{n+1}) = (r(t_{n+1})t_0, ..., r(t_{n+1})t_n, t_{n+1}),$$

(8.50)

where

$$r(t) = \sqrt{1-t^2}, \qquad \Sigma_{i \leqslant n}\, r^2(t_{n+1})t_i^2 = 1 - t_{n+1}^2.$$

This map collapses the upper basis of the cylinder to the 'North pole' $\underline{n}$ of $\mathbb{S}^{n+1}$ (on the last axis), and the lower basis to the 'South pole' $\underline{s}$; it is injective elsewhere, because $r(t)$ only annihilates at $t = \pm 1$.

Therefore, p induces a bijective continuous mapping

$$h: \Sigma(\mathbb{S}^n) \to \mathbb{S}^{n+1},$$

(8.51)

which is a homeomorphism: in fact, $\Sigma(\mathbb{S}^n)$ is a compact space (as a quotient of $\mathbb{S}^n \times [-1, 1]$) and $\mathbb{S}^{n+1}$ is Hausdorff.

*(b) The pointed suspension $\Sigma(\mathbb{S}^n, x_n)$ is the quotient of the cylinder $I(\mathbb{S}^n)$ which *also* collapses the fibre $\{x_n\} \times [-1, 1]$.

For $n = 0$, we are collapsing, in the unpointed suspension $\mathbb{S}^1$, a closed semicircle to a point, as in the following picture, and the quotient is (easily proved to be) homeomorphic to $\mathbb{S}^1$

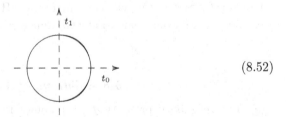

(8.52)

In higher dimension, we still collapse, in the unpointed suspension $\mathbb{S}^{n+1}$, a compact 1-dimensional arc to a base point; the quotient is homeomorphic to $\mathbb{S}^{n+1}$.

An analytic proof requires some work.

8.5.7 Solutions of 5.2.9

(a) Using the representative map $\check{\varphi}: X \to PY$ of the homotopy, we let

$$\varphi_*(x) = [\check{\varphi}(x)] \in \Pi_1(Y)(f(x), g(x)).$$

(8.53)

The naturality condition on an arrow $[a]\colon x \to x'$ in $\Pi_1(X)$ amounts to the commutativity of the following square, in $\Pi_1(Y)$

$$
\begin{array}{ccc}
f(x) & \xrightarrow{\ \varphi_*(x)\ } & g(x) \\
{\scriptstyle [fa]}\big\downarrow & & \big\downarrow{\scriptstyle [ga]} \\
f(x') & \xrightarrow[\ \varphi_*(x')\ ]{} & g(x')
\end{array}
\qquad (8.54)
$$

This follows from the map $\Phi = \varphi(a \times \mathrm{id}\,\mathbb{I})\colon \mathbb{I}^2 \to Y$, whose boundary is represented in the left diagram below

$$(8.55)$$

A convenient map $h\colon \mathbb{I}^2 \to \mathbb{I}^2$ of the square turns Φ into a homotopy with fixed endpoints $\Phi' = \Phi h\colon \mathbb{I}^2 \to Y$, from the path $fa + \check{\varphi}(x')$ to the path $\check{\varphi}(x) + ga$. An interested reader can work out this computation, or see an equivalent procedure in [G8], 3.2.2(c).

(b) Given a homotopy equivalence $f\colon X \rightleftarrows Y\colon g$, with homotopies $gf \simeq 1_X$ and $fg \simeq 1_Y$, we get two functors $f_*\colon \Pi_1(X) \rightleftarrows \Pi_1(Y)\colon g_*$ whose composites are isomorphic to identity functors.

8.5.8 Solutions of 5.5.3

(a) The reflexivity property $f \simeq f$ comes from the sequence of trivial homomorphisms $0\colon A_n \to B_{n+1}$. For symmetry we take the opposite components $-\varphi_n\colon A_n \to B_{n+1}$. For transitivity, we add the components of two homotopies $\varphi\colon f \to g$ and $\psi\colon g \to h$

$$
\varphi + \psi\colon f \to h, \qquad (\varphi + \psi)_n = \varphi_n + \psi_n. \qquad (8.56)
$$

Consistence with composition can be readily verified in form 1.4.5(ii), on the whisker composition of $\varphi\colon f \to g\colon A \to B$ with two chain morphisms $h\colon A' \to A$, $k\colon B \to B'$. Then the sequence of homomorphisms

$$
\psi_n = k_{n+1}\varphi_n h_n\colon A'_n \to B'_{n+1}, \qquad (8.57)
$$

gives a homotopy $\psi\colon kfh \to kgh$

$$\partial_{n+1}\psi_n + \psi_{n-1}\partial_n$$
$$= \partial_{n+1}(k_{n+1}\varphi_n h_n) + (k_n\varphi_{n-1}h_{n-1})\partial_n$$
$$= (k_n\partial_{n+1}\varphi_n h_n) + (k_n\varphi_{n-1}\partial_n h_n) \tag{8.58}$$
$$= k_n(\partial_{n+1}\varphi_n + \varphi_{n-1}\partial_n)h_n = k_n(g_n - f_n)h_n.$$

(b) We have a homotopy $\varphi\colon f \to g\colon A \to B$. For every cycle $z \in Z_n(A)$, the homology class $[z] \in H_n(A)$ gives

$$g_{*n}[z] - f_{*n}[z] = [g_n(z) - f_n(z)] = [\partial_{n+1}\varphi_n(z) + \varphi_{n-1}\partial_n(z)] = 0,$$

because z is an n-cycle and $\partial_{n+1}\varphi_n(z)$ is an n-boundary.

*(c) It is easy to verify that the differential of IA verifies $\partial\partial = 0$.

As to homotopies, we take a morphism $\Phi = [f, \varphi, g]\colon IA \to B$ of chain complexes and analyse the condition $\partial_{n+1}\Phi_{n+1} = \Phi_n\partial_{n+1}$ on a general element $(a, x, b) \in (IA)_{n+1}$

$$\partial_{n+1}\Phi_{n+1}(a, x, b) = \partial_{n+1}(f_{n+1}a + \varphi_n x + g_{n+1}b)$$
$$= f_n\partial_{n+1}a + \partial_{n+1}\varphi_n x + g_n\partial_{n+1}b,$$

$$\Phi_n\partial_{n+1}(a, x, b) = \Phi_n(\partial_{n+1}a - x, -\partial_n x, \partial_{n+1}b + x)$$
$$= f_n\partial_{n+1}a - f_n x - \varphi_{n-1}\partial_n x + g_n\partial_{n+1}b + g_n x.$$

Their equality means that, for all $x \in A_n$

$$\partial_{n+1}\varphi_n(x) + \varphi_{n-1}\partial_n(x) = g_n(x) - f_n(x).$$

8.5.9 Solutions of 5.5.6

(a) For a generator $a \in K_{n+1}$ we have

$$\partial_n\partial_{n+1}(a) = \Sigma_{i,j}(-1)^{i+j}\,\underline{\partial}_i\,\underline{\partial}_j(a)$$
$$(i = 0, ..., n, \ j = 0, ..., n+1), \tag{8.59}$$

where we write $\partial_{ni}\colon K_n \to K_{n-1}$ as $\underline{\partial}_i$, omitting the degree n.

This sum can be split in two parts, for $j \leqslant i$ and $j > i$, respectively. Then we transform the first part using the simplicial relations of the faces, $\underline{\partial}_i\,\underline{\partial}_j = \underline{\partial}_j\,\underline{\partial}_{i+1}$ for $j \leqslant i$ (in (5.63)), and apply a convenient change of variables on the indices

$$\partial_n\partial_{n+1}(a) = \Sigma_{j\leqslant i}(-1)^{i+j}\,\underline{\partial}_i\,\underline{\partial}_j(a) + \Sigma_{j>i}(-1)^{i+j}\,\underline{\partial}_i\,\underline{\partial}_j(a)$$
$$= \Sigma_{j\leqslant i}(-1)^{i+j}\,\underline{\partial}_j\,\underline{\partial}_{i+1}(a) + \Sigma_{j>i}(-1)^{i+j}\,\underline{\partial}_i\,\underline{\partial}_j(a) \tag{8.60}$$
$$= \Sigma_{i<j}(-1)^{i+j+1}\,\underline{\partial}_i\,\underline{\partial}_j(a) + \Sigma_{j>i}(-1)^{i+j}\,\underline{\partial}_i\,\underline{\partial}_j(a) = 0.$$

(b) The proof is similar to the previous one, using the cubical relations of the faces.

(c) There are no mappings $\Delta^n \to \emptyset$. Therefore $\mathrm{Ch}_+(\mathrm{Smp}(\emptyset))$ is the trivial chain complex: all its components are 0, and all homology groups are 0.

As to the singleton, for each $n \geqslant 0$ there is a unique map $a_n \colon \Delta^n \to \{*\}$, and

$$\partial_n(a_n) = \Sigma_{0 \leqslant i \leqslant n} \, (-1)^i \, \partial_{ni}(a) = \Sigma_{0 \leqslant i \leqslant n} \, (-1)^i \, a_{n-1}$$

is 0 for n odd (when the sum has an even number of terms), and a_{n-1} for n even. The singular chain complex of the singleton is isomorphic to

$$\ldots \, \mathbb{Z} \xrightarrow{1} \mathbb{Z} \xrightarrow{0} \mathbb{Z} \xrightarrow{1} \mathbb{Z} \xrightarrow{0} \mathbb{Z}. \tag{8.61}$$

Its homology is $\mathbb{Z}$ in degree 0, and trivial elsewhere.

(d) Using cubical simplices, there are no mappings $\mathbb{I}^n \to \emptyset$ and the first point is as in (c).

For the singleton, there is again a unique mapping $a_n \colon \mathbb{I}^n \to \{*\}$, for every $n \geqslant 0$. Here the differential of the *non-normalised* chain complex always involves an even number of summands

$$\partial_n(a_n) = \Sigma_{i,\alpha} \, (-1)^{i+\alpha} \, \partial_i^\alpha(a) = \Sigma_{i,\alpha} \, (-1)^{i+\alpha} \, a_{n-1} = 0,$$

and all its homology groups are isomorphic to $\mathbb{Z}$.

But we are interested in the normalised chain complex, and every generator a_n is degenerate in positive degree: $a_n = e_i(a_{n-1})$, for $n > 0$ and any i. Thus the cubical chain complex $\mathrm{Ch}_+(\mathrm{Cub}(\{*\}))$ of the singleton is isomorphic to

$$\ldots \, 0 \to 0 \to 0 \to 0 \to \mathbb{Z} \tag{8.62}$$

and its homology is (again, of course) $\mathbb{Z}$ in degree 0, and trivial elsewhere.

(e) For the remaining issues $C_+(X)$ denotes the cubical chain complex $\mathrm{Ch}_+(\mathrm{Cub}(X))$ of 5.5.5.

It is now convenient to use the 'augmented' chain complex

$$\ldots \, C_2(X) \xrightarrow{\partial_2} C_1(X) \xrightarrow{\partial_1} C_0(X) \xrightarrow{\partial_0} \mathbb{Z} \tag{8.63}$$

where we have added a component in degree -1 and the homomorphism $\partial_0(\Sigma \lambda_i x_i) = \Sigma \lambda_i$.

The homomorphism ∂_0 is surjective, because $X \neq \emptyset$, and $\mathrm{Ker} \, \partial_0 = \mathrm{Im} \, \partial_1$, because X is path-connected. Therefore ∂_0 induces a Noether isomorphism on $H_0(X) = C_0(X)/\mathrm{Im} \, \partial_1$

$$H_0(X) \to \mathbb{Z}, \qquad [x] \mapsto 1_{\mathbb{Z}}. \tag{8.64}$$

(f) One extends the previous argument, using an (obvious) homomorphism $\partial_0 \colon C_0(X) \to \mathbb{Z}I$.

(g) By hypothesis, any continuous mapping $a \colon I^n \to X$ is constant, and a degenerate singular cube when $n > 0$. Thus the cubical chain complex $C_+(X) = \mathrm{Ch}_+(\mathrm{Cub}(X))$ is

$$... \; 0 \; \to \; 0 \; \to \; 0 \; \to \; 0 \; \to \; C_0(X). \qquad (8.65)$$

Its homology is the free abelian group $C_0(X) = \mathbb{Z}X$ in degree 0, and trivial elsewhere. (Also $\mathbb{Z}X$ is trivial, if X is empty.)

8.6 Exercises of Chapter 6

8.6.1 Solutions of 6.1.4

(a) For C^2 we use the notation of (1.31).

For every object $x \colon X' \to X''$ of C^2, a cone for the kernel of x is any morphism $h \colon A \to X'$ such that $xh = 0$, which amounts to the following commutative square in C

$$
\begin{array}{ccc}
A & \xrightarrow{\; h \;} & X' \\
\downarrow & & \downarrow{\scriptstyle x} \\
0 & \longrightarrow & X''
\end{array}
\qquad (8.66)
$$

Introducing the functor

$$
J \colon \mathsf{C} \to \mathsf{C}^2, \qquad J(A) = (A \to 0),
$$
$$
J(h \colon A \to B) = (h, 0) \colon (A \to 0) \to (B \to 0),
\qquad (8.67)
$$

the cone h corresponds to an arrow $J(A) \to x$ in C^2, and the kernel of x is a universal arrow $(K(x), \varepsilon(x) \colon JK(x) \to x)$ from the functor J to the object x.

Therefore C has kernels if and only if the functor J has a right adjoint $K \colon \mathsf{C}^2 \to \mathsf{C}$. If this is the case

$$K(x \colon X' \to X'') = \mathrm{Ker}\, x, \qquad (8.68)$$

and the counit has components $\varepsilon(x) = (\ker x, 0) \colon J(\mathrm{Ker}\, x) \to x$.

On a map $f = (f', f'') \colon x \to y$ of C^2, the morphism $Kf \colon \mathrm{Ker}\, x \to \mathrm{Ker}\, y$ is determined in C by the following commutative (solid) diagram

$$
\begin{array}{ccccccc}
\mathrm{Ker}\, x & \xrightarrow{\ker x} & X' & \xrightarrow{\; x \;} & X'' & \dashrightarrow{\mathrm{cok}\, x} & \mathrm{Cok}\, x \\
\downarrow{\scriptstyle Kf} & & \downarrow{\scriptstyle f'} & & \downarrow{\scriptstyle f''} & & \vdots{\scriptstyle Cf} \\
\mathrm{Ker}\, y & \xrightarrow{\ker y} & Y' & \xrightarrow{\; y \;} & Y'' & \dashrightarrow{\mathrm{cok}\, y} & \mathrm{Cok}\, y
\end{array}
\qquad (8.69)
$$

In other words, Kf is the restriction of f' to kernels, and *should not be confused with* $\operatorname{Ker} f$, the kernel of the morphism $f = (f', f'')$ in C^2 (which also exists, in our hypothesis, and is easily constructed on the top of the previous diagram).

(a*) Dually, the embedding

$$J' \colon \mathsf{C} \to \mathsf{C}^2, \qquad J'(A) = (0 \to A),$$
$$J'(h \colon A \to B) = (0, h) \colon (0 \to A) \to (0 \to A), \tag{8.70}$$

has a left adjoint $C \colon \mathsf{C}^2 \to \mathsf{C}$ if and only if the category C has cokernels.

Then $C(x) = \operatorname{Cok} x$, and Cf is induced by f'' on cokernels, as shown in the dashed part of diagram (8.69).

8.6.2 Solutions of 6.2.3

(a) Obviously (i) $\Leftrightarrow$ (ii), and (i) $\Rightarrow$ (iv) $\Rightarrow$ (iii). The implication (iii) $\Rightarrow$ (i) follows from the relations $f_\bullet f^\bullet f_\bullet = f_\bullet$ and $f^\bullet f_\bullet f^\bullet = f^\bullet$. As (i) is selfdual, the proof is complete.

(d) Using the distributivity of lattices we get

$$f_\bullet(x \wedge x') = f_\bullet((x \wedge x') \vee f^\bullet 0) = f_\bullet((x \vee f^\bullet 0) \wedge (x' \vee f^\bullet 0))$$
$$= f_\bullet(f^\bullet f_\bullet x \wedge f^\bullet f_\bullet x') = f_\bullet f^\bullet(f_\bullet x \wedge f_\bullet x')$$
$$= f_\bullet x \wedge f_\bullet x' \wedge f_\bullet 1 = f_\bullet x \wedge f_\bullet x'.$$

8.6.3 Solutions of 6.2.3

(a) It is sufficient to show that the mapping $f \mapsto f^*$ preserves composition (contravariantly) and identities. The first point follows from the pasting property of pullbacks, in 2.3.6(b), the second is obvious.

(b) Take an R-homomorphism $h \colon A \to B$ with kernel $m \colon h^{-1}\{0\} \subset A$ and cokernel $p \colon B \to B/h(A)$.

Then the embedding

$$m_* \colon \operatorname{Sub}(h^{-1}\{0\}) \to \operatorname{Sub}A$$

is the same as the inclusion $\downarrow h^*(0) \subset \operatorname{Sub}A$, while the embedding

$$p^* \colon \operatorname{Sub}(B/h(A)) \to \operatorname{Sub}B$$

can be identified with the inclusion $\uparrow h_*(1) \subset \operatorname{Sub}B$.

8.6.4 Solutions of 6.3.2

(a) The morphism f annihilates ker f, and factorises as $f = hq$ for a unique $h \colon \mathrm{Cok}\,(\ker f) \to B$

$$\mathrm{Ker}\,f \overset{k}{\rightarrowtail} A \xrightarrow{\;f\;} B \overset{c}{\twoheadrightarrow} \mathrm{Cok}\,f \qquad (8.71)$$

with q, h, n and g forming the diagram $\mathrm{Coim}\,f \overset{g}{\to} \mathrm{Im}\,f$.

But $ch = 0$, because

$$(ch)q = cf = 0 \colon A \to \mathrm{Cok}\,f,$$

and q is epi. There is thus a unique morphism g such that $h = ng$.

(b) In the canonical factorisation (8.71) of a monomorphism $f \colon A \to B$, $\ker f = 0 \colon 0 \rightarrowtail A$ and $q = \mathrm{cok}\,(\ker f) = 1_A$. Since g is an isomorphism, f is equivalent to the normal mono n.

(c) The axiom (pex.1) is sufficient to give a pair of antitone mappings

$$\mathrm{cok} \colon \mathrm{Sub}A \rightleftarrows \mathrm{Quo}A \colon \ker. \qquad (8.72)$$

This pair is a contravariant Galois connection

$$m \leqslant \ker\,(\mathrm{cok}\,m), \qquad p \leqslant \mathrm{cok}\,(\ker p), \qquad (8.73)$$

because $(\mathrm{cok}\,m).m = 0$ and $p.(\ker p) = 0$. The connection restricts to an anti-isomorphism between the closed elements, namely the normal subobjects and the normal quotients.

If the category is p-exact, it is now sufficient to apply Exercise (b).

*(d) Essentially, let $f = mp$, where p is a normal epi and m a normal mono. Then $m \sim \ker u$, and $u = nq$ where q is a normal epi and n a normal mono. One verifies that $q \sim \mathrm{cok}\,m = \mathrm{cok}\,f$. Dually, m is a kernel of f.

8.6.5 Solutions of 6.3.5

(a) The canonical factorisations $f = mp$ and $g = nq$ of two consecutive morphisms give a commutative diagram

$$A \xrightarrow{\;f\;} B \xrightarrow{\;g\;} C \qquad (8.74)$$

with p, m through $\mathrm{Im}\,f$ and q, n through $\mathrm{Im}\,g$,

where $m = \mathrm{im}\,f$ and $\ker q = \ker g$.

Therefore the sequence (f, g) is exact if and only if the slanting sequence (m, q) is.

(b) In fact $\operatorname{im}(1_X) = 1_X$ and $\ker(1_X) = 0\colon 0 \rightarrowtail X$.

(c) Follows from (a) and (b).

8.6.6 Solutions of 6.4.5

(d) In Ltc, the biproduct of a finite family (X_i) of lattices is realised as the cartesian product ΠX_i. We use the following projections and injections

$$\Pi X_i \xrightarrow{\ p_i\ } X_i \xrightarrow{\ m_i\ } \Pi X_i \qquad\qquad (8.75)$$

$$m_{i\bullet} \dashv m_i{}^\bullet = \mathrm{pr}_i = p_{i\bullet} \dashv p_i{}^\bullet,$$

where $\mathrm{pr}_i\colon \Pi X_i \to X_i$ is an ordinary cartesian projection; for an element $x \in \Pi X_i$ and $j \neq i$

$$(m_{i\bullet}(x))_i = x, \quad (m_{i\bullet}(x))_j = 0, \qquad (p_i{}^\bullet(x))_i = x, \quad (p_i{}^\bullet(x))_j = 1.$$

Ltc is thus semiadditive, and the sum $f + g$ of two maps $f, g\colon X \to Y$ is defined as in (6.35)

$$X \xrightarrow{\ d\ } X \times X \xrightarrow{\ f \times g\ } Y \times Y \xrightarrow{\ \partial\ } Y \qquad\qquad (8.76)$$

$$d_\bullet(x) = (x, x), \qquad d^\bullet(x, x') = x \wedge x',$$

$$\partial_\bullet(y, y') = y \vee y', \qquad \partial^\bullet(y) = (y, y),$$

$$f + g = \partial.(f \times g).d = (f_\bullet \vee g_\bullet, f^\bullet \wedge g^\bullet).$$

In other words, $f + g$ *is the join* $f \vee g$ with respect to the order relation of Galois connections, defined in (6.10). *The sum is idempotent* ($f \vee f = f$), so that the cancellation law does not hold and our structure is not additive.

Restricting to the category of *complete* lattices and Galois connections we even have arbitrary biproducts of objects, and arbitrary sums of parallel morphisms $\vee f_i$.

8.6.7 Solutions of 6.4.8

(a) Obvious, since $[h, -k].(f, g) = hf - kg\colon A \to B$.

(b) We know (from 2.3.2) that the pullback-object of the cospan (h, k) is the equaliser of the morphisms $hp_1, kp_2\colon X \oplus Y \to B$, that is the subobject

$$\operatorname{Ker}(hp_1 - kp_2) = \operatorname{Ker}[h, -k] \rightarrowtail X \oplus Y.$$

Therefore the square (6.36) is a pullback if and only if $(f, g): A \to X \oplus Y$ is a monomorphism equivalent to the subobject $\ker [h, -k]$ of $X \oplus Y$.

(e) It is sufficient to consider (i), because the exactness condition is selfdual.

We have already noted that the pullback-object of (h, k) is the subobject $\mathrm{Ker}\,[h, -k] \rightarrowtail X \oplus Y$; now, by (b*), the pushout of (f, g) is a pair (h', k') with $\mathrm{Ker}\,[h', -k'] = \mathrm{Im}\,(f, g)$, and its pullback is this subobject of $X \oplus Y$.

Thus the exactness of the square amounts to the coincidence of the two pullbacks.

8.6.8 Solutions of 6.5.4

(c) Plainly, the relation $f \sim_\pi g$ is a congruence of categories and implies $f_* = g_*$. Therefore the canonical projection $P: \mathsf{E} \to \mathrm{Pr}\,\mathsf{E}$ factorises through $\mathsf{E}_\pi = \mathsf{E}/\!\sim_\pi$, as the composite $P = P'' P'$ of two functors

$$P': \mathsf{E} \to \mathsf{E}_\pi, \qquad P'': \mathsf{E}_\pi \to \mathrm{Pr}\,\mathsf{E}. \tag{8.77}$$

Knowing that P preserves and reflects exactness (including the zero object), we easily deduce that P' and P'' also do, whence E_π is p-exact and P', P'' are exact functors.

(d) There are five homomorphisms $f_i \colon \mathbb{Z} \to \mathbb{Z}/5$, determined by the element $f_i(1) = [i]$, for $i = 0, 1, ..., 4$.

There are three equivalence classes modulo $\sim_\pi$, namely

$$[f_0]_\pi, \qquad [f_1]_\pi = [f_4]_\pi, \qquad [f_2]_\pi = [f_3]_\pi.$$

But $\mathbb{Z}/5$ is a cyclic group, whence $X = \mathrm{Sub}\,\mathbb{Z}/5$ is a two-point lattice $\{0, 1\}$ and a join-homomorphism $h \colon X \to X$ can only be null or the identity: the homomorphisms f_i fall in two classes modulo $\sim_S$, namely $[f_0]$ and $[f_1] = ... = [f_4]$.

*(e) Any prime number $p \geqslant 3$ gives a cyclic group $\mathbb{Z}/p$ with p homomorphisms $f_i \colon \mathbb{Z} \to \mathbb{Z}/p$, that form $1 + (p-1)/2$ classes modulo $\sim_\pi$ and two classes modulo $\sim_S$. For $p \geqslant 5$ the outcomes are different.

8.6.9 Solutions of 6.6.2

(a) Obvious, using (6.59).

(b) If u is an injective (resp. surjective) homomorphism, then $u^\sharp u = 1$ (resp. $u u^\sharp = 1$).

(c) The factorisation (6.61) is obvious, using (6.59). Moreover we know that m, n are monorelations, while p, q are epirelations.

It follows that $p m^\sharp$ is an epirelation and $n q^\sharp$ is a monorelation.

(d), (d*) A consequence of (c). (Properties (d) and (d*) are dual to each other, for the involution of relations.)

(e) From (d) and (d*), since $\text{Rel}(R\,\text{Mod})$ is balanced.

(g) A consequence of the existence of epi-mono factorisations in $\text{Rel}(R\,\text{Mod})$: see 2.8.2(ii).

(h) For a non-trivial R-module A, let $\omega, \Omega\colon A \twoheadrightarrow A$ be, respectively, the least and the largest endo-relation (as in (6.60)).

Then there is no relation u with values in A such that $\omega u = \Omega u$. In fact, for every relation $u\colon B \twoheadrightarrow A$, $\text{Val}\,\omega u = \{0\}$ and $\text{Val}\,\Omega u = A$ are different.

8.6.10 Solutions of 6.7.5

(a) In this case, all kernels and cokernels are identities. The normal factorisation of a morphism simply gives $f = (\text{id}B).f.(\text{id}A)$; the morphism f is exact if and only if it is an isomorphism.

(b) The equivalence follows from a covariant Galois connection between subclasses of morphisms and subclasses of objects of E.

To a class $\mathcal{N} \subset \text{Mor}\,\text{E}$ we assign the class $O(\mathcal{N})$ of objects Z such that $1_Z \in \mathcal{N}$; to a class $\mathcal{O} \subset \text{Ob}\,\text{E}$ we assign the ideal $N(\mathcal{O})$ of all morphisms f which factorise through an object of $\mathcal{O}$. These mappings are monotone, and $N \dashv O$:

$$ON(\mathcal{O}) = \{Z \mid 1_Z \text{ factorises through an object of } \mathcal{O}\} \supset \mathcal{O},$$
$$NO(\mathcal{N}) = \{f \mid f \text{ factorises through } Z, \text{ and } 1_Z \in \mathcal{N}\} \subset \mathcal{N}. \tag{8.78}$$

The connection induces a bijective correspondence between the 'fixed' ideals, satisfying $\mathcal{N} = NO(\mathcal{N})$ (called closed ideals), and the 'fixed' classes of objects $\mathcal{O} = ON(\mathcal{O})$. The latter are precisely the classes of objects closed under retracts.

(c) Straightforward computations.

(d) Follows easily from (c). First, every sequence (6.90) is short exact in Top_2. Second, the kernel of a morphism $f\colon (X, A) \to (Y, B)$ is necessarily of the form (C, A), with $A \subset C \subset X$ (inclusions of subspaces), while the cokernel of f is necessarily of the form (Y, D), with $B \subset D \subset Y$.

(e) To prove that the monomorphisms of Top_2 are the injective maps it is sufficient to consider the forgetful functor

$$U\colon \text{Top}_2 \to \text{Set}, \qquad U(X, A) = |X|, \tag{8.79}$$

which is faithful and represented by the pair $(\{*\}, \emptyset)$ (apply 1.6.6(d)). Now, the kernel-object of an injective map of pairs $f\colon (X, A) \to (Y, B)$ is still $(f^{-1}(B), A)$ and need not be null. All the rest is obvious.

8.7 Exercises of Chapter 7

8.7.1 Solutions of 7.1.2

Composing three horizontally consecutive cells, we get

$$X \xrightarrow[g]{\overset{f}{\downarrow\varphi}} Y \xrightarrow[k]{\overset{h}{\downarrow\psi}} Z \xrightarrow[s]{\overset{r}{\downarrow\rho}} W \qquad (8.80)$$

$$\rho\circ(\psi\circ\varphi) = \rho\circ(\psi g.h\varphi) = \rho kg.r\psi g.rh\varphi = (\rho k.r\psi)\circ\varphi = (\rho\circ\psi)\circ\varphi.$$

(b) The following cells give:

$$X \xrightarrow[1]{\overset{1}{\downarrow 1}} X \xrightarrow[g]{\overset{f}{\downarrow\varphi}} Y \xrightarrow[1]{\overset{1}{\downarrow 1}} Y \qquad (8.81)$$

$$\varphi\circ\text{id}(1_X) = \varphi = \text{id}(1_Y)\circ\varphi.$$

(c) In diagram (7.8) we have:

$$X \xrightarrow[h]{\overset{f}{\underset{\downarrow\psi}{\downarrow\varphi}}} Y \xrightarrow[t]{\overset{r}{\underset{\downarrow\tau}{\downarrow\sigma}}} Z \qquad (8.82)$$

$$(\tau.\sigma)\circ(\psi.\varphi) = (\tau.\sigma)h.r(\psi.\varphi) = \tau h.\sigma h.r\psi.r\varphi = \tau h.(\sigma h.r\psi).r\varphi,$$

$$(\tau\circ\psi).(\sigma\circ\varphi) = \tau h.s\psi.\sigma g.r\varphi = \tau h.(s\psi.\sigma g).r\varphi.$$

But $\sigma h.r\psi = s\psi.\sigma g$, by the reduced interchange property (7.6).

8.7.2 Solutions of 7.1.6

(a) Starting from $\varphi\colon F'H \to KF$ and applying both mappings we get a diagram

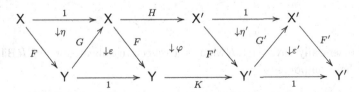

whose 'pasting' gives back φ, by the triangular identities. The other checking is similar.

8.7.3 Solutions of 7.2.8

(a) The special cell a is a natural transformation

$$a\colon u \to v\colon X^{\mathrm{op}} \times Y \to \mathsf{Set},$$

which means a natural transformation of profunctors $a\colon u \to v$, and a cell of their bicategory.

The globular cell b is a natural transformation

$$b\colon \mathrm{Mor}_X \to \mathrm{Mor}_{X'}(f^{\mathrm{op}} \times g)\colon X^{\mathrm{op}} \times X' \to \mathsf{Set},$$

$$b(x,x')\colon X(x,x') \to X'(fx,gx'), \tag{8.83}$$

$$(\xi\colon x \to x') \mapsto b(\xi)\colon fx \to gx',$$

and simply amounts to a natural transformation $b\colon f \to g\colon X \to X'$.

(b) The collage $\mathsf{C}^2 +_{T_*} \mathsf{C}$ consists of the sum $\mathsf{C}^2 + \mathsf{C}$, supplemented with heteromorphisms $\varphi\colon (A,B) \nrightarrow C$ given by C-arrows $\varphi\colon A \otimes B \to C$. The composition of $\mathsf{C}^2 + \mathsf{C}$ is extended in the obvious way

$$(A',B') \xrightarrow{(f,g)} (A,B)$$

$$\Big\downarrow \varphi \qquad\qquad\qquad h.\varphi.(f,g) = h\,\varphi\,(f \otimes g). \tag{8.84}$$

$$C \xrightarrow{\ h\ } C'$$

(c) The tensor product $T\colon R\,\mathsf{Mod}^2 \to R\,\mathsf{Mod}$ gives the profunctor T_*

$$T_*((A,B),C) = R\,\mathsf{Mod}(A \otimes_R B, C) = \mathrm{Bil}(A \times B, C), \tag{8.85}$$

where we identify the linear mappings $A \otimes_R B \to C$ with the bilinear mappings $A \times B \nrightarrow C$. The collage $R\,\mathsf{Bil} = R\,\mathsf{Mod}^2 +_{T_*} R\,\mathsf{Mod}$ has heteromorphisms $(A,B) \nrightarrow C$ given by these arrows.

The universal property of $\eta\colon A \times B \nrightarrow A \otimes_R B$ (in (3.37)) can now be rewritten as a diagram in $R\,\mathsf{Bil}$

$$(A,B) \xrightarrow{\ \eta\ } A \otimes_R B$$

$$\searrow_{\varphi} \qquad \Big\downarrow h \tag{8.86}$$

$$C$$

and essentially says that $R\,\mathsf{Mod}$ is a reflective subcategory of $R\,\mathsf{Bil}$. The unit has components

$$\eta(A,B)\colon (A,B) \nrightarrow U(A \otimes_R B), \qquad \eta A = \mathrm{id}A\colon A \nrightarrow U(A), \tag{8.87}$$

where $U\colon R\,\mathsf{Mod} \to R\,\mathsf{Bil}$ is the embedding.

8.7.4 Solutions of 7.3.4

(a) The unit of the tensor product of R is 0, and the associated forgetful functor is:

$$U = R(0, -): R \to \mathsf{Set},$$

$$U(0) = \{*\}, \qquad U(\lambda) = \emptyset \qquad (\lambda \neq 0).$$

If X is a δ-metric space, the underlying category is the set X, preordered by the relation $x \prec x'$ defined by $d(x, x') = 0$.

References

[AHS]   J. Adámek, H. Herrlich, G.E. Strecker, Abstract and concrete categories. The joy of cats, John Wiley & Sons, New York 1990.

[AnF]   F.W. Anderson, K.R. Fuller, Rings and categories of modules, Springer, 1974.

[AT]   H. Appelgate, M. Tierney, Categories with models, in: Seminar on triples and categorical homology theory, Lecture Notes in Math. Vol. 80, Springer, 1969, pp. 156–244.

[BaL]   J.C. Baez, A.D. Lauda, A prehistory of n-categorical physics, in Deep Beauty: Mathematical Innovation and the Search for an Underlying Intelligibility of the Quantum World, ed. Hans Halvorson, Cambridge University Press, Cambridge 2011, pp. 13–128.
         Available at: https://arxiv.org/abs/0908.2469

[BaP]   J.C. Baez, B.S. Pollard, A Compositional Framework for Reaction Networks, Rev. Math. Phys. 29 (2017), n. 9.
         Available at: https://arxiv.org/abs/1704.02051

[Bar]   M. Barr, Exact categories, in Lecture Notes in Math. Vol. 236, Springer, 1971, pp. 1–120.

[BarB]   M. Barr, J. Beck, Homology and standard constructions, in Seminar on Triples and Categorical Homology Theory, Springer, 1969, pp. 245–335.

[BarW]   M. Barr, C. Wells, Category Theory for Computing Science, Prentice Hall, New York 1990.

[BasE]   A. Bastiani, C. Ehresmann, Multiple Functors I. Limits Relative to Double Categories, Cah. Top. Géom. Diff. 15 (1974), 215–292.

[BBP]   M.A. Bednarczyk, A.M. Borzyszkowski, W. Pawlowski, Generalized congruences - epimorphisms in Cat, Theory Appl. Categ. 5 (1999), No. 11, 266–280.

[Bc]   J.M. Beck, Triples, algebras and cohomology, Repr. Theory Appl. Categ. 2 (2003).

[Be1]   J. Bénabou, Catégories avec multiplication, C. R. Acad. Sci. Paris 256 (1963), 1887–1890.

[Be2]   J. Bénabou, Introduction to bicategories, in: Reports of the Midwest Category Seminar, Lecture Notes in Math. Vol. 47, Springer, 1967, pp. 1–77.

[Be3]   J. Bénabou, Les distributeurs, Inst. de Math. Pure et Appliquée, Université Catholique de Louvain, Rapport n. 33, 1973.

[Bet]   R. Betti, Bicategorie di base, Ist. Mat. Univ. Milano, 2/S (1981).

[Bi]   G. Birkhoff, Lattice theory, 3rd ed., Amer. Math. Soc. Coll. Publ. 25, Providence 1973.

[BKPS]   G.J. Bird, G.M. Kelly, A.J. Power, R.H. Street, Flexible limits for 2-categories, J. Pure Appl. Algebra 61 (1989), 1–27.

[Bo1]   F. Borceux, Handbook of categorical algebra 1, Cambridge University Press, Cambridge 1994.

[Bo2]   F. Borceux, Handbook of categorical algebra 2, Cambridge University Press, Cambridge 1994.

[Bo3]   F. Borceux, Handbook of categorical algebra 3, Cambridge University Press, Cambridge 1994.

[Bo4]   F. Borceux, A survey of semi-abelian categories, in: Galois theory, Hopf algebras, and semiabelian categories, Fields Inst. Commun. Vol. 43, Amer. Math. Soc., Providence 2004, pp. 27–60.

[BoB]   F. Borceux, D. Bourn, Mal'cev, protomodular, homological and semiabelian categories, Kluwer Academic Publishers, Dordrecht 2004.

[BoJ]   F. Borceux, G. Janelidze, Galois theories, Cambridge University Press, Cambridge 2001.

[Bor1]   N. Bourbaki, Algebra I, Chapters 1–3, Hermann and Addison–Wesley, 1974.

[Bor2]   N. Bourbaki, General Topology I, Chapters 1–4, Springer, 1989.

[Bor3]   N. Bourbaki, General Topology II, Chapters 5–10, Springer, 1989.

[Bou1]   D. Bourn, Normalization equivalence, kernel equivalence and affine categories, in: Category theory (Como, 1990), Lecture Notes in Math. Vol. 1488, Springer, 1991, pp. 43–62.

[Bou2]   D. Bourn, Moore normalization and Dold–Kan theorem for semiabelian categories, in: Categories in algebra, geometry and mathematical physics, 105–124, Amer. Math. Soc., Providence 2007.

[Bra]   H. Brandt, Über eine Verallgemeinerung des Gruppenbegriffes, Math. Ann. 96 (1927), 360–366.

[Bri]   H.B. Brinkmann, Relations for exact categories, J. Algebra 13 (1969), 465–480.

[BriP]   H.B. Brinkmann, D. Puppe, Abelsche und exacte Kategorien, Korrespondenzen, Lecture Notes in Math. Vol. 96, Springer, 1969.

[Bro1]   R. Brown, Elements of modern topology, McGraw–Hill, New York 1968.

[Bro2]   R. Brown, Topology and groupoids, Third edition of Elements of modern topology, BookSurge, LLC, Charleston SC 2006.

[BroH1]   R. Brown, P.J. Higgins, On the algebra of cubes, J. Pure Appl. Algebra 21 (1981), 233–260.

[BroH2]   R. Brown, P.J. Higgins, Tensor products and homotopies for w-groupoids and crossed complexes, J. Pure Appl. Algebra 47 (1987), 1–33.

[BruMM]   R. Bruni, J. Meseguer, U. Montanari, Symmetric monoidal and Cartesian double categories as a semantic framework for tile logic, Math. Structures Comput. Sci. 12 (2002), 53–90.

[Bu1]   D.A. Buchsbaum, Exact categories and duality, Trans. Amer. Math. Soc. 80 (1955), 1–34.

[Bu2]   D.A. Buchsbaum, Appendix, in: H. Cartan, S. Eilenberg, Homological algebra, Princeton University Press, Princeton 1956.

[BunGS]   M.C. Bunge, F. Gago, A.M. San Luis, Synthetic Differential Topology, Cambridge University Press, Cambridge 2018.

[Ca]   F. Cagliari, Right adjoint for the smash product functor, Proc. Amer. Math. Soc. 124 (1996), 1265–1269.

[CaG]   A. Carboni, M. Grandis, Categories of projective spaces, J. Pure Appl. Algebra 110 (1996) 241–258.

[CaJKP]   A. Carboni, G. Janelidze, G.M. Kelly, R. Paré, On localization and stabilization for factorization system, Appl. Categ. Structures 5 (1997), 1–58.

[CE]   H. Cartan, S. Eilenberg, Homological algebra, Princeton University Press, Princeton 1956.

[ClP]   A.H. Clifford, G.B. Preston, The algebraic theory of semigroups, Vol. 1, Math. Surveys of the Amer. Math. Soc. 7, Providence 1961.

[CoL]   J.R.B. Cockett, S. Lack, Restriction categories I: categories of partial maps, Theor. Comput. Sci. 270 (2002), 223–259.

[Coh]   P.M. Cohn, Universal algebra, Harper & Row, New York 1965.

[CoC]   A. Connes, C. Consani, Homological algebra in characteristic one, High. Struct. 3 (2019), 155–247.

[Cop]   L. Coppey, Algèbres de décompositions et précategories, Diagrammes 3 Suppl. (1980), 1–139.

[Cr]   R.L. Crole, Categories for Types, Cambridge University Press, Cambridge 1993.

[Da1]   G. Darbo, Aspetti algebrico-categoriali della teoria dei dispositivi, in: Symposia Mathematica, Vol. IV, INDAM, Roma 1968/69. Academic Press, London 1970, pp. 303–336.

[Da2]   G. Darbo, Un'algebra locale reticolata che interviene nella teoria dei sistemi, Rend. Sem. Mat. Univ. Padova 60 (1978), 115–139.

[DaK]   B.J. Day, G.M. Kelly, On topological quotient maps preserved by pullbacks or products, Proc. Cambridge Philos. Soc. 67 (1970), 553–558.

[Dg]   J. Dugundij, Topology, Allyn & Bacon, Boston 1966.

[Ds]   J. Duskin, Variations on Beck's tripleability criterion, Lecture Notes in Math. Vol. 106, Springer, 1969, pp. 74–129.

[EE]   A. Ehresmann, C. Ehresmann, Multiple functors II. The monoidal closed category of multiple categories, Cah. Top. Géom. Diff. 19 (1978), 295–333.

[Eh1]   C. Ehresmann, Catégories des foncteurs types, Rev. Un. Mat. Argentina 20 (1962), 194–209.

[Eh2]   C. Ehresmann, Catégories structurées, Ann. Sci. Ecole Norm. Sup. 80 (1963), 349–425.

[Eh3]   C. Ehresmann, Catégories et structures, Dunod, Paris 1965.

[Eh4]   C. Ehresmann, Oeuvres complètes et commentées, Partie II-1, Structures locales. Editée et commentée par Andrée C. Ehresmann, Amiens 1982.
          http://ehres.pagesperso-orange.fr/C.E.WORKS_fichiers/C.E_Works.htm

[Eh5]   C. Ehresmann, Oeuvres complètes et commentées, Partie II.2, Editée et commentée par Andrée C. Ehresmann, Amiens 1982.
          http://ehres.pagesperso-orange.fr/C.E.WORKS_fichiers/C.E_Works.htm

[EiM]   S. Eilenberg, S. Mac Lane, General theory of natural equivalences, Trans. Amer. Math. Soc. 58 (1945), 231–294.

[EiK]   S. Eilenberg, G.M. Kelly, Closed categories, in Proc. Conf. Categorical Algebra, La Jolla 1965, Springer, 1966, pp. 421–562.

[EiS]   S. Eilenberg, N. Steenrod, Foundations of algebraic topology, Princeton University Press, Princeton 1952.

[FGR1]   L. Fajstrup, E. Goubault, M. Raussen, Detecting deadlocks in concurrent systems, in: CONCUR'98 (Nice), 332–347, Lecture Notes in Comput. Sci., 1466, Springer, 1998.

[FGR2]   L. Fajstrup, E. Goubault, M. Raussen, Algebraic topology and concurrency, Theor. Comput. Sci. 357 (2006), 241–178. (Revised version of a preprint at Aalborg, 1999.)

[Fo] B. Fong, The Algebra of Open and Interconnected Systems, Ph.D. thesis, Department of Computer Science, University of Oxford, 2016.

[Fr1] P. Freyd, Abelian categories, An introduction to the theory of functors, Harper & Row, New York 1964. Republished in: Reprints Theory Appl. Categ. 3 (2003).

[Fr2] P. Freyd, Stable homotopy, in: Proc. Conf. Categ. Algebra, La Jolla, 1965, Springer, 1966, pp. 121–176.

[Fr3] P. Freyd, On the concreteness of certain categories, in: Symposia Mathematica, Vol. IV, INDAM, Roma 1968/69. Academic Press, London 1970, pp. 431–456.

[Fr4] P. Freyd, Homotopy is not concrete, in: The Steenrod algebra and its applications, Battelle Memorial Inst., Columbus, OH, 1970, Lecture Notes in Math. Vol. 168, Springer, 1970, pp. 25–34.

[FrK] P. Freyd, G.M. Kelly, Categories of continuous functors, J. Pure Appl. Algebra 2 (1972), 1–18.

[FrS] P.J. Freyd, A. Scedrov, Categories, allegories, North-Holland Publ. Co., Amsterdam 1990.

[GaZ] P. Gabriel, M. Zisman, Calculus of fractions and homotopy theory, Springer, 1967.

[Ga] R. Garner, A homotopy-theoretic universal property of Leinster's operad for weak ω-categories, Math. Proc. Cambridge Philos. Soc. 147 (2009), 615–628.

[GiS] G. Gierz, K.H. Hofmann, K. Keimel, J.D. Lawson, M. Mislove, D.S. Scott, A compendium of continuous lattices, Springer, 1980.

[G1] M. Grandis, Transfer functors and projective spaces, Math. Nachr. 118 (1984), 147–165.

[G2] M. Grandis, Cohesive categories and manifolds, Ann. Mat. Pura Appl. 157 (1990), 199–244.
Available at: https://link.springer.com/journal/10231/157/1

[G3] M. Grandis, Cubical monads and their symmetries, in: Proceedings of the Eleventh International Conference on Topology, Trieste 1993, Rend. Ist. Mat. Univ. Trieste 25 (1993), 223–262.

[G4] M. Grandis, Weak subobjects and the epi-monic completion of a category, J. Pure Appl. Algebra 154 (2000), 193–212.

[G5] M. Grandis, Finite sets and symmetric simplicial sets, Theory Appl. Categ. 8 (2001), No. 8, 244–252.

[G6] M. Grandis, On the monad of proper factorisation systems in categories, J. Pure Appl. Algebra 171 (2002), 17–26.

[G7] M. Grandis, The role of symmetries in cubical sets and cubical categories (On weak cubical categories, I), Cah. Topol. Géom. Différ. Catég. 50 (2009), 102–143.

[G8] M. Grandis, Directed Algebraic Topology, Models of non-reversible worlds, Cambridge University Press, Cambridge 2009.
Available at: http://www.dima.unige.it/~grandis/BkDAT_page.html

[G9] M. Grandis, Homological Algebra, The interplay of homology with distributive lattices and orthodox semigroups, World Scientific Publishing Co., 2012.

[G10] M. Grandis, Homological Algebra in strongly non-abelian settings, World Scientific Publishing Co., 2013.

[G11] M. Grandis, Higher Dimensional Categories, From Double to Multiple Categories, World Scientific Publishing Co., 2020.

[G12]   M. Grandis, Manifolds and local structures, A general theory, World Scientific Publishing Co., in printing.

[GM]   M. Grandis, L. Mauri, Cubical sets and their site, Theory Appl. Categ. 11 (2003), No. 8, 185–211.

[GP1]   M. Grandis, R. Paré, Limits in double categories, Cah. Topol. Géom. Différ. Catég. 40 (1999), 162–220.

[GP2]   M. Grandis, R. Paré, Adjoint for double categories, Cah. Topol. Géom. Différ. Catég. 45 (2004), 193–240.

[GP3]   M. Grandis, R. Paré, Kan extensions in double categories (On weak double categories, III), Theory Appl. Categ. 20 (2008), No. 8, 152–185.

[GP4]   M. Grandis, R. Paré, Lax Kan extensions for double categories (On weak double categories, Part IV), Cah. Topol. Géom. Différ. Catég. 48 (2007), 163–199.

[GT]   M. Grandis, W. Tholen, Natural weak factorisation systems, Archivum Mathematicum (Brno) 42 (2006), 397–408.

[Gr1]   G. Grätzer, Universal algebra, Van Nostrand Co., Princeton 1968.

[Gr2]   G. Grätzer, General lattice theory, Academic Press, New York 1978.

[Gra]   J.W. Gray, The existence and construction of lax limits, Cah. Topol. Géom. Différ. 21 (1980), 277–304.

[Grt]   P.A. Grillet, Exact categories and categories of sheaves, in Lecture Notes in Math. Vol. 236, Springer, 1971, pp. 121–222.

[Gt]   A. Grothendieck, Sur quelques points d'algèbre homologique, Tôhoku Math. J. 9 (1957), 119– 221.

[Ha]   A. Hatcher, Algebraic topology, Cambridge University Press, Cambridge 2002.
          Available at: https://www.math.cornell.edu/~hatcher/AT/ATpage.html

[He]   H. Herrlich, Topologische Reflexionen und Coreflexionen, Lecture Notes in Math. 78, Springer, 1968.

[HeS]   H. Herrlich, G.E. Strecker, Category theory, an introduction, Allyn and Bacon, Boston 1973.

[Hi]   P.J. Hilton, Correspondences and exact squares, in: Proc. of the Conf. on Categorical Algebra, La Jolla 1965, Springer, 1966, 255–271.

[HiW]   P.J. Hilton, S. Wylie, Homology theory: An introduction to algebraic topology, Cambridge University Press, Cambridge 1960.

[Ho]   J.M. Howie, An introduction to semigroup theory, Academic Press, London 1976.

[HyP]   M. Hyland, J. Power, The category theoretic understanding of universal algebra: Lawvere theories and monads. In: Computation, meaning, and logic: articles dedicated to Gordon Plotkin, Elsevier Sci. B. V., Amsterdam 2007, pp. 437–458.

[In]   H. Inassaridze, Non-abelian homological algebra and its applications, Kluwer Academic Publishers, Dordrecht 1997.

[JaMT]   G. Janelidze, L. Márki, W. Tholen, Semi-abelian categories, in: Category theory 1999 (Coimbra), J. Pure Appl. Algebra 168 (2002), 367–386.

[Je]   T. Jech, Set theory, Academic Press, New York 1978.

[Jo1]   P.T. Johnstone, The point of pointless topology, Bull. Amer. Math. Soc. 8 (1983), 41–53.

[Jo2]   P.T. Johnstone, The art of pointless thinking: a student's guide to the category of locales, in: Category theory at work (Bremen, 1990) Heldermann, Berlin 1991, pp. 85–107.

[Jo3]   P.T. Johnstone, Sketches of an elephant: a topos theory compendium, Vol. 1, Oxford University Press, New York 2002.

[Jo4]   P.T. Johnstone, Sketches of an elephant: a topos theory compendium, Vol. 2, Oxford University Press, New York 2002.

[K1]   D.M. Kan, Abstract homotopy I, Proc. Nat. Acad. Sci. U.S.A. 41 (1955), 1092–1096.

[K2]   D.M. Kan, Adjoint functors, Trans. Amer. Math. Soc. 87 (1958), 294–329.

[Ke]   J.L. Kelley, General topology, Van Nostrand, New York 1955.

[Kl1]   G.M. Kelly, On Mac Lane's conditions for coherence of natural associativities, commutativities, etc., J. Algebra 1 (1964), 397–402.

[Kl2]   G.M. Kelly, Basic concepts of enriched category theory, Cambridge University Press, Cambridge 1982.

[Kl3]   G.M. Kelly, Elementary observations on 2-categorical limits, Bull. Austral. Math. Soc. 39 (1989), 301–317.

[KlS]   G.M. Kelly, R. Street, Review of the elements of 2-categories, in: Category Seminar, Sydney 1972/73, Lecture Notes in Math. Vol. 420, Springer, 1974, pp. 75–103.

[Ko]   A. Kock, Synthetic differential geometry, Cambridge University Press, Cambridge 1981.

[KrT]   M. Korostenski, W. Tholen, Factorization systems as Eilenberg–Moore algebras, J. Pure Appl. Algebra 85 (1993), 57–72.

[LaS]   J. Lambek, P.J. Scott, Introduction to higher order categorical logic, Cambridge University Press, Cambridge 1986.

[Law]   M.V. Lawson, Inverse semigroups, The theory of partial symmetries, World Scientific Publishing Co., 1998.

[Lw1]   F.W. Lawvere, Metric spaces, generalized logic and closed categories, Rend. Sem. Mat. Fis. Univ. Milano 43 (1974), 135–166. Republished in: Reprints Theory Appl. Categ. 1 (2002). http://www.tac.mta.ca/tac/reprints/articles/1/tr1.pdf

[Lw2]   F.W. Lawvere, Functorial semantics of algebraic theories and some algebraic problems in the context of functorial semantics of algebraic theories, Reprinted with comments from two papers of 1963 and 1968, Reprints Theory Appl. Categ. 5 (2004). http://www.tac.mta.ca/tac/reprints/articles/5/tr5.pdf

[LwR]   F.W. Lawvere, R. Rosebrugh, Sets for mathematics, Cambridge University Press, Cambridge 2003.

[LwS]   F.W. Lawvere, S.H. Schanuel, Conceptual mathematics. A first introduction to categories. Second edition. Cambridge University Press, Cambridge 2009.

[Le]   T. Leinster, Higher operads, higher categories, Cambridge University Press, Cambridge 2004.

[LoR]   E. Lowen-Colebunders, G. Richter, An elementary approach to exponential spaces, Appl. Categ. Structures 9 (2001), 303–310.

[M1]   S. Mac Lane, An algebra of additive relations, Proc. Nat. Acad. Sci. USA 47 (1961), 1043–1051.

[M2]   S. Mac Lane, Homology, Springer, 1963.

[M3]   S. Mac Lane, Natural associativity and commutativity, Rice University Studies 49 (1963), 28–46.

[M4]   S. Mac Lane, Categories for the working mathematician, Springer, 1971.

[MaM]   S. Mac Lane, I. Moerdijk, Sheaves in geometry and logic. A first introduction to topos theory, Springer, 1994.

[MaP]   S. Mac Lane, R. Paré, Coherence for bicategories and indexed categories, J. Pure Appl. Algebra 37 (1985), 59–80.

[MakR]   M. Makkai, G.E. Reyes, First order categorical logic. Model-theoretical methods in the theory of topoi and related categories, Lecture Notes in Math. Vol. 611, Springer, 1977.

[Mas]   W.S. Massey, Singular homology theory, Springer, 1980.

[Mat]   M. Mather, Pull-backs in homotopy theory, Can. J. Math. 28 (1976), 225–263.

[MayS]   J.P. May, J. Sigurdsson, Parametrized homotopy theory, American Mathematical Society, Providence 2006.

[Me]   J. Meisen, Relations in regular categories, in: Localization in group theory and homotopy theory and related topics, Lecture Notes in Math. Vol. 418, Springer, 1974, pp. 96–102.

[Mit]   B. Mitchell, Theory of categories, Academic Press, New York 1965.

[Mo]   K. Morita, Duality for modules and its applications to the theory of rings with minimum condition, Sci. Rep. Tokyo Kyoiku Daigaku, Sect. A 6 (1958), 83–142.

[Pa]   R. Paré, Simply connected limits, Canad. J. Math. 42 (1990), 731–746.

[Pi]   C. Pisani, Convergence in exponentiable spaces, Theory Appl. Categ. 5 (1999), No. 6, 148–162.

[Pu]   D. Puppe, Korrespondenzen in abelschen Kategorien, Math. Ann. 148 (1962), 1–30.

[Qu]   D. Quillen, Higher algebraic K-theory, I, in: Lecture Notes in Math. Vol. 341, Springer, 1973, pp. 85–147.

[Ro]   K. Rosenthal, The theory of quantaloids, Longman, Harlow 1996.

[Se]   Z. Semadeni, Banach spaces of continuous functions, Polish Sci. Publ., Warszawa 1971.

[St1]   R. Street, Fibrations and Yoneda's lemma in a 2-category, in: G.M. Kelly Ed., Category Seminar, Sydney 1972/73, Lecture Notes in Math. Vol. 420, Springer, 1974, pp. 104–133.

[St2]   R. Street, Limits indexed by category-valued 2-functors, J. Pure Appl. Alg. 8 (1976), 149–181.

[St3]   R. Street, Cosmoi of internal categories, Trans. Amer. Math. Soc. 258 (1980), 271–318.

[T1]   M.S. Tsalenko, Correspondences over a quasi exact category, Dokl. Akad. Nauk SSSR 155 (1964), 292–294.

[T2]   M.S. Tsalenko, Correspondences over a quasi exact category, Mat. Sbornik 73 (1967), 564–584.

[Vi]   J.W. Vick, Homology theory. An introduction to algebraic topology. Second edition, Springer, 1994.

[Wa]   R.F.C. Walters, Sheaves and Cauchy-complete categories, Cah. Top. Géom. Diff. 22 (1981), 282–286.

[We]   W. Wechler, Universal algebra for computer scientists, Springer, 1992.

[ZS]   O. Zariski, P. Samuel, Commutative Algebra, Vol. II, Van Nostrand, New York 1960.

Index

Printed in the United States
by Baker & Taylor Publisher Services